(continued)

[7.4] The solutions to $ax^2 + bx + c = 0$ are given by

$$x = \frac{-b \pm \sqrt{b^2 - 4ac}}{2a} \quad (a \neq 0)$$

[8.3] The equation of the line through point $(x_1, y_1)$ with slope $m$ is

$$y - y_1 = m(x - x_1)$$

[10.3] (Cramer's rule) The solution to the system

$$ax + by = e$$
$$cx + dy = f$$

is given by

$$x = \frac{\begin{vmatrix} e & b \\ f & d \end{vmatrix}}{\begin{vmatrix} a & b \\ c & d \end{vmatrix}} \qquad y = \frac{\begin{vmatrix} a & e \\ c & f \end{vmatrix}}{\begin{vmatrix} a & b \\ c & d \end{vmatrix}} \quad (ad - bc \neq 0)$$

[11.1] $\log_b N = x$ means $b^x = N$

[11.4] $\log xy = \log x + \log y$

$$\log \frac{x}{y} = \log x - \log y$$

$$\log x^r = r(\log x)$$

$$10^x = \text{antilog } x$$

[11.5] $\log_b N = \dfrac{\log N}{\log b}$

[12.3] For an arithmetic progression $a_1, a_2, a_3, \ldots$, with common difference $d$:

$$a_n = a_1 + (n - 1)d$$

$$S_n = \frac{n}{2}(a_1 + a_n)$$

[12.4] For a geometric progression $a_1, a_2, a_3, \ldots$, with common ratio $r$:

$$a_n = a_1 r^{n-1}$$

$$S_n = \frac{a_1(r^n - 1)}{r - 1}$$

[12.5] For an infinite geometric progression $a_1, a_2, a_3, \ldots$, with common ratio $r$, $|r| < 1$:

$$S_\infty = \frac{a_1}{1 - r}$$

[12.6] The Binomial Theorem for a positive integer $n$:

$$(x + y)^n = x^n + nx^{n-1}y + \frac{n(n - 1)}{1 \cdot 2}x^{n-2}y^2 + \cdots$$

# INTERMEDIATE ALGEBRA

# INTERMEDIATE ALGEBRA

Howard A. Silver
Department of Mathematics
Chicago State University

PRENTICE-HALL, INC., ENGLEWOOD CLIFFS, NEW JERSEY 07632

*Library of Congress Cataloging in Publication Data*

SILVER, HOWARD A (date)
Intermediate algebra.

Includes index.
1. Algebra. I. Title.
QA154.2.S52 512.9 80-23302
ISBN 0-13-469411-2



Printed in the United States of America

10 9 8 7 6 5

*Editorial/production supervision by Kathleen M. Lafferty*
*Interior design by Wanda Lubelska*
*Cover design by Miriam Recio*
*Manufacturing buyer: John B. Hall*

Prentice-Hall International, Inc., *London*
Prentice-Hall of Australia Pty. Limited, *Sydney*
Prentice-Hall of Canada, Ltd., *Toronto*
Prentice-Hall of India Private Limited, *New Delhi*
Prentice-Hall of Japan, Inc., *Tokyo*
Prentice-Hall of Southeast Asia Pte. Ltd., *Singapore*
Whitehall Books Limited, *Wellington, New Zealand*

To my parents

# CONTENTS

## Chapter 9 FUNCTIONS AND SECOND-DEGREE RELATIONS 291



## Chapter 10 SYSTEMS OF EQUATIONS AND INEQUALITIES 321



## Chapter 11 LOGARITHMS 367

# PREFACE

This text, *Intermediate Algebra,* is written for today's college student. It is tailored to the changes in the mathematical preparation of students, the shifts in their major fields of study, and new ideas in presenting the material to college students.

This text squarely faces the fact that many of the college students in an intermediate algebra course have weak mathematical backgrounds or are older students who have not had a mathematics course in many years. Frankly, this text is *not* written for the mathematics major or the mathematically elite. Rather, it is written for the business major, the health science major, the life science major, the social science major, the art major, and so on. It is written for the student with gaps in his or her background, years since his or her last math course, poor reading ability, or "math anxiety."

The material in this text is the set of topics standard to most intermediate algebra courses. It is the presentation of these topics that is compelling and unique. The following special features are molded together to form a strong, integrated text with which the students will feel comfortable.

EXAMPLES The text contains over 350 worked examples. But more than mere quantity, the emphasis is on the quality of the examples and their presentation. Every example is worked out slowly, with the steps being explained by a graphic flow chart to the left of the problem. Also, each example is carefully chosen to demonstrate a skill needed to work a group of exercises in the problem set.

PROBLEM SET The text contains over 3300 exercises. About 55 percent of the answers appear in the back of the text. Each problem set is a carefully graded set of exercises that are keyed to the examples. For instance, in the margin next to Problems 10–17 there might appear a box which reads, "See Example 22." In addition to providing a ready reference for the student when working the problems, this grouping of the exercises will also help both the instructor and the student to organize the material of the chapter as the various problem-types are clearly separated.

PROCEDURES In addition to the worked-out examples, explicit step-by-step procedures are stated and boxed for student reference.

APPLICATIONS Students taking intermediate algebra often ask, "How does this relate to my major?" This text presents applications to a wide variety of fields, such as business, health, engineering, life science, social science, physical science, and art. These applications appear at the end of almost every problem set to provide maximum impact and reinforcement of skills.

READABILITY This text is written in a crisp, plain-English style that does not drown the presentation in a sea of excessive verbiage. The striking simplicity of the flow-charted explanations has far more visual impact than paragraphs of exposition. The open white-space layout of the text makes it very appealing to the eye and non-threatening to the student.

SKILLS ORIENTED Since very few of the students using this text will pursue advanced mathematics courses, the level of abstraction and the proving of theorems is kept to a minimum. Set notation is used very discreetly.

HAND CALCULATOR Since almost every college student now seems to own a hand calculator, it is only natural to integrate usage of the calculator into the text wherever possible. For instance, the calculator is used in the discussion of polynomial evaluation, exponents, logarithms, square roots, and checking solutions to linear systems of equations.

YES/NO BOXES Experience has shown that students seem to make the same errors term after term. These Yes/No boxes appear as an effort to preempt these common mistakes by showing the correct procedure (Yes) and by negating the procedure with a giant × drawn through it (No).

REVIEW Every chapter ends with a thorough review including important words and phrases, important properties, and important procedures. In addition, there is a chapter test on all the material of the chapter.

I gratefully acknowledge the help of the following reviewers during the various stages in the preparation of the manuscript: Terry Czerwinski, University of Illinois at Chicago Circle; Suzanna Hesson, San Antonio College; and Thomas Hill, University of Oklahoma.

I would like to thank the great Prentice-Hall staff for all their help and encouragement: Harry Gaines, Bob Sickles, Beth Elderd, Kathleen Lafferty, and all the others whose names I never learned. I would like to thank my typist, Mary Lou Leggett, who did a beautiful job with the manuscript. Finally, I would like to thank my friends and colleagues for their input and feedback along the way; specifically, let me thank Bob Bernhardt, Mike Sullivan, Jackie Blackmon, and Jeni Morison.

A final note to the faculty and students using this text: I would love to hear your comments on the text; write to me at Chicago State University or through Prentice-Hall.

Howard A. Silver

# 1
# SETS AND REAL NUMBERS

**Algebra** is a study of numbers, different ways to express numbers, and relationships between numbers. In this chapter, we discuss and review some of the basic ideas and notation that will be used throughout.

## 1.1 SETS AND SET OPERATIONS

A **set** is a collection of some kind. It might be a collection of coins, women, or odd numbers between 2 and 50. Sets are written with braces and noted by capital letters, such as

$$A = \{1, 2, 3, 4, 5, 6\}$$
$$B = \{\text{Americans with IQs over } 130\}$$

These two sets show two different ways that we can write sets. Set $A$ uses the **list method,** in which we simply list every single element of the set. Set $B$ uses the **description method,** in which we describe the elements by a common property.

Two sets are **equal** if they have the exact same elements, whether they are listed or described. The order in which the elements are listed is not important. For example, $\{1, 2, 3\} = \{2, 1, 3\} = \{2, 3, 1\}$.

**EXAMPLE 1** Write $A = \{\text{first 10 counting numbers}\}$ using the list method.

**SOLUTION** We simply list all the elements using the given description. Therefore,

$$A = \{1, 2, 3, 4, 5, 6, 7, 8, 9, 10\}$$

**EXAMPLE 2** Write $B = \{10, 12, 14, 16, 18\}$ using the description method.

**SOLUTION** There are two ways to write out the set. They are

$$B = \{\text{even numbers between 10 and 18, inclusive}\}$$
$$B = \{x \mid x \text{ is an even number between 10 and 18, inclusive}\}$$

We use the notation $x \in A$ to mean that $x$ is an **element** of the set $A$. Also, $x \notin A$ means that $x$ is *not* an element of $A$.

**EXAMPLE 3** Let $A = \{a, b, c, d, e\}$. Then

**(a)** $a \in A$ and $c \in A$

**(b)** $f \notin A$ and $27 \notin A$

A set is called **finite** if its elements can be counted and there is a last element. A set is called **infinite** if there is no last element. A set is **empty,** written $\varnothing$**.** if it has no elements at all. As examples,

$$A = \{1, 2, 3, 4\} \text{ is finite}$$

$$B = \{\text{even numbers}\} \text{ is infinite}$$

$$C = \{\text{living Civil War veterans}\} \text{ is empty}$$

We say that $A$ is a **subset** of $B$, written $A \subset B$, if all the elements of the set $A$ are contained in the set $B$. For example,

$$\{1, 2, 3\} \subset \{1, 2, 3, 4, 5\}$$

As illustrated in Fig. 1.1, we can form new sets from old sets. The **union** of sets $A$ and $B$, written $A \cup B$, is the set formed by joining all the elements in $A$ or $B$, or both.

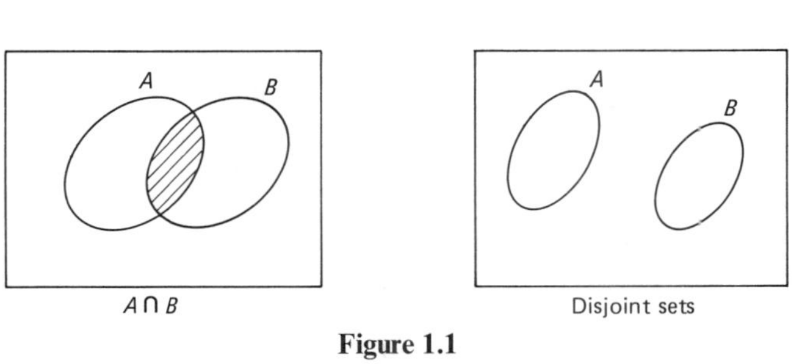


**Figure 1.1**

The **intersection** of sets $A$ and $B$, written $A \cap B$, is the set formed by taking the overlap or elements in both $A$ and $B$. When two sets have no elements in common, we say the sets are **disjoint.** Thus, $A$ and $B$ are disjoint if $A \cap B = \varnothing$.

**EXAMPLE 4** Let $A = \{x, y, z\}$, $B = \{t, u, v, w, x\}$, and $C = \{s, t, u\}$. Then

**(a)** $A \cup B = \{t, u, v, w, x, y, z\}$

**(b)** $A \cup C = \{s, t, u, x, y, z\}$

**(c)** $B \cup C = \{s, t, u, v, w, x\}$

**(d)** $A \cap B = \{x\}$

**(e)** $A \cap C = \varnothing$ (Thus, $A$ and $C$ are disjoint.)

## PROBLEM SET 1.1

Write the following sets using the list methods.

*See Example 1*

1. {even numbers between 2 and 12, inclusive}
2. {first seven counting numbers}
3. {states of the union that start with the letter A}
4. {months of the year that start with the letter J}
5. {U.S. senators from your home state}
6. {houses of Congress}
7. {whole numbers greater than 10 and less than 6}
8. {odd numbers greater than 4 and less than 12}

Write the following sets using the description method.

*See Example 2*

9. {1, 3, 5, 7, 9, 11, 13}
10. {1, 2, 3, 4, 5, 6, 7, 8, 9}
11. {10, 20, 30, 40, 50, 60}
12. {20, 22, 24, 26, 28, 30, 32}
13. {Washington, Adams, Jefferson, Madison, Monroe}
14. {Lincoln, Garfield, McKinley, Kennedy}

15. {Monday, Tuesday, Wednesday, Thursday, Friday}

16. {General Motors, Ford Motors, Chrysler Motors, AMC}

Let $A = \{1, 2, 3, 4, 5\}$ and $B = \{3, 5, 7, 9\}$. Indicate which of the following statements are true and which are false.

*See Example 3*

17. $1 \in A$ 18. $7 \notin A$ 19. $4 \notin B$

20. $3 \notin B$ 21. $4 \in B$ 22. $4 \in A$

Let $A = \{1, 2, 3, 4, 5, 6\}$, $B = \{2, 4, 6, 8, 10\}$, $C = \{1, 3, 5\}$, $D = \{5, 6, 7, 8\}$, and $E = \{1, 4, 7, 10, 13\}$.

*See Example 4*

23. Find $A \cup B$. 24. Find $B \cup C$. 25. Find $A \cup D$.

26. Find $B \cup D$. 27. Find $C \cup E$. 28. Find $D \cup E$.

29. Find $A \cap B$. 30. Find $A \cap C$. 31. Find $D \cap E$.

32. Find $C \cap D$. 33. Find $B \cap C$. 34. Find $B \cap D$.

*Health application*

In a study of 680 addicts (tobacco, alcohol, and drugs), the subjects were classified as shown in Fig. 1.2. For each of the following sets, describe the set in words and give the total number in the set using the illustration.

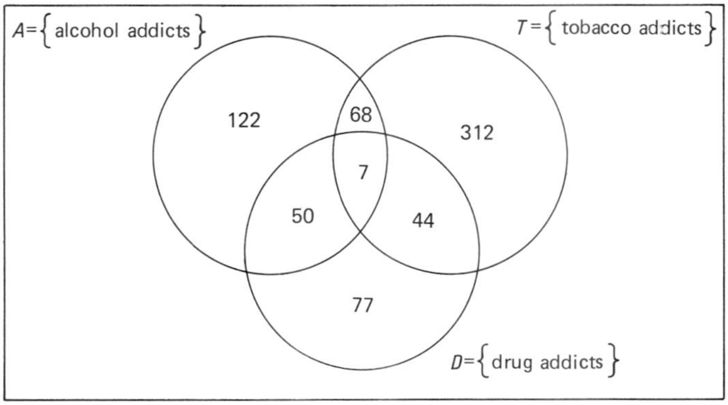


**Figure 1.2**

35. $A$ 36. $T$ 37. $D$

38. $A \cap D$ 39. $D \cap T$ 40. $A \cap D \cap T$

41. $A \cup T$ 42. $D \cup T$ 43. $A \cup D \cup T$

*Business application*

In a recent consumer study, 460 shoppers listed certain major purchases they would make within the year: automobile, television, and bedroom furniture.

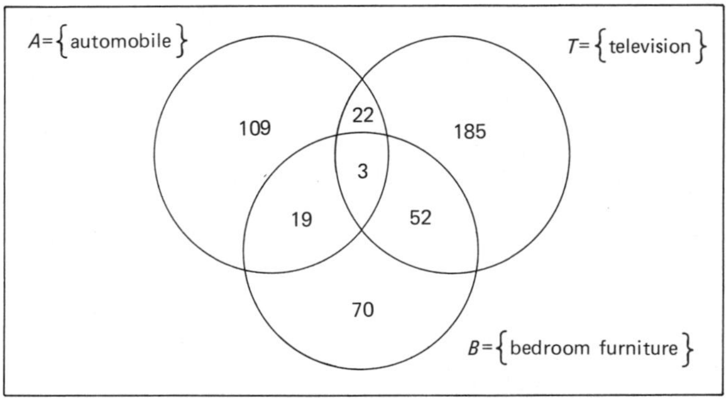


**Figure 1.3**

The data are illustrated in Fig. 1.3. For each of the following sets, describe the set in words and give the total number in the set using the illustration.

44. $A$
45. $T$
46. $B$
47. $A \cap B$
48. $B \cap T$
49. $A \cap B \cap T$
50. $A \cup T$
51. $A \cup B$
52. $A \cup B \cup T$

## 1.2 SETS OF NUMBERS; EQUALITY AND INEQUALITY

In algebra, we encounter various sets of numbers. The following are the sets we use throughout the book, together with their common symbols.

1. **Natural numbers** (or **counting numbers**): **N** $= \{1, 2, 3, 4, 5, \ldots\}$
2. **Whole numbers: W** $= \{0, 1, 2, 3, 4, 5, \ldots\}$
3. **Integers: I** $= \{\ldots, -3, -2, -1, 0, 1, 2, 3, \ldots\}$
4. **Rational numbers: Q** $= \left\{\begin{matrix}\text{fractions, } \frac{a}{b}\text{, with } b \neq 0;\ a, b \in I; \\ \text{terminating or repeating decimals}\end{matrix}\right\}$
5. **Irrational numbers: H** $= \left\{\begin{matrix}\text{nonterminating, nonrepeating} \\ \text{decimals}\end{matrix}\right\}$
6. **Real numbers: R** $= \left\{\begin{matrix}\text{all decimal numbers, or coordinates} \\ \text{of points on the number line}\end{matrix}\right\}$

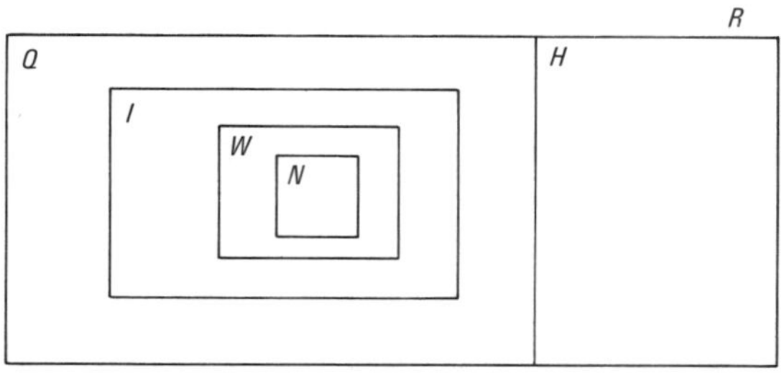


**Figure 1.4**

These sets obey the following rules:

$$N \subset W \subset I \subset Q \subset R \quad \text{and} \quad H \subset R$$

Because of this, if a number is in one of the sets, it is also in all the larger sets. Figure 1.4 shows these relations pictorially. Notice that $Q \cup H = R$ and $Q \cap H = \varnothing$.

**EXAMPLE 5** **(a)** $-10 \in I$; also, $-10 \in Q$ and $-10 \in R$

**(b)** $1.73 = \frac{173}{100} \in Q$; also, $1.73 \in R$

**(c)** $0 \in W$; also, $0 \in I$, $0 \in Q$, and $0 \in R$

**(d)** $\pi = 3.14159\ldots \in H$; also, $\pi \in R$

**(e)** $0.3333\ldots = \frac{1}{3} \in Q$; also, $0.3333\ldots \in R$

**(f)** $\sqrt{3} \in H$; also $\sqrt{3} \in R$ (The square roots of numbers that are not perfect squares are always irrational.)

Notice that all these numbers are in the set of real numbers. As mentioned in the description of real numbers, they can be very nicely pictured on the number line. Each number on the line is called the **coordinate** of the point.

**EXAMPLE 6** Find the coordinate of each of the points on the following number line.

**SOLUTION** We often have to approximate the coordinates for points between the hatch marks. Here we have $A = 2$, $B = -3$, $C = 3.6$, $D = -0.3$. The point $E = 0$ is also called the **origin.**

Two expressions are **equal** if they name the same numbers. For instance, 10, $6 + 4$, $2 \cdot 5$, and $\frac{30}{3}$ all name the same number. We have the following laws for equality.

For any real numbers $a$, $b$, and $c$:

| | |
|---|---|
| **Reflexive law:** | $a = a$. |
| **Symmetric law:** | If $a = b$, then $b = a$. |
| **Transitive law:** | If $a = b$, and $b = c$, then $a = c$. |
| **Substitution law:** | If $a = b$, then $a$ can replace $b$, and vice versa, in a statement without affecting the statement's truth. |

**EXAMPLE 7** The following use the laws for equality.

**(a)** $12 = 12$. (reflexive law)

**(b)** If $x = 7$, then $7 = x$. (symmetric law)

**(c)** If $x = y$ and $y = 9$, then $x = 9$. (transitive law)

**(d)** If $x = 7$ and $y = x^2 + x - 4$, then $y = 7^2 + 7 - 4$. (substitution law)

An **inequality** is a statement that two numbers are not equal: for example, $5 \neq 8$. We also have symbols for one number being larger or smaller than the other.

| SYMBOL | MEANING |
|---|---|
| $a < b$ | $a$ is less than $b$ |
| $a \leq b$ | $a$ is less than or equal to $b$ |
| $a > b$ | $a$ is greater than $b$ |
| $a \geq b$ | $a$ is greater than or equal to $b$ |

**EXAMPLE 8** The following use the inequality symbols.

**(a)** $6 < 9$ **(b)** $-10 \leq 3$ **(c)** $12 \leq 12$

**(d)** $8 > 7.9$ **(e)** $2 \geq -8$ **(f)** $13 \geq 13$

Inequalities can be pictured on the number line. For example, $-3 < 1$ can be viewed as $-3$ *is to the left of* 1, as shown at the top of the next page.

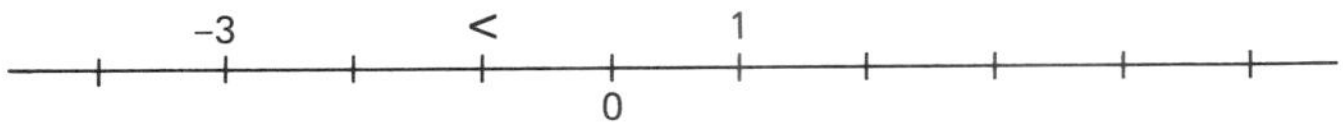


We can graph sets of real numbers on the number line.

**EXAMPLE 9** Graph the set $\{x \mid x \leq 6\}$.

**SOLUTION** $\{x \mid x \leq 6\}$ means the set of real numbers to the left of or equal to 6. We graph this as shown. (We put a heavy dot on the 6 since it is in the set.)

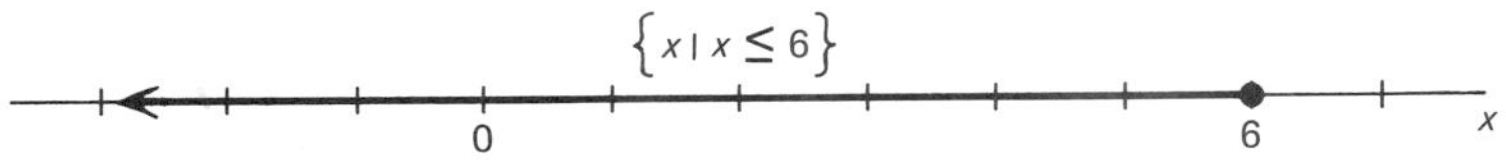


**EXAMPLE 10** Graph the set $\{x \mid x > -2\}$.

**SOLUTION** $\{x \mid x > -2\}$ means the set of real numbers to the right of (but not equal to) $-2$. When graphing, we use an open circle for $-2$ to show that it is not in the set.

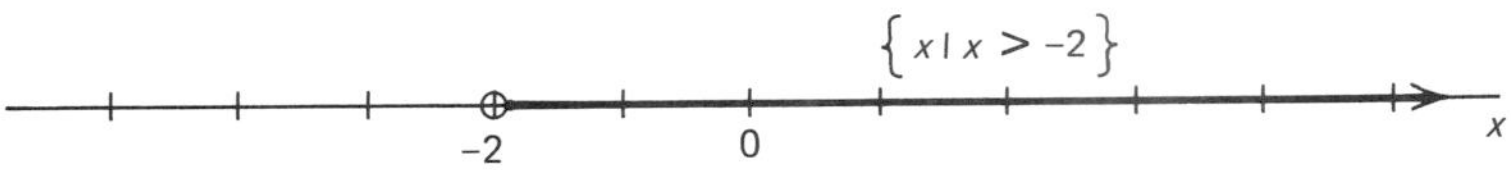


**EXAMPLE 11** Graph the set $\{x \mid -1 < x < 8\}$.

**SOLUTION** $\{x \mid -1 < x < 8\}$ means the set of real numbers to the right of $-1$ and to the left of 8; in other words, $x$ is between $-1$ and 8.

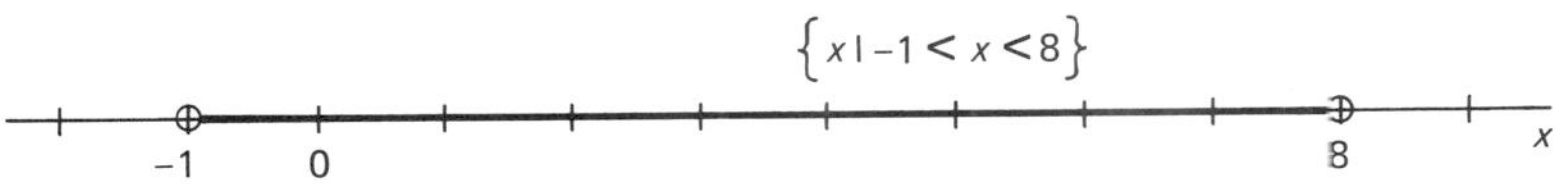


We finally state two properties about inequalities.

> For any real numbers $a$, $b$, and $c$:
>
> **Trichotomy law:** Exactly one of the following is true: $a < b$, $a = b$, or $a > b$.
>
> **Transitive law:** If $a < b$ and $b < c$, then $a < c$.

**EXAMPLE 12** **(a)** Either $x < 10$, $x = 10$, or $x > 10$. (trichotomy law)

**(b)** If $2 < x$ and $x < y$, then $2 < y$. (transitive law)

## PROBLEM SET 1.2

Indicate whether the following statements are true or false.

*See Example 5*

1. $\frac{2}{3} \in Q$
2. $-8 \in R$
3. $2.9 \in I$
4. $0 \in N$
5. $-8.02 \in Q$
6. $6.12 \in H$
7. $18 \in W$
8. $1.2493726\ldots \in R$
9. $0.833333\ldots \in Q$
10. $-6 \in W$
11. $\sqrt{2} \in Q$
12. $-65 \in I$

Estimate the coordinates of the points on the following number line.

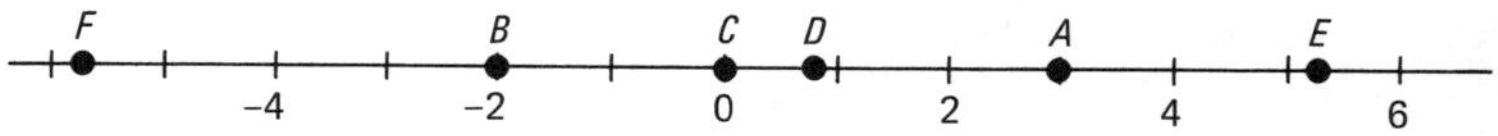


*See Example 6*

13. Point $A$
14. Point $B$
15. Point $C$
16. Point $D$
17. Point $E$
18. Point $F$

On a number line, locate the following points.

*See Example 6*

19. $M = 10$
20. $N = -7$
21. $P = 3\frac{3}{8}$
22. $Q = -4.2$
23. $R = 0.1$
24. $\pi = 3.1415\ldots$

State the equality or inequality law that makes the following statements true.

*See Examples 7 and 12*

25. If $100 < T$ and $T < R$, then $100 < R$.
26. If $K = a + b$, then $a + b = K$.
27. Either $x = 200$, $x < 200$, or $x > 200$.
28. If $x = 2$ and $y = x^2 + 7$, then $y = 2^2 + 7$.
29. If $a = b$ and $b = p + q$, then $a = p + q$.
30. $12 = 12$.

Indicate which of the following statements are true and which are false.

*See Example 8*

31. $6 < 9$
32. $-10 \le 3$
33. $0 > 7$
34. $41 \le 41$
35. $-11 > -8$
36. $14 \ge -7$

37. $-8 \leq 0$
38. $15 \neq -15$
39. $16 \geq 16$
40. $-4 \leq -7$
41. $0 \geq -3$
42. $-5 < -2$

Graph the following inequality sets.

*See Examples 9 and 10*

43. $\{x \mid x < 7\}$
44. $\{x \mid x \leq 5\}$
45. $\{x \mid x > 10\}$
46. $\{x \mid x < -3\}$
47. $\{x \mid x \geq -6\}$
48. $\{x \mid x \leq -8\}$
49. $\{x \mid 3 < x < 8\}$
50. $\{x \mid 4 \leq x \leq 10\}$
51. $\{x \mid -2 \leq x \leq 12\}$
52. $\{x \mid -5 < x < 5\}$
53. $\{x \mid 3 < x \leq 6\}$
54. $\{x \mid -4 \leq x < 2\}$

*Meterology application*

55. Let $e$ be the actual vapor pressure, $E$ the saturation vapor pressure in air, and $E_w$ the saturation vapor pressure in water. If $e < E$ and $E < E_w$, what is the relation between $e$ and $E_w$?

*Engineering application*

56. In the study of materials under stress, it is known that a column will support a weight if

$$\frac{P/A}{\sigma_P} + \frac{M/Z}{\sigma_M} \leq 1$$

Substitute the values $P/A = 6800$, $M/Z = 9000$, $\sigma_P = 15{,}000$, and $\sigma_M = 20{,}000$ to see if the column will support this load.

*Health application*

57. A certain adult male needs at least 70 grams of protein but less than 2000 calories per day. A dietician wants to use hamburger, bread, peanut butter, and milk to fulfill these needs, according to the following relations:

$$\text{Protein:} \quad 7h + 2b + 5p + 8m \geq 70$$

$$\text{Calories:} \quad 80h + 55b + 100p + 160m \leq 2000$$

Substitute the following values into the relations to see if both requirements are met: $h = 5$ ounces of hamburger, $b = 4$ slices of bread, $p = 2$ tablespoons of peanut butter, and $m = 3$ cups of milk.

*Business application*

58. ABC Motion Co. produces cars, trucks, and locomotives. The company is limited in the amount of steel and labor it can put intc these and uses the following relations:

$$\text{Steel:} \quad 1.5C + 8T + 20L \leq 150$$

$$\text{Labor:} \quad 100C + 500T + 1500L \leq 9000$$

Substitute the following values into the relations to see if these conditions are met: $C = 50$ cars, $T = 4$ trucks, and $L = 2$ locomotives.

## 1.3 PROPERTIES OF REAL NUMBERS

The two basic operations on the real numbers are addition and multiplication (subtraction and division are really only the inverse operations of addition and multiplication). Following is a list of the 11 properties of the basic operations on real numbers.

For any real numbers $a$, $b$, and $c$:

**Closure properties:** $a + b$ is a real number
$a \cdot b$ is a real number

**Commutative properties:** $a + b = b + a$
$a \cdot b = b \cdot a$

**Associative properties:** $(a + b) + c = a + (b + c)$
$(a \cdot b) \cdot c = a \cdot (b \cdot c)$

**Identity properties:** There are numbers 0 and 1 such that
$a + 0 = 0 + a = a$
$a \cdot 1 = 1 \cdot a = a$

**Inverse properties:** For every real number $a$, there is a number $(-a)$ such that $a + (-a) = 0$.
For every real number $a$ (except 0), there is a number $\frac{1}{a}$ such that $a \cdot \frac{1}{a} = 1$.

**Distributive property:** $a \cdot (b + c) = a \cdot b + a \cdot c$

Since this text is aimed at teaching algebraic skills, we will not prove any of the standard theorems of algebra but simply state them.

***THEOREM 1.1*** ***(Addition Property of Equality)*** For any real numbers $a$, $b$, and $c$:

If $a = b$, then $a + c = b + c$

| |
|---|
| ***THEOREM 1.2*** **(*Multiplication Property of Equality*)** For any real numbers $a$, $b$, and $c$:<br><br>If $a = b$, then $a \cdot c = b \cdot c$ |

| |
|---|
| ***THEOREM 1.3*** **(*Zero Multiplier Theorem*)** For any real number $a$:<br><br>$a \cdot 0 = 0$ |

| |
|---|
| ***THEOREM 1.4*** **(*Zero Product Theorem*)** For any real numbers $a$ and $b$:<br><br>If $a \cdot b = 0$, then either $a = 0$, $b = 0$, or both. |

| |
|---|
| ***THEOREM 1.5*** **(*Double Negative Theorem*)** For any real number $a$:<br><br>$-(-a) = a$ |

**EXAMPLE 13**

**(a)** $7 + 0 = 7$ (identity property)

**(b)** $8(a + 4) = 8a + 32$ (distributive property)

**(c)** $12 \cdot 0 = 0$ (zero multiplier theorem)

**(d)** $6 \cdot x$ is a real number. (closure property)

**(e)** $-(-2) = 2$ (double negative theorem)

**(f)** $6 \cdot 3 = 3 \cdot 6$ (commutative property)

**(g)** $4 + (10 + 2) = (4 + 10) + 2$ (associate property)

**(h)** If $a = b$, then $a + 5 = b + 5$ (addition property of equality)

**(i)** $6 \cdot (\frac{1}{6}) = 1$ (inverse property)

The inverse properties are very important. The number $-a$ is called the **opposite** of $a$. As examples, $-6$ is the opposite of 6, and 12 is the opposite of $-12$. Also, the number $\frac{1}{a}$ is called the **reciprocal** of $a$. As examples, $\frac{1}{2}$ is the reciprocal of 2, and $\frac{3}{5}$ is the reciprocal of $\frac{5}{3}$.

**EXAMPLE 14** The following table shows opposites and reciprocals.

| NUMBER | OPPOSITE | RECIPROCAL |
|---|---|---|
| $2$ | $-2$ | $\frac{1}{2}$ |
| $\frac{2}{3}$ | $-\frac{2}{3}$ | $\frac{3}{2}$ |
| $-7$ | $7$ | $\frac{-1}{7}$ |
| $0$ | $0$ | (none) |
| $-\frac{4}{9}$ | $\frac{4}{9}$ | $-\frac{9}{4}$ |

## PROBLEM SET 1.3

Tell the property of the real numbers that makes the following statements true. (All letters are real numbers.)

*See Example 13*

1. $7 \cdot a = a \cdot 7$
2. $0 + x = x$
3. If $x = y$, then $8x = 8y$.
4. $\left(\frac{a}{b}\right) \cdot \left(\frac{b}{a}\right) = 1$
5. $5 \cdot (4 \cdot 3) = (5 \cdot 4) \cdot 3$
6. $12 \cdot y$ is a real number.
7. $3 + (2 + 9) = (3 + 2) + 9$
8. $-(-3) = 3$
9. $x \cdot 0 = 0$
10. $6 + x = x + 6$
11. $4 + t$ is a real number.
12. $14 + (-14) = 0$
13. $3 \cdot 1 = 3$
14. $3(a + 7) = 3a + 21$
15. If $(a - 5) \cdot (a - 8) = 0$, then either $a - 5 = 0$ or $a - 8 = 0$.
16. If $x^2 + x = 2$, then $x^2 + x + (-2) = 2 + (-2)$.
17. $a + (b + c) = (b + c) + a$
18. $x + y + (-x - y) = 0$
19. $5 \cdot 6 \cdot 7 \cdot 8 \cdot 9 \cdot 10 \cdot 11 \cdot 0 = 0$
20. $8 \cdot (x - 3y) = 8x - 24y$

Use the property given in parentheses to fill in the blank and make the statement true. (All letters are real numbers.)

*See Example 13*

21. $6 \cdot x =$ ______ $\cdot 6$ (commutative property)
22. $8 +$ ______ $= 8$ (identity property)

23. If $m = n$, then $m + 10 = n +$ _______ (addition property of equality)

24. $\frac{2}{3} \cdot$ _______ $= 1$ (inverse property)

25. $8 \cdot (2 \cdot 5) = ($_______ $\cdot 2) \cdot 5$ (associative property)

26. $18 + x$ is a _______ (closure property)

27. $4 + (6 + 3) = (4 + 6) +$ _______ (associate property)

28. $-(-12) =$ _______ (double negative theorem)

29. $(x^2 + x + 12) \cdot 0 =$ _______ (zero multiplier theorem)

30. $a + 11 = 11 +$ _______ (commutative property)

31. $14 \cdot t$ is a _______ (closure property)

32. $8 +$ _______ $= 0$ (inverse property)

33. $p \cdot$ _______ $= p$ (identity property)

34. $7(k + 9) = 7k +$ _______ (distributive law)

35. If $(x - 2)(x + 3) = 0$, then either $x - 2 = 0$ or _______ $= 0$ (zero product theorem)

36. If $p = q$, then $8p =$ _______ $\cdot q$ (multiplication property of equality)

37. $8x + 16y =$ _______ $\cdot (x + 2y)$ (distributive property)

38. $a - b +$ _______ $= 0$ (inverse property)

Complete the following table.

*See Example 14*

| | Number | Opposite | Reciprocal |
|---|---|---|---|
| 39. | $8$ | | |
| 40. | $-6$ | | |
| 41. | $\frac{2}{5}$ | | |
| 42. | $\frac{-1}{2}$ | | |
| 43. | | $3$ | |
| 44. | | $-9$ | |
| 45. | | | $\frac{1}{3}$ |
| 46. | | | $5$ |

## 1.4 ADDITION AND SUBTRACTION OF REAL NUMBERS

Before we discuss the operations on the real numbers, let us define the **absolute value.**

**DEFINITION 1.1** For any real number $a$,

$$|a| = \begin{cases} a & \text{if } a \geq 0 \\ -a & \text{if } a < 0 \end{cases}$$

This means that if $a$ is positive, leave it positive to compute $|a|$; if $a$ is negative, change it to a positive number to compute $|a|$.

Is $-a$ a negative number? Not necessarily. If $a = -5$, then $-a = -(-5) = 5$. Thus, $-a$ can be positive.

**EXAMPLE 15** **(a)** $|6| = 6$ **(b)** $|-3| = 3$ **(c)** $|0| = 0$

**EXAMPLE 16** **(a)** $-|10| = -10$ **(b)** $-|-1| = -1$ **(c)** $-|0| = 0$

Notice in Example 16 that we first found the absolute value and then took the negative of the number. We can also picture the absolute value as the distance from the point to the origin on the number line (Fig. 1.5).

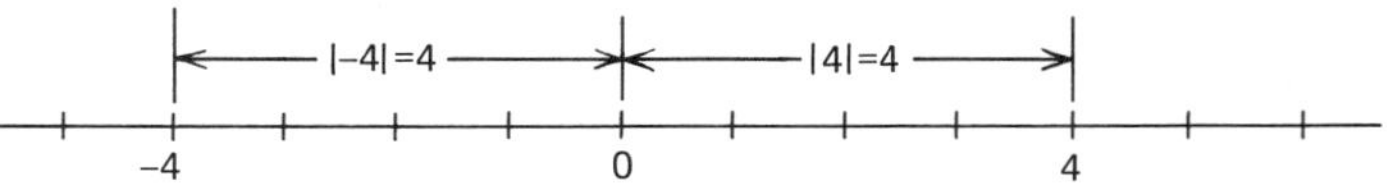


**Figure 1.5**

We are now ready to discuss addition. We have two cases when we add real numbers.

### Same Signs

Notice on the number lines in Fig. 1.6 what happens when we add numbers of the same sign.

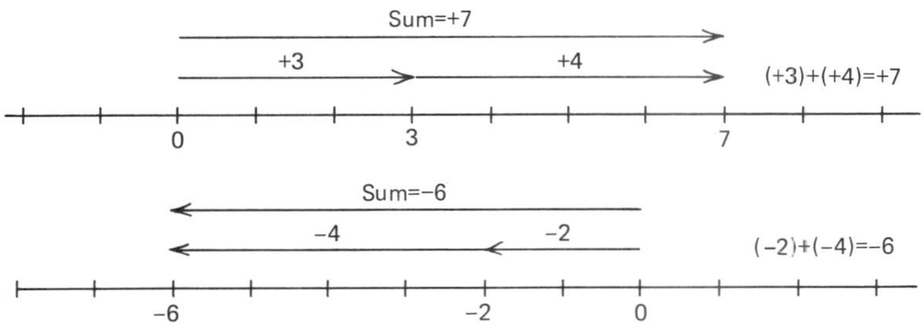


**Figure 1.6**

> To add real numbers with the same sign:
>
> 1. Add the absolute values.
> 2. Give the answer the common sign.

**EXAMPLE 17** The following real numbers have the same sign, so we add the absolute value and give the answer the common sign.

**(a)** $(+10) + (+6) = +16$

**(b)** $(+4) + (+1) = +5$

**(c)** $(-5) + (-6) = -11$

**(d)** $(-20) + (-30) = -50$

## Opposite Signs

Notice on the number line in Fig. 1.7 what happens when we add numbers of opposite signs (one positive, one negative).

> To add real numbers of opposite signs:
>
> 1. Subtract their absolute values (larger − smaller).
> 2. Give the answer the sign of the number with the larger absolute value.

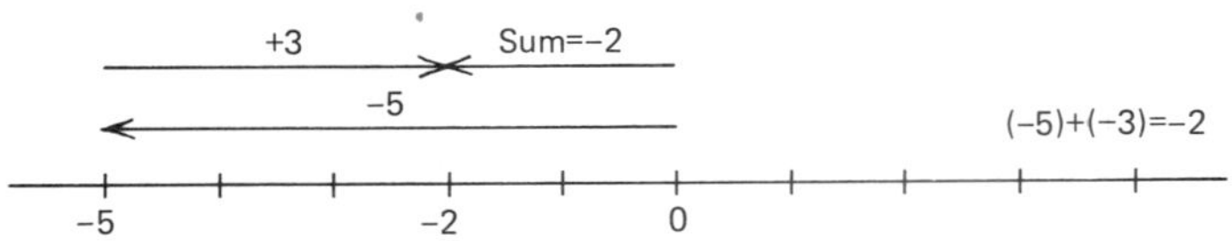


**Figure 1.7**

**EXAMPLE 18** The following real numbers have opposite signs, so we subtract the absolute values and give the answer the sign of the larger absolute value.

**(a)** $(-10) + (+6) = -4$

**(b)** $(-2) + (+8) = +6$

**(c)** $(+12) + (-3) = +9$

**(d)** $(+5) + (-11) = -6$

We see from Example 18(c) that $12 + (-3) = 9$ just as $12 - 3 = 9$. This suggests that subtracting 3 is the same as adding $-3$. In other words, we add the opposite when subtracting.

**DEFINITION 1.2** For any real numbers $a$ and $b$,

$$a - b = a + (-b)$$

To subtract real numbers:

1. Replace the number to be subtracted by its opposite. (Change the sign of the number to be subtracted.)
2. Add, using the rules to add real numbers.

**EXAMPLE 19** In the following cases, we change subtraction to addition of the opposite.

**(a)** $4 - 9 = 4 + (-9) = -5$

**(b)** $-3 - 16 = -3 + (-16) = -19$

**(c)** $-5 - (-3) = -5 + (+3) = -2$

To simplify a sequence of additions and subtractions:

1. Change all subtractions to additions of opposites.
2. Group all the positives together.
3. Group all the negatives together.
4. Simplify this.

**EXAMPLE 20** Simplify $4 - (-3) - 5 + (-9) + 8 - (-10) + (-2) - 1$.

**SOLUTION** We first change all the subtractions to additions.

| | |
|---|---|
| Change all subtractions to additions | $4 + (+3) + (-5) + (-9) + 8 + (+10) + (-2) + (-1)$ |
| Group all positives, negatives | $= (4 + 3 + 8 + 10) + [(-5) + (-9) + (-2) + (-1)]$ |
| Simplify | $= 25 + (-17) = 8$ |

## PROBLEM SET 1.4

Simplify the following absolute values.

*See Example 15*

1. $|10|$ 2. $|-8|$ 3. $|-2.5|$
4. $|0|$ 5. $|6.9|$ 6. $|-1.01|$

*See Example 16*

7. $-|7|$ 8. $-|-3|$ 9. $-(-|6|)$
10. $-(-|-9|)$ 11. $-(-|0|)$ 12. $-(-|-4|)$

Add the following real numbers.

*See Example 17*

13. $(+4) + (+3)$ 14. $(+10) + (+13.2)$ 15. $(+8.1) + (10.7)$
16. $(-9) + (-6)$ 17. $(-12) + (-8.6)$ 18. $(-9.3) + (-7.9)$

*See Example 18*

19. $3 + (-2)$ 20. $8 + (-11)$ 21. $14 + (-10)$
22. $-14 + 9$ 23. $-8 + 13$ 24. $-12 + 20$

*See Examples 17 and 18*

25. $4 + (-6)$ 26. $-5 + (-8)$ 27. $-11 + 2$
28. $-15 + (-12)$ 29. $3 + (-10)$ 30. $-19 + 6$

Subtract the following real numbers.

*See Example 19*

31. $17 - 3$ 32. $14 - 19$ 33. $21 - (-17)$
34. $-4 - 6$ 35. $-12 - (-8)$ 36. $11 - (-3)$

37. $15 - (-10)$ 38. $-7 - 18$ 39. $-5 - (-11)$

40. $8 - 21$ 41. $-9 - 3$ 42. $-9 - (-2)$

Simplify each of the following to a single real number.

*See Example 20*

43. $-8 - 6 + 3 - (-5) - 7 + (-10)$
44. $-10 + (-9) - 8 - (-7) - 6 + 5 - 4 - (-3)$
45. $12 - (-3) + 7 - 2 - (-8) - 11 + 6 + (-5) - 1$
46. $-11 - 10 + 16 - (-3) - 4 - 5 + 7 + (-8) - 15$
47. $17 - 21 - 15 - 3 - (-7) - (-9) + (-8) - 6 + 2$
48. $-6 - 19 - (-23) - (-14) - 6 + 15 + (-17) - 8 + 3 - (-10)$

*Business applications*

In accounting, we have the relation

$$\textit{Net profit} = \textit{Sales} - \textit{Variable costs} - \textit{Fixed costs}$$

Complete the following table for net profit.

| | Sales | Variable Costs | Fixed Costs | Net Profit |
|---|---|---|---|---|
| 49. | 100,000 | 40,000 | 20,000 | ? |
| 50. | 60,000 | 32,000 | 33,000 | ? |
| 51. | 250,000 | 140,000 | 110,000 | ? |
| 52. | 93,000 | 65,000 | 46,000 | ? |

In business, we also have the formula

$$\textit{Change in profit} = \textit{Change in revenue} - \textit{Change in costs}$$

Complete the following table.

| | Change in Revenue | Change in Cost | Change in Profit |
|---|---|---|---|
| 53. | 10,000 | 12,000 | ? |
| 54. | 25,000 | 18,000 | ? |
| 55. | −11,000 | −8,000 | ? |
| 56. | −52,000 | −73,000 | ? |

# 1.5 MULTIPLICATION AND DIVISION OF REAL NUMBERS

Multiplication and division of real numbers are fairly easy to do.

> To multiply (or divide) two real numbers:
>
> 1. Multiply (or divide) the absolute values.
>
> 2a. If numbers have the same sign, the answer is positive.
>
> 2b. If numbers have opposite signs, the answer is negative.

**EXAMPLE 21** In the following multiplication examples, we simply multiply the absolute values, and make the answers + if the signs are the same, and − if the signs are different.

Same sign means +

Opposite signs mean −

**(a)** $4 \cdot 5 = 20$

**(b)** $(-3) \cdot (-7) = 21$

**(c)** $2 \cdot (-9) = -18$

**(d)** $(-8) \cdot 7 = -56$

**EXAMPLE 22** Division is just like multiplication as far as the signs are concerned.

Same sign means +

Opposite signs mean −

**(a)** $\frac{28}{7} = 4$

**(b)** $\frac{-36}{-4} = 9$

**(c)** $\frac{-16}{2} = -8$

**(d)** $\frac{20}{-10} = -2$

**EXAMPLE 23** Simplify $\frac{(2)(-6)(-10)}{(-4)(5)}$.

**SOLUTION** To simplify, we simply multiply and divide everything in pairs, according to the rules just given.

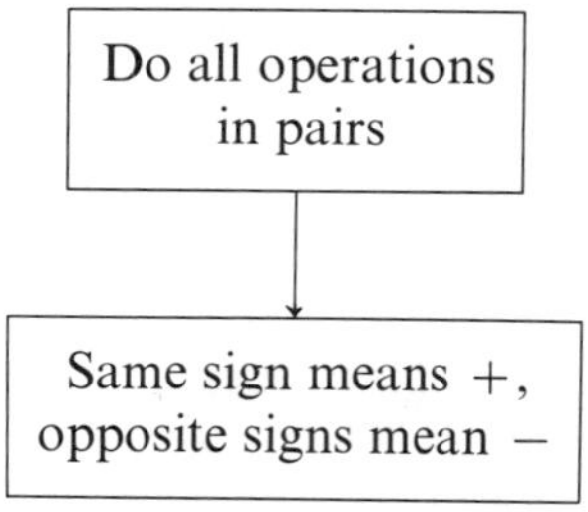


$$\frac{(2)(-6)(-10)}{(-4)(5)} = \frac{(-12)(-10)}{-20}$$

$$= \frac{120}{-20}$$

$$= -6$$

Exponents can be thought of as "repeated multiplication."

> **DEFINITION 1.3** For any real number $b$ and natural number $n$,
>
> $$b^n = \underbrace{b \cdot b \cdot b \cdot \ldots \cdot b}_{n \text{ factors}}$$

In the expression $b^n$, we call $b$ the **base** and $n$ the **power** or **exponent.** For example,

$$5^2 = 5 \cdot 5 = 25 \qquad \text{and} \qquad 2^5 = 2 \cdot 2 \cdot 2 \cdot 2 \cdot 2 = 32$$

This shows us that $b^n$ is not (usually) equal to $n^b$. In Chapter 6, we study zero, negative, and fractional exponents.

**EXAMPLE 24** **(a)** $a^6 = a \cdot a \cdot a \cdot a \cdot a \cdot a$

**(b)** $(1.05)^3 = (1.05) \cdot (1.05) \cdot (1.05) = 1.157625$

**(c)** $(\frac{1}{2})^4 = \frac{1}{2} \cdot \frac{1}{2} \cdot \frac{1}{2} \cdot \frac{1}{2} = \frac{1}{16}$

**(d)** $(-6)^2 = (-6) \cdot (-6) = 36$

**(e)** $(x + y)^5 = (x + y) \cdot (x + y) \cdot (x + y) \cdot (x + y) \cdot (x + y)$

**(f)** $(-3)^3 = (-3) \cdot (-3) \cdot (-3) = -27$

| YES | NO |
|---|---|
| $-6^2 = -36$ | ~~$-6^2 = 36$~~ |
| $(-6)^2 = (-6)(-6) = 36$ | ~~$(-6)^2 = -36$~~ |

A negative sign within the parentheses is a part of the base. When there are no parentheses, the negative sign is *not* a part of the base.

**EXAMPLE 25** Compute $(1.06)^7$ on a hand calculator.

**SOLUTION** This would be very difficult to do without a hand calculator. The obvious method is just to multiply 1.06 as a factor seven times. Here is an easier trick that works on most calculators:

We enter 1.06, punch [×], and punch [=] 6 times.

| PUNCH | DISPLAY | MEANING |
|---|---|---|
| [C] | 0. | Clear |
| [1] [·] [0] [6] | 1.06 | Enter base, 1.06 |
| [×] | 1.06 | Start multiplying |
| [=] | 1.1236 | $(1.06)^2$ |
| [=] | 1.191016 | $(1.06)^3$ |
| [=] | 1.262477 | $(1.06)^4$ |
| [=] | 1.3382256 | $(1.06)^5$ |
| [=] | 1.4185191 | $(1.06)^6$ |
| [=] | 1.5036303 | $(1.06)^7$ = answer |

> To find $B^n$ on most hand calculators:
>
> 1. Enter $B$.
> 2. Punch [×].
> 3. Punch [=] $(n - 1)$ times.

For calculators with a [$y^x$] key:

| PUNCH | DISPLAY | MEANING |
|---|---|---|
| [C] | 0. | Clear |
| [1] [·] [0] [6] | 1.06 | Enter base |
| [$y^x$] [7] | 7. | Enter exponent |
| [=] | 1.5036303 | Answer |

## PROBLEM SET 1.5

Simplifying the following.

*See Example 21*

1. $2(-10)$
2. $(-6)(-5)$
3. $3 \cdot 7$
4. $(-7)(8)$
5. $(-10)(-9)$
6. $15(-14)$
7. $(2.5)(4.2)$
8. $(-3)11$
9. $(-5)(-3.3)$

10. $\frac{1}{2}\left(-\frac{1}{3}\right)$ 11. $\left(\frac{1}{9}\right)\left(\frac{3}{5}\right)$ 12. $\left(\frac{4}{11}\right)\left(\frac{2}{3}\right)$

*See Example 22*

13. $\frac{26}{13}$ 14. $\frac{-18}{9}$ 15. $\frac{-60}{-10}$

16. $\frac{-20}{2}$ 17. $\frac{32}{-4}$ 18. $\frac{100}{5}$

19. $\frac{-80}{16}$ 20. $\frac{-90}{-15}$ 21. $\frac{200}{-10}$

22. $\frac{5.1}{0.3}$ 23. $\frac{-240}{0.06}$ 24. $\frac{-4000}{-0.005}$

*See Example 23*

25. $\frac{(-6)(-8)}{2}$ 26. $\frac{4(-10)}{-5}$ 27. $\frac{(-11)(-30)(2)}{-6}$

28. $\frac{30}{(-2)(-3)}$ 29. $\frac{-100}{(-5)2(-2)}$

30. $\frac{16}{(-2)(-2)(-2)}$ 31. $\frac{(-1)(-2)(-3)(-4)}{-6}$

32. $\frac{(-8)(-6)}{2(-12)}$ 33. $\frac{7(-4)(-2)}{(-1)(-14)}$

34. $\frac{20(-2)(-3)4}{(-1)(-6)(-10)}$ 35. $\frac{(-100)(-10)}{(-2)(-5)(-2)(-2)}$

36. $\frac{(-20)(-30)(-40)}{(-10)2(50)}$

Express each of the following as a single number.

*See Example 24*

37. $2^4$ 38. $8^2$ 39. $10^3$

40. $3^5$ 41. $(-4)^3$ 42. $(-7)^2$

43. $(-10)^4$ 44. $-(1.7)^2$ 45. $(0.6)^2$

46. $\left(\frac{1}{3}\right)^4$ 47. $\left(\frac{-2}{5}\right)^2$ 48. $\left(\frac{-1}{10}\right)^5$

Write each of the following in exponent form.

*See Example 24*

49. $x \cdot x \cdot x \cdot x \cdot x \cdot x$ 50. $b \cdot b \cdot b \cdot b$

51. $(p + q)(p + q)$ 52. $(r - 2s)(r - 2s)(r - 2s)(r - 2s)$

53. $\left(\frac{1}{a}\right)\left(\frac{1}{a}\right)\left(\frac{1}{a}\right)$ 54. $\left(\frac{m}{n}\right)\left(\frac{m}{n}\right)\left(\frac{m}{n}\right)\left(\frac{m}{n}\right)$

Use a hand calculator to evaluate the following expressions.

*See Example 25*

55. $(1.07)^5$ 56. $(2.6)^3$ 57. $(1.25)^8$
58. $(1.065)^{10}$ 59. $(18.32)^4$ 60. $(4.53)^9$
61. $(298.93)^2$ 62. $(89.477)^4$

*Chemical application*

In chemistry, elements and compounds have valences or oxidation numbers, which combine as signed numbers do. Complete the following table.

| | Element (Valence) | Element (Valence) | Compound | Net Valence |
|---|---|---|---|---|
| | Al (+3) | O (−2) | $Al_2O_3$ | $2(+3) + 3(-2) = 6 + (-6) = 0$ |
| | N (−3) | H (+1) | $NH_4$ | $1(-3) + 4(+1) = -3 + 4 = +1$ |
| 63. | H (+1) | O (−2) | $H_2O$ | ? |
| 64. | N (−3) | O (−2) | $NO_2$ | ? |
| 65. | Na (+1) | Cl (−1) | NaCl | ? |
| 66. | Fe (+3) | O (−2) | $Fe_2O_3$ | ? |
| 67. | Cu (+2) | Cl (−1) | $CuCl_2$ | ? |
| 68. | Cr (+6) | O (−2) | $Cr_2O_7$ | ? |

*Business application*

69. There is a 90% (= 0.9) chance that a student will complete a data-processing training course. If 10 students are in such a program, the chance that all 10 will complete it is $(0.9)^{10}$. Compute this.

*Health application*

70. The chance that a drug will not have a harmful side effect on a patient is 98% (= 0.98). If the drug is given to 20 patients, the chance that none of them will experience a harmful side effect is $(0.98)^{20}$. Compute this.

## 1.6 ORDER OF OPERATIONS

Suppose that we are confronted by an expression such as

$$3 - 4(2 + (3 + 1) \cdot 5) - 8 \cdot (6 - 2)^3$$

There is a standard procedure and a simple memory device to help remember it. Consider the sentence, "Please excuse my dear Aunt Sally." Look at the first letters.

| | | |
|---|---|---|
| 1. **P**lease | → | 1. **P**arentheses[1] (and fraction bars) |
| 2. **E**xcuse | → | 2. **E**xponents |
| 3. **M**y | → | 3. **M**ultiplication } |
| 4. **D**ear | → | 4. **D**ivision } |
| 5. **A**unt | → | 5. **A**ddition } |
| 6. **S**ally | → | 6. **S**ubtraction } |

To simplify any numerical expression:

1. Work all expressions within parentheses (inside to outside) and above or below fraction bars.
2. Do all exponents.
3. Do all multiplication and division, left to right.
4. Do all addition and subtraction, left to right.
   Skip any step that is not needed.

**EXAMPLE 26** Simplify $2 + 6 \cdot 3$.

**SOLUTION** We work these in order of priority.

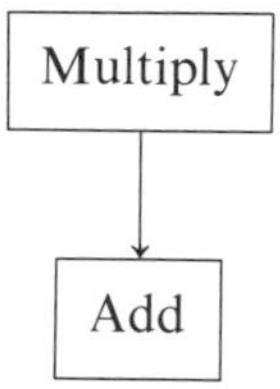


$$2 + 6 \cdot 3 = 2 + 18$$

$$= 20$$

| YES | NO |
|---|---|
| $2 + 6 \cdot 3 = 2 + 18 = 20$ | ~~$2 + 6 \cdot 3 = 8 \cdot 3 = 24$~~ |

We multiply first; then add.

[1] By parentheses, we will mean any symbols of inclusion, such as parentheses ( ), brackets [ ], and braces { }.

**EXAMPLE 27** Simplify $225 - 3 \cdot 7^2$.

**SOLUTION** We work these in order of priority.

| | |
|---|---|
| Do exponent ↓ | $225 - 3 \cdot 7^2 = 225 - 3 \cdot 49$ |
| Multiply ↓ | $= 225 - 147$ |
| Subtract | $= 78$ |

**EXAMPLE 28** Simplify $(3 + 4) \cdot 6 + 4 \cdot (5 - 3)^4$.

**SOLUTION** Here we do the operations within the parentheses first.

| | |
|---|---|
| Parentheses first ↓ | $(3 + 4) \cdot 6 + 4 \cdot (5 - 3)^4 = 7 \cdot 6 + 4 \cdot 2^4$ |
| Do exponent ↓ | $= 7 \cdot 6 + 4 \cdot 16$ |
| Multiply ↓ | $= 42 + 64$ |
| Add | $= 106$ |

Notice that we carefully simplify, using the priority rules.

**EXAMPLE 29** Simplify $\dfrac{2 + 5^2}{3^2 - 6} - \dfrac{6^2 - 3}{3 \cdot 2^2 - 1}$.

**SOLUTION** Here we have fraction bars that act like parentheses. Therefore, we simplify the expressions above and below the bars, first using the usual priorities: exponents first, and so on.

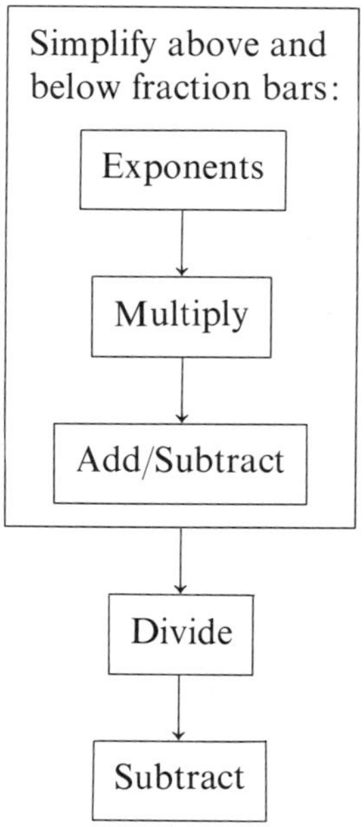


$$\frac{2+5^2}{3^2-6}-\frac{6^2-3}{3\cdot 2^2-1}=\frac{2+25}{9-6}-\frac{36-3}{3\cdot 4-1}$$

$$=\frac{2+25}{9-6}-\frac{36-3}{12-1}$$

$$=\frac{27}{3}-\frac{33}{11}$$

$$=9-3$$

$$=6$$

**EXAMPLE 30** Simplify $3-4\cdot[2+(5-1)^2\cdot 6]$.

**SOLUTION** Here we have parentheses and brackets. We simplify the inner parentheses first, then the outer.

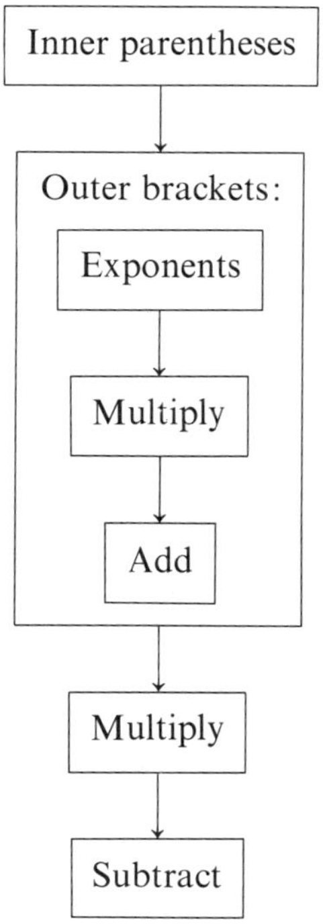


$$3-4\cdot[2+(5-1)^2\cdot 6]=3-4\cdot[2+4^2\cdot 6]$$

$$=3-4\cdot[2+16\cdot 6]$$

$$=3-4\cdot[2+96]$$

$$=3-4\cdot 98$$

$$=3-392$$

$$=-389$$

## PROBLEM SET 1.6

Simplify each of the following expressions to a single number.

*See Example 26*

1. $10 - 2 \cdot 3$
2. $8 + 4 \cdot 5$
3. $2 \cdot 7 - 3$
4. $5 \cdot 8 - 1$
5. $10 - 3 \cdot 2 - 1$
6. $7 + 4 \cdot 9 - 2$
7. $4 \cdot 10 - 3 \cdot 7$
8. $9 \cdot 3 + 8 \cdot 6 + 4 \cdot 12$

*See Example 27*

9. $6 + 2^3$
10. $25 - 4^2$
11. $2 \cdot 5^3$
12. $4 \cdot 2^5$
13. $3 + 8 \cdot 3^3$
14. $1000 - 6 \cdot 2^4$
15. $6 \cdot 3^2 + 4 \cdot 3 + 1$
16. $2 \cdot 4^3 - 5 \cdot 4^2 + 8 \cdot 4 - 3$

*See Example 28*

17. $8 \cdot (4 + 1)$
18. $7 \cdot (10 - 2)$
19. $2 \cdot (6 + 3)^2$
20. $8 \cdot (9 - 4)^2$
21. $(4 + 7) \cdot 8 - 2 \cdot (1 + 3)^2$
22. $(1 + 5)^2 - 2 \cdot (2 + 3)^3 - 6$
23. $4 \cdot (6 - 1)^2 - 7 \cdot (6 - 1) - 3$
24. $5(10 - 6)^3 - 7(10 - 6)^2 + 4(10 - 6) + 1$

*See Example 29*

25. $\dfrac{6 + 10}{8 - 4}$
26. $\dfrac{19 + 2}{10 - 3}$
27. $\dfrac{2 + 3 \cdot 6}{6 + 2 \cdot 2}$
28. $\dfrac{18 - 3 \cdot 5}{12 + 6 \cdot 3}$
29. $\dfrac{2 - 3 \cdot 4^3 - 10}{2 - 3 \cdot 4^3}$
30. $\dfrac{100 - 2 \cdot 3^3 - 4}{2 \cdot 5 + 8 \cdot 2^2}$
31. $\dfrac{2 + (5 - 1)^3 + 19}{8 + 2 \cdot (2 + 1)^4}$
32. $\dfrac{3 + (9 - 2)^2 + 23}{6 - 3 \cdot (5 - 2)^3}$
33. $\dfrac{8 - 3}{4 + 2} + \dfrac{2 + 1}{3 \cdot 2^2}$
34. $\dfrac{2^3 - 1}{2^4 - 6} - \dfrac{2^2 - 1}{5 \cdot 2^2}$

*See Example 30*

35. $2 + [3 - 2 \cdot (2 + 1)]$
36. $30 - [46 - 4 \cdot (5 + 1)]$
37. $[2 + (8 - 5) \cdot 6] \cdot [3 + 5]$
38. $[4 + 10] \cdot [4 + 6 \cdot (5 - 2)]$
39. $3^2 - [6 + 2 \cdot (2 + 3)^3]$
40. $2^4 + [2 \cdot 3^2 + 5 \cdot (6 - 4)^3]$
41. $2 \cdot 3^2 + [45 - 4 \cdot (6 - 3)^2]^2$
42. $8 \cdot 5^2 + [10^2 - 2 \cdot (5 + 2)^2]^3$
43. $4 + (2 - 3 \cdot [5 + 6 \cdot (3 + 2)^2])$
44. $20^2 - (10 + 6 \cdot [7 - 2 \cdot (8 - 4)^2])$

*Business applications*

45. In a certain industry, the formulas for total cost (TC) and total revenue (TR) are given by

$$\text{TC} = (Q - 60)^3 + 2000Q + 200{,}000$$

$$\text{TR} = 300{,}000 - 20(Q - 20)^2$$

Evaluate TC and TR if $Q = 100$ items produced.

*Health applications*

46. If there is a 30% chance of a person contracting a certain contagious disease, the chance that at least one out of a group of five people will contact it is $1 - (0.7)^5$. Compute this.

47. The rate of cancer deaths in a year is given by

$$R = 1 - \frac{A_0\left[1 + \dfrac{A_x}{A_0 + A_c}\right] + \dfrac{1}{2}D_0\left[1 + \dfrac{D_x + L}{D_0 + D_c}\right]}{A + D + L}$$

Compute this if

$$\begin{aligned}
A &= \text{people alive (total)} = 2573\\
A_0 &= \text{people alive (no cancer)} = 1981\\
A_c &= \text{people live (with cancer)} = 83\\
A_x &= \text{people alive (cancer unknown)} = 509\\
L &= \text{untraced people} = 25\\
D &= \text{people dead (total)} = 112\\
D_0 &= \text{people dead (no cancer)} = 12\\
D_c &= \text{people dead (with cancer)} = 78\\
D_x &= \text{people dead (cancer unknown)} = 13
\end{aligned}$$

*Consumer applications*

48. A car depreciates according to the formula $V = P(0.75)^N$. Compute this value ($V$) if the new price $P = 7000$ and the age $N = 6$ years.

49. A bank account grows according to the formula $B = P(1 + r)^N$. Evaluate the balance $B$ if the original deposit $P = 500$, the rate $r = 0.06$, and the number of years $N = 8$.

*Electronics application*

50. In television technology, we have the formula

$$R_k = \frac{\frac{1}{g_m} \cdot Z_0}{\frac{1}{g_m} - Z_0}$$

Evaluate the cathode resistance $R_k$ if the transconductance $g_m = 0.01$ and the output impedance $Z_0 = 72$.

*Optical applications*

51. The fraction of light reflected at the surface of the lens is

$$R = \frac{(n - 1)^2}{(n + 1)^2}$$

Find the reflection $R$ if the index of refraction for glass $n = 1.5$.

The percentage of light intensity ($I$) away from the center of a camera lens depends on the focal length ($f$) and the distance from the center ($x$). The formula is

$$I = 100 \frac{f^4}{(f^2 + x^2)^2}$$

Evaluate the following.

52. *Normal lens:* $f = 56$ mm; $x = 28$ mm
53. *Wide-angle lens:* $f = 28$ mm; $x = 28$ mm
54. *Telephoto lens:* $f = 135$ mm; $x = 28$ mm

# CHAPTER 1 SUMMARY

## Important Words and Phrases

absolute value (1.4)
addition property of equality (1.3)
algebra (introduction)
associative properties (1.3)
base (1.5)
closure properties (1.3)
commutative properties (1.3)
coordinate (1.2)
descriptive method (1.1)
disjoint (1.1)
distributive property (1.3)
double negative theorem (1.3)

element (1.1)
empty set (1.1)
equal numbers (1.2)
equal sets (1.1)
exponent (1.5)
finite set (1.1)
identity properties (1.3)
inequality (1.2)
infinite set (1.1)
integers (1.2)
intersection (1.1)
inverse properties (1.3)
irrational numbers (1.2)
list method (1.1)
multiplication property of equality (1.3)
natural numbers (1.2)
opposite (1.3)
origin (1.2)
power (1.5)
rational numbers (1.2)
real numbers (1.2)
reciprocal (1.3)
reflexive law (1.2)
set (1.1)
subset (1.1)
substitution law (1.2)
symmetric law (1.2)
transitive law (1.2)
trichotomy law (1.2)
union (1.1)
whole numbers (1.2)
zero multiplier theorem (1.3)
zero product theorem (1.3)

## Important Properties, Theorems, and Definitions

For real numbers $a$, $b$, and $c$:

$a = a$.

If $a = b$, then $b = a$.

If $a = b$ and $b = c$, then $a = c$.

If $a = b$, then $a$ can replace $b$, and vice versa, in a statement without affecting the statement's truth.

Exactly one of the following is true:

$a < b$, $a = b$, or $a > b$.

If $a < b$ and $b < c$, then $a < c$.

| | |
|---|---|
| $a + b$ is a real number | $a \cdot b$ is a real number |
| $a + b = b + a$ | $a \cdot b = b \cdot a$ |
| $(a + b) + c = a + (b + c)$ | $(a \cdot b) \cdot c = a \cdot (b \cdot c)$ |
| $a + 0 = 0 + a = a$ | $a \cdot 1 = 1 \cdot a = a$ |
| $a + (-a) = 0$ | $a \cdot \frac{1}{a} = 1$ |

$a \cdot (b + c) = a \cdot b + a \cdot c$

If $a = b$, then $a + c = b + c$.

If $a = b$, then $a \cdot c = b \cdot c$.

$a \cdot 0 = 0$

If $a \cdot b = 0$, then either $a = 0$, $b = 0$, or both.

$-(-a) = a$

$$|a| = \begin{cases} a & \text{if } a \geq 0 \\ -a & \text{if } a < 0 \end{cases}$$

$b^n = b \cdot b \cdot \ldots \cdot b$, with $n$ factors

## Important Procedures

*To subtract real numbers:*

1. Replace the number to be subtracted by its opposite (change the sign of the number to be subtracted).
2. Add, using the rules for signed numbers.

*To simplify a sequence of additions and subtractions:*

1. Change all subtractions to additions.
2. Group all positives together.
3. Group all negatives together.
4. Simplify.

*To simplify a numerical expression:*

1. Work all expressions within parentheses (inside to outside) and above or below fraction bars.
2. Do all exponents.
3. Multiply and divide (left to right).
4. Add and subtract (left to right).
   Use the memory device: **P**lease **E**xcuse **M**y **D**ear **A**unt **S**ally.

## CHAPTER 1 REVIEW EXERCISES

1. Write $A = \{\text{integers between } -4 \text{ and } 5\text{, inclusive}\}$ using the list method.
2. Write $B = \{100, 200, 300, 400, 500\}$ using the description method.

For problems 3 to 6, $A = \{a, b, c\}$ and $B = \{b, c, d, e\}$.

3. Is $b \in A$?
4. Is $e \notin B$?
5. Find $A \cup B$.
6. Find $A \cap B$.

For problems 7 and 8, indicate whether the statement is true or false.

7. $1.2 \in Q$

8. $-7.98473\ldots \in R$

Find the coordinates of the points on the following line.

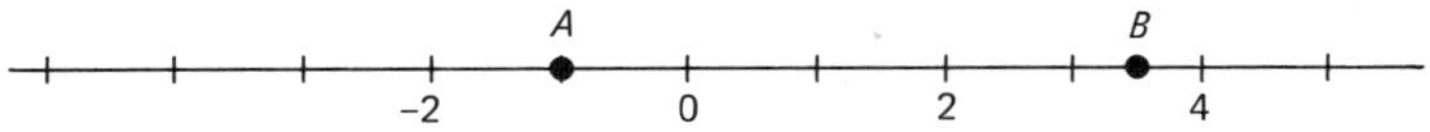


9. Point $A$

10. Point $B$

Locate the following points on the number line.

11. $C = 5\frac{1}{2}$

12. $D = -3$

State whether the following is true or false.

13. $-5 > 0$

14. $18 \leq 18$

15. Graph the set $\{x \mid -4 \leq x \leq 6\}$.

State the property of equality, inequality, or real-number operations that make the following statements true.

16. If $50 < R$ and $R < S$, then $50 < S$.

17. If $x = 5$ and $a = x^2 + x + 1$, then $a = 5^2 + 5 + 1$.

18. $8 \cdot (4 \cdot a) = (8 \cdot 4) \cdot a$

19. $6 + r = r + 6$

20. $8(a + 2b) = 8a + 16b$

21. If $(x + 7)(x - 3) = 0$, then either $x + 7 = 0$ or $x - 3 = 0$.

Use the property in the parentheses to fill in the blank.

22. $\frac{5}{9} \cdot (\underline{\qquad\qquad}) = 1$ (inverse property)

23. $(x^{12} + x^{11} + x^8 + x + 1) \cdot 0 = \underline{\qquad\qquad}$ (zero multiplier theorem)

Simplify each of the following.

24. $|-8.2|$

25. $-|-3|$

26. $-5 + (-8)$

27. $-19 + 6$

28. $-12 - 8$

29. $-9 - (-5)$

30. $12 - (-7) - 5 - 11 + 10 + (-8) - 9 - (-13)$

31. $(-3)10$

32. $(-8)(-10)$

33. $\dfrac{-35}{7}$

34. $\dfrac{-60}{-6}$

35. $\dfrac{(-10)(-4)}{(-2)(-2)(-2)}$

36. $(-5)^3$

37. $10 + 5 \cdot 4$

38. $100 - 2 \cdot 4^3$

39. $(6 + 2)^2 - 5(6 + 2) + 1$

40. $\dfrac{3 + 2^2 \cdot 3^2}{2^4 - 3}$

41. $4 \cdot 3^3 + [8^2 - 3 \cdot (6 - 4)^4]$

42. What is the opposite of 4?

43. What is the opposite of $-8$?

44. What is the reciprocal of 5?

45. What is the reciprocal of $\frac{2}{3}$?

# 2

# LINEAR EQUATIONS AND INEQUALITIES IN ONE VARIABLE

| | |
|---|---|
| $2 + 3 = 5$ | This is a true statement. |
| $3 + 5 = 10$ | This is a false statement. |
| $x + 2 = 6$ | Is this true or false? That depends on $x$. If $x = 4$, it is true; if $x \neq 4$, it is false. |

An equation that is neither true or false is called an **open equation.** Any number that makes an open equation true is called a **solution.** The set of all solutions is called the **solution set.**

In the following three examples, we have written each solution in the two different forms in which solutions may be written.

| EQUATION | SOLUTION | SOLUTION SET |
|---|---|---|
| $x + 2 = 6$ | $x = 4$ | $\{4\}$ |
| $x^2 = 4$ | $x = +2$ or $-2$ | $\{+2, -2\}$ |
| $x = x + 2$ | No solution | $\varnothing$ |

Notice from the third example that an equation may not have a solution at all.

## 2.1 SOLVING LINEAR EQUATIONS

Equations such as $x + 2 = 6$ are easy to solve, but most others are not that simple. Therefore, we need a set of rules and procedures to help us. First, let us state the goal in all equation solving.

> **GOAL** Isolate the unknown.

This means that we want to end up with the unknown by itself on one side of the equation.

To help us do this, we have the following two properties. First, we have the **Addition Property of Equality.**

**Figure 2.1**

> Addition Property of Equality.
>
> Let $a$, $b$, and $c$ be real numbers.
> If $a = b$, then $a + c = b + c$

In words, *we can add the same number to both sides of an equation, and the results are still equal.* Since subtraction is the addition of an opposite, this property also holds for subtraction.

Also, we have the **Multiplication Property of Equality.**

> Multiplication Property of Equality.
>
> Let $a$, $b$, and $c$ be real numbers.
> If $a = b$, then $ac = bc$

In words, *we can multiply both sides of an equation by the same number, and the results are still equal.* Since division is multiplication by a reciprocal, this property also holds for division (by a non-zero number).

The student should realize that an equation is like a balance: whatever we do to one side, we must do to the other side to maintain the balance (Fig. 2.1).

**EXAMPLE 1** Solve $x + 9 = 14$.

**SOLUTION** Since 9 was added to $x$, we "undo" this and isolate $x$ by subtracting 9 from sides.

| | |
|---|---|
| Given | $x + 9 = 14$ |
| ↓ Subtract 9 from both sides | $x + 9 \boxed{-9} = 14 \boxed{-9}$ |
| ↓ Simplify | $x = 5$ |

We can say that the solution is $x = 5$ or that the solution set is $\{5\}$. It is easy to check $5 + 9 = 14$.

**EXAMPLE 2** Solve $2x + 11 = 3$.

**SOLUTION** We isolate $x$ by "undoing" the operations that were done on it.

| | |
|---|---|
| Given | $2x + 11 = 3$ |
| Subtract 11 from both sides | $2x + 11 - 11 = 3 - 11$ |
| Simplify | $2x = -8$ |
| Divide both sides by 2 | $\frac{2x}{2} = \frac{-8}{2}$ |
| Simplify | $x = -4$ |

Sometimes, we have more than one $x$-term in an equation. We must simplify this to one single $x$-term. We do this by using the distributive law. For example,

$$6x - 3x + 8x = (6 - 3 + 8)x = 11x$$

$$4x - 5 - 7x - 3 = 4x - 7x - 5 - 3 = -3x - 8$$

This process of simplifying is called **combining terms.**

**EXAMPLE 3** Solve $7t - 5 = 4t + 7$.

**SOLUTION** Recall that our goal is to isolate $t$. To do this, we must put all the $t$-terms on one side of the equation and all the numbers on the other.

| | |
|---|---|
| Given | $7t - 5 = 4t + 7$ |
| Subtract $4t$ from both sides | $3t - 5 = 7$ |
| Add 5 to both sides | $3t = 12$ |
| Divide both sides by 3 | $t = 4$ |

We can check the solution, $t = 4$, as follows:

$$\text{Left side} = 7(4) - 5 = 28 - 5 = 23$$

$$\text{Right side} = 4(4) + 7 = 16 + 7 = 23$$

It checks, so $t = 4$ is the solution.

**EXAMPLE 4** Solve $5x + 2 + 3x = 7x - 7 - 2x + 5$.

**SOLUTION** We first combine terms, and then isolate $x$.

| | |
|---|---|
| Given | $5x + 2 + 3x = 7x - 7 - 2x + 5$ |
| Combine terms | $8x + 2 = 5x - 2$ |
| Subtract $5x$ from both sides | $3x + 2 = -2$ |
| Subtract 2 from both sides | $3x = -4$ |
| Divide both sides by 3 | $x = \dfrac{-4}{3}$ |

Thus, the solution is the fraction $\dfrac{-4}{3}$. The student should check this solution.

> To solve linear equations for an unknown, such as $x$:
>
> 1. Use the distributive law to remove all parentheses.
> 2. Simplify by combining terms on each side of the equation.
> 3. Move all $x$-terms to one side of the equation.
> 4. Move all number terms to the other side.
> 5. Multiply or divide to isolate $x$.
>
> (*Note:* If any of these steps is not needed, skip it.)

**EXAMPLE 5** Solve $5(x - 5) + 7 = 3(2x + 3) - 4x$.

**SOLUTION** We follow the rules given above.

| | |
|---|---|
| Given | $5(x - 5) + 7 = 3(2x + 3) - 4x$ |
| ↓ Distributive law | $5x - 25 + 7 = 6x + 9 - 4x$ |
| ↓ Combine terms | $5x - 18 = 2x + 9$ |
| ↓ Subtract $2x$ from both sides | $3x - 18 = 9$ |
| ↓ Add 18 to both sides | $3x = 27$ |
| ↓ Divide both sides by 3 | $x = 9$ |

| YES | NO |
|---|---|
| $5(x - 5) + 7 = 5x - 25 + 7$ | ~~$5(x - 5) + 7 = 5x - 25 + 35$~~ |

The distributive law only applies to the terms inside the parentheses.

**EXAMPLE 6** Solve $x + 5 = x + 7$.

**SOLUTION** We start to solve this as usual.

| | |
| --- | --- |
| Given | $x + 5 = x + 7$ |
| Subtract $x$ from both sides | $5 = 7$ ???? |

The equation $5 = 7$ is false. Since we did nothing illegal, this simply means that there is *no solution* to the equation; that is, the solution set is $\varnothing$.

## PROBLEM SET 2.1

Solve the following equations for the unknowns. (*Be careful:* Some equations may not have a solution—see Example 6.)

*See Example 1*

1. $x - 7 = 11$
2. $a - 4.2 = 5.3$
3. $r + 12 = 3$
4. $k + 6.2 = 8.1$
5. $5t = 40$
6. $0.7m = 2.1$
7. $\frac{z}{8} = 9$
8. $\frac{u}{1.5} = 6$

*See Example 2*

9. $2x - 7 = 3$
10. $8a + 2 = 50$
11. $4q + 6 = 2$
12. $10b - 1 = 14$
13. $-3k + 7 = 19$
14. $6m + 13 = 25$
15. $9s - 5 = 14$
16. $-5r - 4 = 23$
17. $-8y + 3 = 16$
18. $14t - 6 = -17$

*See Example 3*

19. $4x - 2 = 8x - 18$
20. $6t + 5 = 9t - 4$
21. $5k - 9 = 2k - 15$
22. $10a + 4 = 3a + 39$
23. $2m + 7 = 4m + 21$
24. $9r - 11 = 8r - 10$
25. $5z - 3 = 5z - 6$
26. $15b + 1 = 7b - 4$
27. $4x - 13 = -7x + 3$
28. $3y - 10 = -3y + 11$

*See Example 4*

29. $8x - 5x - 4 = x + 7 - 5$
30. $10a - 8 + 4a = 7 + 3a + 8a$
31. $4m - 10 - 6m = m - 8 + 1$
32. $2y - 6 - 4y = 5y - 9 - 8y$

33. $8q - 10q - 5 = 5 - q - 6q$

34. $12b - 5b - 6 = 8b + 3 - b$

35. $6k - 7 + 4k = 8k - 7 + k$

36. $4w + 9 - 6w = 5w - 12 + 5$

37. $9t - 3t - 4t = 6 + 5t - t$

38. $7x - x - 2 - 9x = 4x + 7 - x - 2$

*See Example 5*

39. $5(x + 1) + 2 = 8(x - 1) + 3$

40. $6(y + 4) - 9 = 4(2y + 3) + 1$

41. $4(n + 5) - 2n = 5(2n + 7) - 3n$

42. $8(2b + 4) - 3b = 7(b + 8) + 12$

43. $6(q - 4) - 3q = 8 - 4(q + 1)$

44. $8(r + 7) - 12 = 4r - 6(2r - 1)$

45. $8(u + 3) - 5 = 2(4u - 3) + 6$

46. $6(3p - 1) - p = 8 - 2(5 - 2p)$

47. $4(x - 5) - 3(2x + 1) = 8(x + 2) + 3$

48. $5(z - 2) - 4(2z + 3) = 7 - 2(3 - z)$

*Business applications*

49. In Illinois, the income tax that a family pays is given by $T = 0.025(I - 1000E)$. Find the income $I$ if the tax $T = 500$ and the number of exemptions $E = 4$.

50. The cost $C$ to produce a certain line of stereos is given by

$$C = 100{,}000 + 120x$$

If a firm has a cost budget $C = \$550{,}000$, find $x$, the number of stereos that they can produce.

*Physical applications*

51. The formula for an object thrown in the air is $h = vt + \frac{1}{2}at^2$. Find the initial velocity $v$ if the time $t = 3$, the acceleration $a = -32$, and the height $h = 87$.

52. The relation between Celsius and Fahrenheit temperature is given by $C = \frac{5}{9}(F - 32)$. Find the temperature Fahrenheit $F$ if the temperature Celsius $C = 10°$.

*Consumer applications*

53. Judy Morison earned \$11,000 in a year when the Consumer Price Index (CPI) was 134. If the CPI has increased to 180, find the income $x$ that Judy must earn just to keep up with inflation by solving

$$\frac{x}{11{,}000} = \frac{180}{134}$$

54. The true interest of a loan is given approximately by

$$\text{APR} = \frac{24I}{B(n+1)}$$

Find the interest charge $I$ if APR $= 0.135$, the amount borrowed $B = \$2000$, and the number of payments $n = 24$.

*Population application*

55. The population $P$ of a town is growing by the formula $P = 8000 + 700t$. In how many years $t$ will the population $P = 11{,}500$?

*Health application*

56. To get 1 milligram of thiamine, a patient must take $x$ ounces of prune juice and $y$ ounces of orange juice. These satisfy the equation

$$0.05x + 0.08y = 1$$

If the patient gets $x = 4$ ounces of prune juice, how much orange juice does he need?

## 2.2 LITERAL EQUATIONS

A **literal equation** is an open equation with numbers and other letters in addition to the unknown. For example, $ax + b = 5$ is a literal equation since it contains an unknown, $x$, and other letters, $a$ and $b$.

To solve a literal equation:

1. Identify the unknown letter (perhaps with a circle).
2. Use the distributive law to remove all parentheses.
3. Move all terms with the unknown to one side of the equation.
4. Move all terms without the unknown to the other side of the equation.
5. Use the distributive law to combine the unknown terms.
6. Divide or multiply to isolate the unknown.

These rules are almost identical to those of the last sections, except that we have to sort the unknown from the other letters.

**EXAMPLE 7** Solve $ax + b = 5$, for $x$.

**SOLUTION** Since $x$ is the unknown, we move all the other terms to the other side. In this example, we circle $x$ so that we do not lose it.

| | |
|---|---|
| Identify unknown | $a\textcircled{x} + b = 5$ |
| ↓ Subtract $b$ from both sides | $a\textcircled{x} = 5 - b$ |
| ↓ Divide both sides by $a$ | $\textcircled{x} = \dfrac{5 - b}{a}$ |

That is the solution: $x = \dfrac{5 - b}{a}$.

**EXAMPLE 8** Solve $m(n - x) = p(m + x) - 7$, for $x$.

**SOLUTION** We again circle all the $x$-terms so that we do not lose them.

| | |
|---|---|
| Identify unknown | $m(n - \textcircled{x}) = p(m + \textcircled{x}) - 7$ |
| ↓ Distributive law | $mn - m\textcircled{x} = pm + p\textcircled{x} - 7$ |
| ↓ Add $m\textcircled{x}$ to both sides | $mn = pm + p\textcircled{x} + m\textcircled{x} - 7$ |
| ↓ Subtract $pm$ from both sides | $mn - pm = p\textcircled{x} + m\textcircled{x} - 7$ |
| ↓ Add 7 to both sides | $mn - pm + 7 = p\textcircled{x} + m\textcircled{x}$ |
| ↓ Distributive law | $mn - pm + 7 = (p + m)\textcircled{x}$ |
| ↓ Divide both sides by $p + m$ | $\dfrac{mn - pm + 7}{p + m} = \textcircled{x}$ |

Therefore, the solution is $x = \dfrac{mn - pm + 7}{p + m}$.

**EXAMPLE 9** The average, $A$, of three numbers, $p$, $q$, and $r$, is given by the formula

$$A = \tfrac{1}{3}(p + q + r)$$

Solve this for $r$.

**SOLUTION** To isolate $r$ it will be faster to first multiply by 3.

| | |
|---|---|
| Identify unknown | $A = \tfrac{1}{3}(p + q + \textcircled{r})$ |
| ↓ Multiply both sides by 3 | $3A = p - q + \textcircled{r}$ |
| ↓ Subtract $p$ and $q$ from both sides | $3A - p - q = \textcircled{r}$ |

| YES | NO |
|---|---|
| $a\textcircled{x} = b\textcircled{x} + 5$<br>$(a - b)\textcircled{x} = 5$<br>$\textcircled{x} = \dfrac{5}{a - b}$ | ~~$ax = bx - 5$~~<br>~~$x = \dfrac{bx + 5}{a}$~~ |

We must get all $x$-terms on *one* side.

## PROBLEM SET 2.2

Solve the following equations for $x$.

*See Example 7*

1. $ax = 7$
2. $xy = 8$
3. $gx - 4 = 11$
4. $kx + 5 = 9$
5. $px - 8 = q$
6. $mx + 6 = n$
7. $h + kx = j$
8. $rx - s = t$

*See Example 8*

9. $ax = bx - 5$
10. $mx - s = nx + 7$
11. $px - qx = 8 - rx$
12. $xyz - 2x = 10 - uwx$
13. $6(a - x) = 8(b - x)$
14. $7(p - x) = 10(q + x)$

15. $9(2x - m) = 3(n + x) + 2$

16. $4(3x - a) = 9(b - x) + 4x$

17. $5(u - vx) - 6(wx - 5) = 7(yx - 4)$

18. $2(a - bx) - 3(cx + d) = 4(ax - c)$

*See Example 9*

The following formulas come from various fields of study. Solve each for the indicated variable.

*Geometry applications*

19. $C = 2\pi r$, for $r$

20. $A = lw$, for $w$

21. $S = 2\pi r^2 + 2\pi rh$, for $h$

22. $A = \frac{1}{2}h(B + b)$, for $h$

23. $A = \frac{1}{2}h(B + b)$, for $b$

24. $P = 2l + 2w$, for $l$

*Business applications*

25. $I = prt$, for $r$

26. $B = P(1 + r)$, for $r$

27. $C = 10{,}000 + 400x$, for $x$

28. $T = \dfrac{C}{x}$, for $C$

29. $V = 50{,}000 - 4000n$, for $n$

30. $Q + 1000P = 6000$, for $P$

*Physical applications*

31. $E = mc^2$, for $m$

32. $E = \frac{1}{2}mv^2 + mgh$, for $m$

33. $V = iR$, for $i$

34. $pV = nRT$, for $T$

35. $uv = fu + fv$, for $f$

36. $R_1R_2 = RR_1 + RR_2$, for $R_2$

*Health applications*

37. $(F + a)V = (F_0 - F)B$, for $F$

38. $C = \dfrac{W}{70} \cdot A$, for $W$

39. $0.05x + 0.08y = 1$, for $x$

40. $165x + 90y = 300$, for $y$

## 2.3 LINEAR INEQUALITIES

Recall from Chapter 1 that an **inequality** is a statement that one number is less than (or greater than) another number: for example, $5 < 12$. Just as we have open equations, we can have an **open inequality,** which is an inequality that may be true or false, depending on the variable. For example, $x > 6$ is true if $x$ is 6.1, 7, 9.8, 57.3, $100\frac{1}{2}$, and so on. Notice that the open inequality has many (in fact, infinitely many) solutions. Usually, these solution sets are shown as entire portions of the real line, as illustrated in Fig. 2.2.

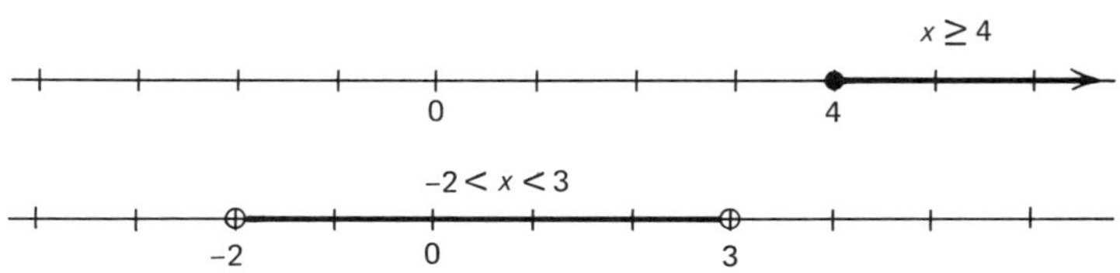


**Figure 2.2**

Solving inequalities is very similar to solving equations. We have the following properties.

> **Addition Property of Inequality.**
>
> Let $a$, $b$ and $c$ be real numbers.
> If $a < b$, then $a + c < b + c$.
> (This relation also holds true for $\leq$, $>$, and $\geq$.)

> **Multiplication Property of Inequality.**
>
> Let $a$, $b$, and $c$ be real numbers.
>
> 1. If $a < b$, and $c$ is positive, then $ac < bc$.
> 2. If $a < b$, and $c$ is negative, then $ac > bc$.
>
> (These relations also hold true for $\leq$, $>$, and $\geq$.)

Notice that multiplying (or dividing) by a *negative* number *reverses* the inequality symbol.

We will not try to prove these statements, but will let the sketches in Fig. 2.3 illustrate the principles.

> To solve linear inequalities:
>
> 1. The goal is to isolate the unknown.
> 2. We use the inequality properties and the same procedures used in solving linear equations (see page 42).
> 3. Remember that multiplying or dividing *by* a *negative* number *reverses* the inequality symbol.

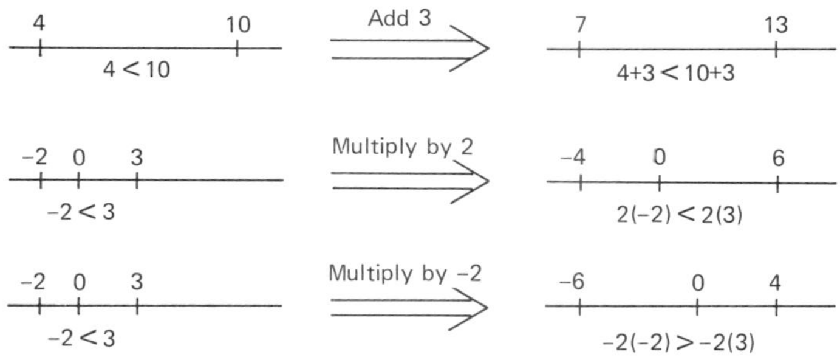


**Figure 2.3**

**EXAMPLE 10** Solve $x + 5 < 12$, and graph the solution set.

**SOLUTION** We solve this as we would an equation.

| | |
|---|---|
| Given | $x + 5 < 12$ |
| ↓ Subtract 5 from both sides | $x < 7$ |

This solution is written $\{x \mid x < 7\}$ and looks as follows.

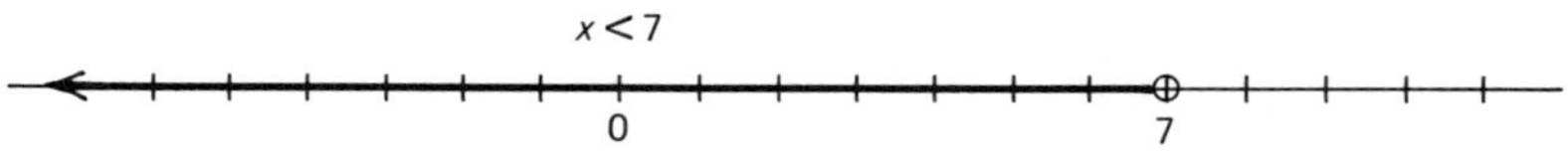


**EXAMPLE 11** Solve $-3k + 6 \geq 21$, and graph the solution set.

**SOLUTION**

| | |
|---|---|
| Given | $-3k + 6 \geq 21$ |
| ↓ Subtract 6 from both sides | $-3k \geq 15$ |
| ↓ Divide both sides by $-3$, reverse inequality sign | $k \leq -5$ |

Notice that dividing by $-3$ changed the $\geq$ to $\leq$. The solution set is $\{k \mid k \leq -5\}$ and is graphed as follows.

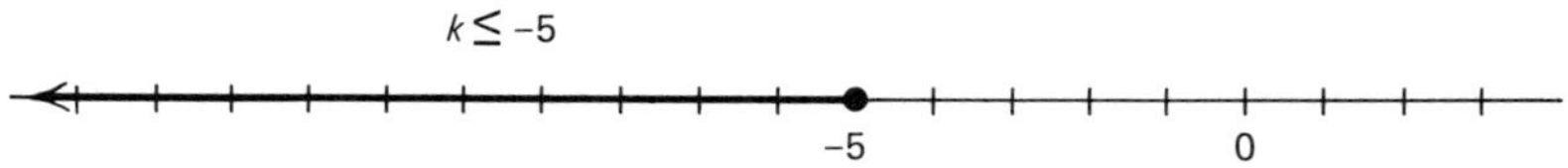


**EXAMPLE 12** Solve $3x + 12 \leq 5x + 7$, and graph the solution set.

**SOLUTION** We treat this just as we would an equation: isolate $x$.

| | |
|---|---|
| Given | $3x + 12 \le 5x + 7$ |
| Subtract $3x$ from both sides | $12 \le 2x + 7$ |
| Subtract 7 from both sides | $5 \le 2x$ |
| Divide both sides by 2 | $\frac{5}{2} \le x$ |

The solution set is $\{x \mid \frac{5}{2} \le x\}$ and looks as follows.

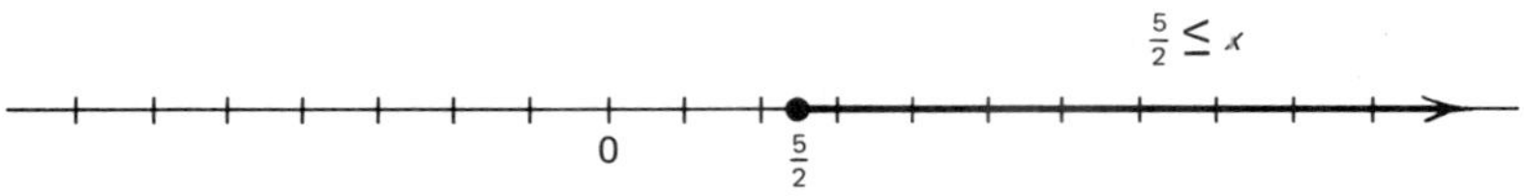


**EXAMPLE 13** Solve $2(a - 3) - 4 > 7(a - 5) - 2a$, and graph the solution set.

**SOLUTION** We begin by clearing the parentheses and then proceed to isolate $a$.

| | |
|---|---|
| Given | $2(a - 3) - 4 > 7(a - 5) - 2a$ |
| Distributive Law | $2a - 6 - 4 > 7a - 35 - 2a$ |
| Combine terms | $2a - 10 > 5a - 35$ |
| Subtract $2a$ from both sides | $-10 > 3a - 35$ |
| Add 35 to both sides | $25 > 3a$ |
| Divide both sides by 3 | $8\frac{1}{3} = \frac{25}{3} > a$ |

The solution set is $\{a \mid 8\frac{1}{3} > a\}$ and appears as follows.

$8\frac{1}{3} > a$

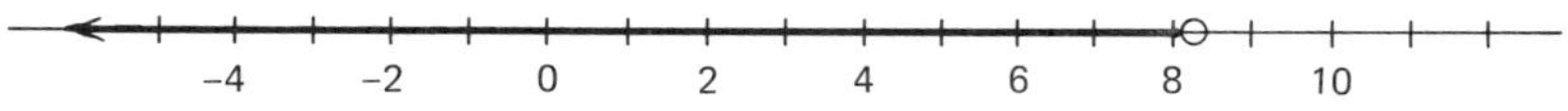

**EXAMPLE 14** Solve $4 < x + 3 < 10$, and graph the solution set.

**SOLUTION** In some courses, especially calculus and statistics, we see a double inequality such as this. This is really an abbreviation for *two* inequalities: $4 < x + 3$ *and* $x + 3 < 10$. However, we can use the same rules as before and solve them both at one time.

| | |
|---|---|
| Given | $4 < x + 3 < 10$ |
| ↓ Subtract 3 from all sides | $1 < x < 7$ |

Thus, $x$ is between 1 and 7, but not including the endpoints. This is written $\{x \mid 1 < x < 7\}$ and appears as follows.

$1 < x < 7$

**EXAMPLE 15** Solve $-2 \le \dfrac{x - 60}{12} \le 2$, and graph the solution set.

**SOLUTION** We isolate $x$ in the middle.

| | |
|---|---|
| Given | $-2 \le \dfrac{x - 60}{12} \le 2$ |
| ↓ Multiply all sides by 12 | $-24 \le x - 60 \le 24$ |
| ↓ Add 60 to all sides | $36 \le x \le 84$ |

The solution set is $\{x \mid 36 \le x \le 84\}$ and looks as follows.

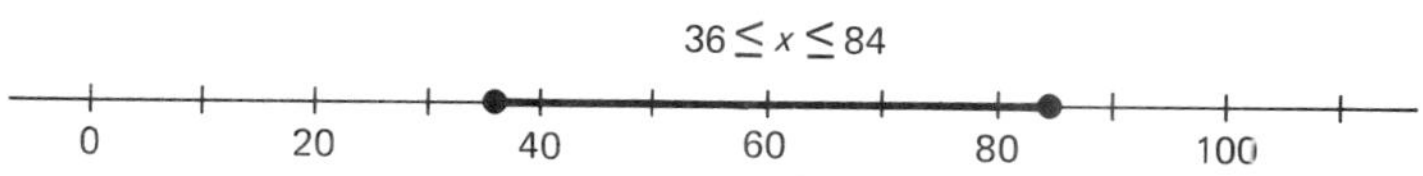


**EXAMPLE 16** Solve $2x + 5 < 11$ and $3x + 9 > 12$ together.

**SOLUTION** Here we solve each inequality separately and then take the intersection or overlap of their solution sets.

| | |
|---|---|
| Given | $2x + 5 < 11$ |
| Subtract 5 from both sides | $2x < 6$ |
| Divide both sides by 2 | $x < 3$ |

And

| | |
|---|---|
| Given | $3x + 9 > 12$ |
| Subtract 9 from both sides | $3x > 3$ |
| Divide both sides by 3 | $x > 1$ |

We now put these on the same graph (Fig. 2.4) and see that the common solution set is $\{x \mid 1 < x < 3\}$.

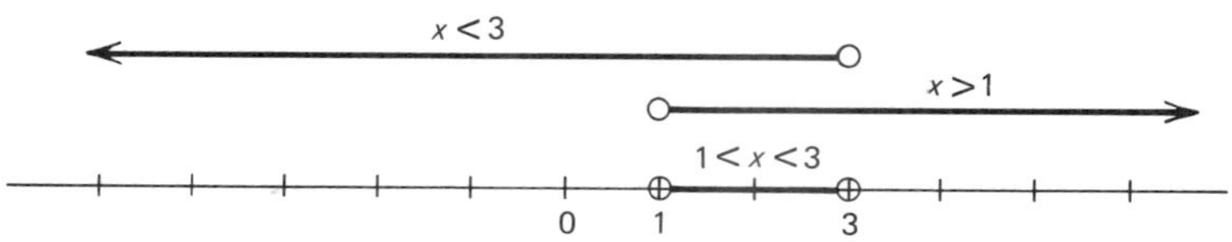


**Figure 2.4**

| YES | NO |
|---|---|
| $2x < -10$<br>$x < -5$ | ~~$2x < -10$~~<br>~~$x > -5$~~ |

We only reverse the inequality when we divide or multiply *by* a negative number, not *into* a negative number.

## PROBLEM SET 2.3

Solve the following inequalities, and graph their solution sets.

*See Examples 10 and 11*

1. $x + 7 > 10$
2. $a + 9 < 4$
3. $k - 4 \leq 6$
4. $b - 8 < -3$
5. $5w > 30$
6. $-6u \leq 42$
7. $-12 > -4p$
8. $10r < -20$
9. $\frac{x}{7} \leq 3$
10. $5 < \frac{t}{4}$
11. $10 \geq \frac{r}{-6}$
12. $\frac{z}{-8} > 5$

*See Example 11*

13. $3x + 1 > 13$
14. $8a - 2 \leq -10$
15. $5m - 4 \geq 21$
16. $4k - 7 < 1$
17. $-6y + 15 < 9$
18. $-4t + 11 \geq 3$
19. $8 \leq 2f - 3$
20. $-5m - 2 > -11$
21. $12 > -10p + 5$
22. $-13 < 9q - 3$

*See Example 12*

23. $8x - 1 < 6x + 7$
24. $5m + 2 > 12m - 19$
25. $4g + 2 \leq 7g + 4$
26. $9n - 5 \geq -15n + 19$
27. $-3t + 6 - t \geq 9t - 7$
28. $4z - 6 - 5z < 9z - 5$
29. $8r - 2 - 2r > 10r - 6$
30. $7p - 4 - 2p \leq -4p + 12$
31. $-10k - 4 < 15k - 8 - k$
32. $6q - 10 > -8q + 10 + 3q$

*See Example 13*

33. $2(x + 2) < 20$
34. $4(a - 6) \geq 12$
35. $3(3y - 4) > y - 3$
36. $5(4p - 3) \leq p$

37. $-6(2w - 7) \leq 17w + 42$

38. $8(u - 3) > 3(u - 5)$

39. $4(2m + 5) > 5(m + 7)$

40. $-3(2x - 7) < 2(5x - 3)$

41. $-10(k - 2) < 2(2k - 4) - 7$

42. $7(2c - 6) \geq 3(c + 8) - 5c$

*See Examples 14 and 15*

43. $6 < x + 3 < 10$

44. $45 \geq y + 10 \geq 25$

45. $-4 \leq a + 6 \leq 4$

46. $-15 < z - 100 < 15$

47. $13 \leq 3p - 2 \leq 25$

48. $5 \leq 4q + 1 \leq 17$

49. $8 < 4t + 7 < 20$

50. $11 < -5k - 9 < 16$

51. $-3 < -2u + 5 < 7$

52. $61 > 8h - 3 > 21$

53. $-3 \leq \dfrac{x - 5}{10} \leq 3$

54. $-2 \leq \dfrac{m - 100}{20} \leq 2$

55. $-1.5 < \dfrac{a + 50}{20} \leq 1.5$

56. $2 < \dfrac{x + 10}{5} \leq 6$

*See Example 16*

57. $x \geq 4$ and $x \leq 10$

58. $a \leq 12$ and $a \geq 2$

59. $x - 2 \geq 6$ and $x + 1 \leq 10$

60. $t + 3 < 9$ and $t + 4 > 2$

61. $2r + 1 < 7$ and $4r - 3 > 1$

62. $5u - 3 > 12$ and $2u - 5 < 17$

63. $2x + 1 < 9$ and $3x + 1 > 8$

64. $4k + 1 < 5$ and $3k - 7 > 2$

*Business applications*

65. A manufacturing firm has 240 labor-hours to spend producing a certain desk and chair. The desk requires 6 labor-hours and the chair 2. These satisfy the relation $6x + 2y \leq 240$. If $x = 20$ desks, find $y$, the number of chairs that can be made.

66. A retail store needs at least 3000 shirts for its inventory. The three warehouses can supply quantities $x$, $y$, and $z$ which satisfy the relation

$$x + y + z \geq 3000$$

Find $x$ if $y = 1200$ and $z = 560$.

*Health applications*

67. A patient must ingest less than 300 milligrams of cholesterol. His diet of eggs and liver satisfies the relation $165x + 90y \leq 300$. If $x = 1$ ounce of eggs, find $y$, the amount of liver the patient can have.

68. An adult should have at least 55 grams of protein. On a restricted diet of bread, peanut butter, and milk, the relation is $2b + 5p + 8m \geq 55$. If the patient gets $b = 3$ slices of bread and $p = 3$ tablespoons of peanut butter, how many glasses of milk $m$ must he drink?

*Statistics applications* In many fields, it is of interest to find a range in which a quantity occurs. Solving the following inequalities will give a range for 95% of the data involved.

69. *Light-bulb wattage:* $-1.96 \leq \dfrac{x - 75}{1.5} \leq +1.96$

70. *IQ scores:* $-1.96 \leq \dfrac{x - 100}{15} \leq +1.96$

71. *Human pH:* $-1.96 \leq \dfrac{x - 7.2}{0.15} \leq +1.96$

72. *Plant height:* $-1.96 \leq \dfrac{x - 30}{5} \leq +1.96$

## 2.4 ABSOLUTE-VALUE EQUATIONS

Recall from Chapter 1 that the absolute value is given by

$$|a| = \begin{cases} a & \text{if } a \geq 0 \\ -a & \text{if } a < 0 \end{cases}$$

For example, $|7| = 7$ and $|-10| = 10$.

Notice that $|a|$ might be $a$ or $-a$. We will use this fact to solve **absolute-value equations,** which are equations with the unknown inside absolute-value bars.

> To solve an absolute-value equation, such as $|P| = c$, where $P$ is some expression:
>
> 1. Split this into two equations: $P = c$ or $P = -c$.
> 2. Solve $P = c$ and $P = -c$ separately.
>
> *Beware:* If $c$ is negative, then $|P| = c$ has no solution because $|P|$ is always positive or zero.

**EXAMPLE 17** Solve $|x| = 5$.

**SOLUTION** We split this into two equations.

| | |
|---|---|
| Given | $\lvert x \rvert = 5$ |
| Split into 2 equations | $x = 5$ or $x = -5$ |

Thus, the solution set is $\{5, -5\}$.

**EXAMPLE 18** Solve $\lvert x - 7 \rvert = 10$.

**SOLUTION** We split this into two equations.

| | |
|---|---|
| Given | $\lvert x - 7 \rvert = 10$ |
| Split into 2 equations | $x - 7 = 10$ or $x - 7 = -10$ |
| Add 7 to both sides | $x = 17$ or $x = -3$ |

The solution set is $\{17, -3\}$. Notice that we can solve both equations with the same steps.

**EXAMPLE 19** Solve $\lvert 3x - 8 \rvert = 1$.

**SOLUTION** We again split and solve two equations.

| | |
|---|---|
| Given | $\lvert 3x - 8 \rvert = 1$ |
| Split into 2 equations | $3x - 8 = 1$ or $3x - 8 = -1$ |
| Add 8 to both sides | $3x = 9$ or $3x = 7$ |
| Divide both sides by 3 | $x = 3$ or $x = \frac{7}{3}$ |

The solution set is $\{3, \frac{7}{3}\}$. Again, we used the same steps to solve both equations.

**EXAMPLE 20** Solve $|5x - 7| = -3$.

**SOLUTION** An absolute value is always positive or zero and cannot be $-3$; thus, this equation has no solution, or we can say that the solution set is $\varnothing$.

### PROBLEM SET 2.4

Solve the following absolute-value equations. (*Be careful:* Some equations may not have a solution—see Example 20.)

*See Example 17*

1. $|x| = 6$
2. $|k| = 100$
3. $|a| = \frac{1}{2}$
4. $|r| = 0$
5. $|w| = -4$
6. $|u| = 0.1$
7. $|m| = 12$
8. $|p| = -\frac{2}{3}$

*See Example 18*

9. $|x - 1| = 12$
10. $|a - 3| = 8$
11. $|k + 10| = 7$
12. $|n + 4| = 5$
13. $|p + 7| = -4$
14. $|x - 6| = 0$
15. $|m + 1| = 3$
16. $|g + 5| = -13$
17. $|u - 4| = 0$
18. $|t - 11| = 11$

*See Example 19*

19. $|3x - 1| = 2$
20. $|5a - 3| = 7$
21. $|2k + 10| = 4$
22. $|7n - 4| = 9$
23. $|5p + 7| = -3$
24. $|-8x - 7| = 1$
25. $|7m - 1| = 0$
26. $|9g + 5| = -8$
27. $|-10u - 4| = 12$
28. $|3t - 11| = 0$

## 2.5 ABSOLUTE-VALUE INEQUALITIES

**Absolute-value inequalities** are inequalities with the unknown inside absolute-value bars. These are a bit harder than absolute-value equations. Recall from Chapter 1 that $|x|$ can be thought of as the distance from $x$ to 0 on the number line. Therefore, $|x| < 5$ means that the distance from $x$ to 0 is less than 5. We can see that

$$|x| < 5 \quad \text{is the same as} \quad -5 < x < 5$$

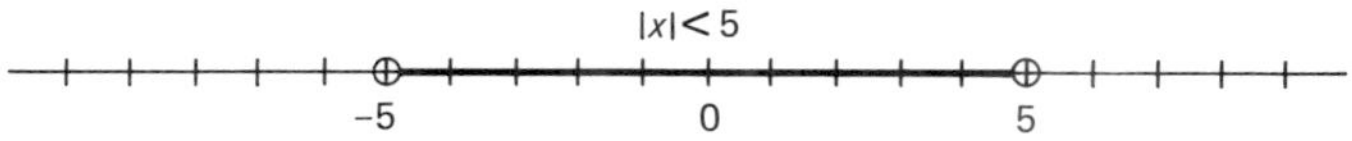


We use this to solve some of the absolute-value inequalities.

> To solve inequalities of the form $|P| < c$, where $P$ is some expression:
>
> 1. Rewrite $|P| < c$ as $-c < P < c$.
> 2. Solve this.
>
> *Note:* These rules also hold for $|P| \leq c$.

**EXAMPLE 21** Solve $|x - 7| < 2$, and graph the solution set.

**SOLUTION** We write this as two inequalities.

| | |
|---|---|
| Given | $|x - 7| < 2$ |
| Write as $-c < P < c$ | $-2 < x - 7 < 2$ |
| Add 7 to all sides | $5 < x < 9$ |

The solution set is $\{x \mid 5 < x < 9\}$ and can be drawn as follows.

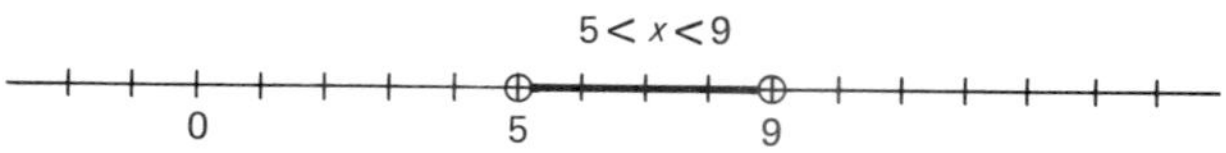


**EXAMPLE 22** Solve $|5x - 2| \leq 7$.

**SOLUTION** We write this as two inequalities.

| | |
|---|---|
| Given | $\lvert 5x - 2\rvert \le 7$ |
| ↓ Write as $-c \le P \le c$ | $-7 \le 5x - 2 \le 7$ |
| ↓ Add 2 to all sides | $-5 \le 5x \le 9$ |
| ↓ Divide all sides by 5 | $-1 \le x \le \frac{9}{5}$ |

The solution set is $\{x \mid -1 \le x \le \frac{9}{5}\}$ and can be graphed as follows.

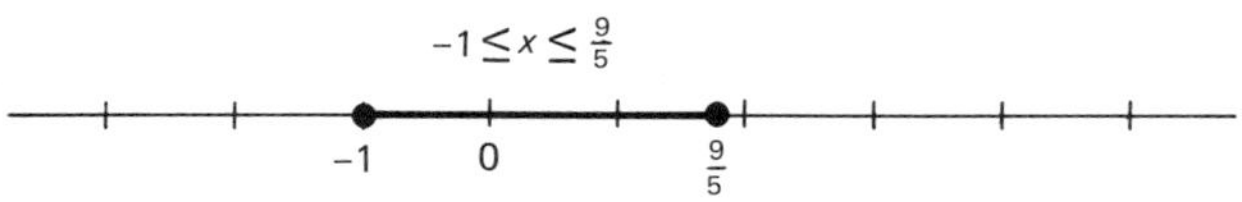


The other type of absolute-value inequality is like $\lvert x\rvert > 3$. This says that the distance from $x$ to 0 is more than 3. Pictorially, we see this as two inequalities: $x < -3$ or $x > 3$.

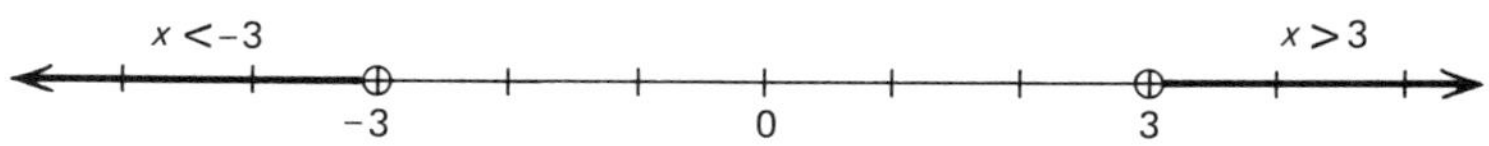


> To solve inequalities of the form $\lvert P\rvert > c$, where $P$ is some expression:
>
> 1. Split the inequality into two inequalities: $P > c$ or $P < -c$.
> 2. Solve both of these inequalities.
>
> *Note:* These rules also hold for $\lvert P\rvert \ge c$.

**EXAMPLE 23** Solve $\lvert x + 9\rvert > 4$.

**SOLUTION** We split this into two inequalities.

| | |
|---|---|
| Given | $\lvert x + 9 \rvert > 4$ |
| Split into $P < -c$ or $P > c$ | $x + 9 < -4$ or $x + 9 > 4$ |
| Subtract 9 from both sides | $x < -13$ or $x > -5$ |

The solution set is $\{x \mid x < -13 \text{ or } x > -5\}$ and can be graphed as follows.

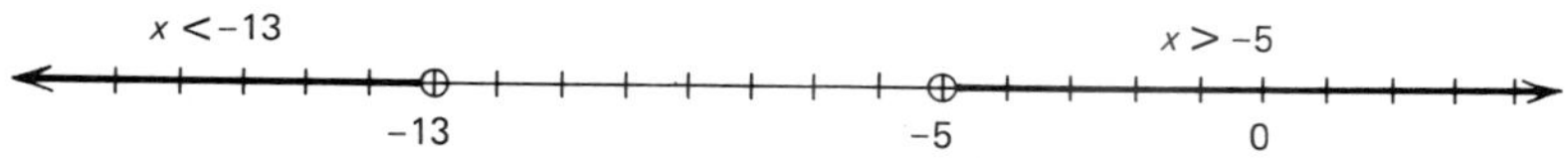


**EXAMPLE 24** Solve $\lvert 4x - 7 \rvert \geq 5$.

**SOLUTION** We split this into two inequalities.

| | |
|---|---|
| Given | $\lvert 4x - 7 \rvert \geq 5$ |
| Split into $P \leq -c$ or $P \geq c$ | $4x - 7 \leq -5$ or $4x - 7 \geq 5$ |
| Add 7 to both sides | $4x \leq 2$ or $4x \geq 12$ |
| Divide both sides by 4 | $x \leq \frac{1}{2}$ or $x \geq 3$ |

The solution set is $\{x \mid x \leq \frac{1}{2} \text{ or } x \geq 3\}$ and can be graphed as follows.

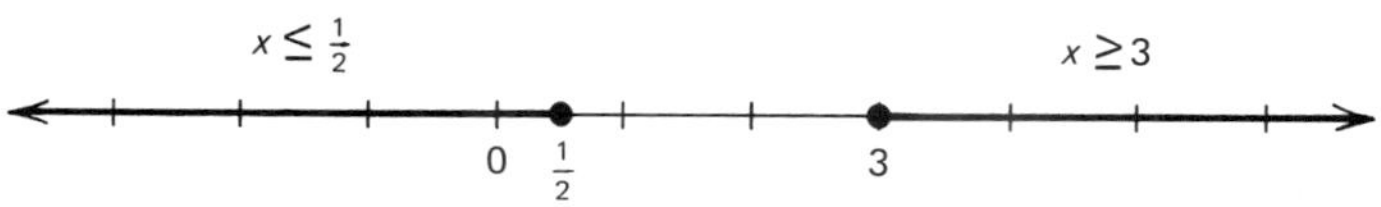

| YES | NO |
|---|---|
| $\lvert x\rvert \geq 8$ <br> $x \leq -8$ or $x \geq 8$ | $\lvert x\rvert \geq 8$ <br> $x \geq -8$ or $x \geq 8$ |

$|P| \geq c$ splits as $P \leq -c$ or $P \geq c$.

| YES | NO |
|---|---|
| 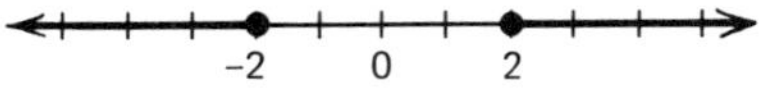 <br> $\{x < -2$ or $x > 2\}$ | 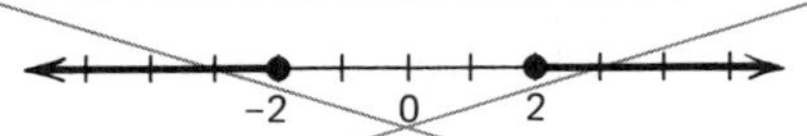 <br> $\{-2 > x > 2\}$ |

There are two separate sections to the graph. We must write this as two inequalities. The expression on the right would say that $-2 > 2$, which is wrong!

## PROBLEM SET 2.5

Solve the following absolute-value inequalities, and graph the solution sets.

*See Example 21*

1. $|x| < 4$ 2. $|p| \leq 7$
3. $|m| \leq \frac{1}{2}$ 4. $|r| < 6.2$
5. $|u - 9| < 3$ 6. $|a + 6| \leq 10$
7. $|t + 11| \leq 5$ 8. $|n - 8| < 13$
9. $|b - 2| < 8$ 10. $|c + 13| \leq 12$

*See Example 22*

11. $|6y - 3| < 9$ 12. $|11s + 2| \leq 13$
13. $|7z + 4| \leq 5$ 14. $|13t - 2| < 3$
15. $|3p - 10| < 8$ 16. $|4w + 12| \leq 2$
17. $|9q + 1| \leq 6$ 18. $|8v - 15| < 12$

*See Example 23*

19. $|x| > 10$ 20. $|u| > 3$
21. $|a| \geq \frac{2}{3}$ 22. $|b| > 1.5$
23. $|r + 7| > 2$ 24. $|p - 4| \geq 6$
25. $|s - 8| \geq 10$ 26. $|q + 20| > 9$
27. $|v + 1| > 5$ 28. $|m - 2| \geq 12$

*See Example 24*

29. $|3r - 2| \geq 7$ 30. $|8m + 1| > 9$
31. $|5k + 6| > 10$ 32. $|10n - 3| \geq 12$

33. $|11x - 7| \geq 1$ 34. $|2z + 9| > 8$

35. $|6y + 4| > 5$ 36. $|3w - 15| \geq 15$

*Engineering application* 37. The width of a certain flange has a tolerance of 0.001 and satisfies $|w - 2.5| \leq 0.001$. Solve this for the range of $w$.

## 2.6 WORD PROBLEMS

Here we solve equations that are stated in words. There are different types of problems, and we discuss each by use of an example.

### Translation Problems

These problems can be "translated" directly from English into mathematics. To do this we will use Table 2.1, a translation table of common English phrases and their mathematical equivalences.

**Table 2.1**

| English | Mathematics |
|---|---|
| Sum of $a$ and $b$; $a$ plus $b$; $a$ more than $b$ | $a + b$ |
| $b$ less than $a$; $a$ minus $b$; the difference between $a$ and $b$ | $a - b$ |
| $a$ times $b$; $a$ of $b$ (with fractions and percents); product of $a$ and $b$ | $ab$ |
| $a$ out of $b$; $a$ divided by $b$; quotient of $a$ and $b$ | $\frac{a}{b}$ |
| is, was, represents, should be | $=$ |
| what, some number | $x$ |
| three consecutive integers | $a, a + 1, a + 2$ |
| three consecutive odd (or even) integers | $a, a + 2, a + 4$ |

> To solve translation problems:
>
> 1. Translate the English information into mathematical relations using Table 2.1.
> 2. Be sure to identify the unknown.
> 3. Solve the equation or inequality for the unknown.

**EXAMPLE 25** Seven-fifths of a number is 10 more than the number. Find the number.

**SOLUTION** We translate and solve.

| | |
|---|---|
| Identify unknown | Let $x$ = unknown number |
| Given | $\frac{7}{5}$ of a number is 10 more than the number |
| Translate | $\frac{7}{5} \cdot x = 10 + x$ |
| Subtract $x$ from both sides | $\frac{2}{5}x = 10$ |
| Multiply both sides by $\frac{5}{2}$ | $x = \frac{5}{2} \cdot 10$ |
| Simplify | $x = 25$ |

We can check: $\frac{7}{5}$ of 25 is 35, which is 10 more than 25.

**EXAMPLE 26** The sum of three consecutive even integers is 48. Find the integers.

**SOLUTION** We translate and solve.

| | |
|---|---|
| Identify unknowns | Let $a$, $a + 2$, and $a + 4$ be three consecutive even integers |
| Given | Sum of 3 consecutive even integers is 48 |
| Translate | $a + a + 2 + a + 4 = 48$ |
| Combine terms | $3a + 6 = 48$ |
| Subtract 6 from both sides | $3a = 42$ |
| Divide both sides by 3 | $a = 14$ |
| Find other unknowns | $a + 2 = 16$<br>$a + 4 = 18$ |

We can check: the consecutive even integers 14, 16, and 18 add to $14 + 16 + 18 = 48$.

## Interest Problems

Another type of word problem is that of different amounts of money invested at different rates of interest. The keys to solving these problems are the following formulas.

$$I = r \cdot P$$

***FORMULA 2.1*** $\left[\begin{matrix}\text{Amount of}\\\text{interest}\end{matrix}\right] = \left[\begin{matrix}\text{Rate (\%) of}\\\text{interest}\end{matrix}\right] \cdot \left[\begin{matrix}\text{Principal}\\\text{invested}\end{matrix}\right]$

***FORMULA 2.2*** $\left[\begin{matrix}\text{Interest}\\\text{from I}\end{matrix}\right] + \left[\begin{matrix}\text{Interest}\\\text{from II}\end{matrix}\right] + \text{etc.} = \left[\begin{matrix}\text{Total}\\\text{interest}\end{matrix}\right]$

> To solve interest problems:
>
> 1. Identify the unknown amounts.
> 2. Use Formula 2.1 to determine the different amounts of interest.
> 3. Use Formula 2.2 to equate the sum of all investment interests to the total interest.
> 4. Solve for the unknowns.

**EXAMPLE 27** The Martins invest \$12,000, some at 7% interest and the rest at 5%. They earn \$660 in total interest. How much did they invest at each interest rate?

**SOLUTION** They have \$12,000 to invest. If they invest $x$ at 7%, then they must invest $12{,}000 - x$ at 5%. We use the notation $I_{5\%}$ for the interest from the 5% investment and $I_{7\%}$ for the interest from the 7% investment.

| Step | |
|---|---|
| Identify unknowns | Let $x$ = amount invested at 7%<br>$12{,}000 - x$ = amount invested at 5% |
| Use Formula 2.1 ($I = rP$) to determine different interests | $I_{7\%} = 0.07x$<br>$I_{5\%} = 0.05(12{,}000 - x)$ |
| Formula 2.2: Sum of all interests = total | $I_{7\%} + I_{5\%} = 660$ |
| Substitute | $0.07x + 0.05(12{,}000 - x) = 660$ |
| Distributive law | $0.07x + 600 - 0.05x = 660$ |
| Combine terms | $0.02x + 600 = 660$ |
| Subtract 600 from both sides | $0.02x = 60$ |
| Divide both sides by 0.02 | $x = 3000$ |
| Find other amount | $12{,}000 - x = 9000$ |

Therefore, they invested \$3000 at 7% (for \$210 interest) and \$9000 at 5% (for \$450 interest). The total interest checks as \$660.

## Mixture Problems

Mixture problems are very much like interest problems. In these problems, we are mixing various mixtures with different percents of pure mixture within them. For instance, we might be mixing a solution that is 20% alcohol with one that is 50% alcohol. We use the following formulas.

$$A = P \cdot B$$

***FORMULA 2.3*** $\begin{bmatrix}\text{Amount of}\\ \text{substance}\end{bmatrix} = \begin{bmatrix}\text{Percent of}\\ \text{substance}\end{bmatrix} \cdot \begin{bmatrix}\text{Total}\\ \text{base}\end{bmatrix}$

Compare this formula to Formula 2.1. Also, we have

***FORMULA 2.4*** $\begin{bmatrix}\text{Pure}\\ \text{Substance}\\ \text{from I}\end{bmatrix} + \begin{bmatrix}\text{Pure}\\ \text{Substance}\\ \text{from II}\end{bmatrix} + \text{etc.} = \begin{bmatrix}\text{Pure}\\ \text{Total}\\ \text{Substance}\end{bmatrix}$

Compare this formula to Formula 2.2.

To solve mixture problems:

1. Identify the unknown amounts.
2. Use Formula 2.3 to determine the different amounts of pure substance used.
3. Use Formula 2.4 to equate the sum of all the pure substances used to the total.
4. Solve for the unknowns.

**EXAMPLE 28** How many pounds of a nut mixture that is 80% cashews must be blended with a mixture that is 20% cashews to produce 50 pounds of a mixture that is 38% cashews?

**SOLUTION** The sketch in Fig. 2.5 might help. The trick is to compute the amount of cashews in each and then use Formula 2.4.

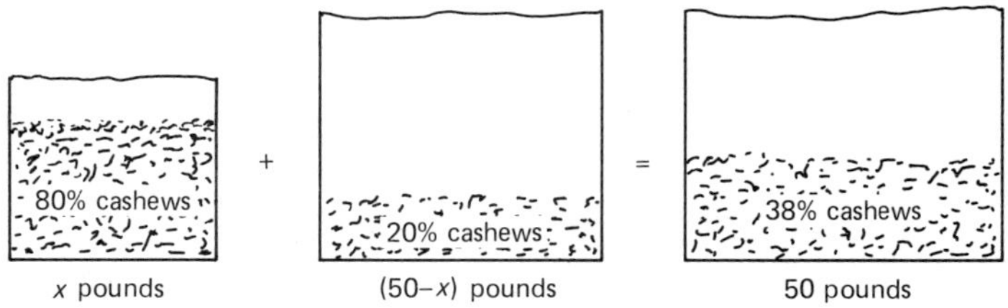


**Figure 2.5**

| | |
|---|---|
| Identify unknowns | Let $x$ = pounds of 80% mix<br>$50 - x$ = pounds of 20% mix |
| Use Formula 2.3 ($A = PB$) to find different amount of cashews | Cashews (80% mix) = $0.80x$<br>Cashews (20% mix) = $0.20(50 - x)$<br>Cashews (38% mix) = $0.38(50) = 19$ |
| Formula 2.4: Sum of cashews = total | Cashews (80%) + Cashews (20%) = Cashews (38%) |
| Substitute | $0.80x + 0.20(50 - x) = 19$ |
| Distributive law | $0.80x + 10 - 0.20x = 19$ |
| Combine terms | $0.60x + 10 = 19$ |
| Subtract 10 from both sides | $0.60x = 9$ |
| Divide both sides by 0.60 | $x = 15$ |
| Find other unknown | $50 - x = 35$ |

Therefore, we need 15 pounds of the 80% mix (= 12 pounds of cashews) and 35 pounds of the 20% mix (= 7 pounds of cashews). This is a total of 19 pounds of cashews out of a total of 50 pounds. This is 38%.

## Inequalities

The last type of word problem deals with inequalities.

> To solve inequality word problems:
>
> 1. Translate the problem as you would an equation.
> 2. Solve for the unknown.

**EXAMPLE 29** A grade of C in Professor Johnson's class is 70% to 80%. Nancy's grades on the first three tests are 74, 72, and 79. What range of scores on the last test will give her an average that is a grade of C?

**SOLUTION** Recall that the average of four scores is the sum of the scores divided by 4.

| | |
|---|---|
| Identify unknown | Let $x$ = score on fourth test |
| ↓ Given range for C | $70 \le \text{average} \le 80$ |
| ↓ Substitute | $70 \le \dfrac{74 + 72 - 79 + x}{4} \le 80$ |
| ↓ Multiply both sides by 4 | $280 \le 74 + 72 + 79 + x \le 320$ |
| ↓ Combine terms | $280 \le 225 + x \le 320$ |
| ↓ Subtract 225 from all sides | $55 \le x \le 95$ |

Therefore, Nancy can score between 55 and 95 and still get a grade of C.

## PROBLEM SET 2.6

*See Example 25*

1. One number is 7 more than another, and their sum is 41. What are the numbers?
2. One number is twice another, and their sum is 63. What are the numbers?
3. Five times a number is 24 more than 3 times the number. What is the number?
4. Four times a number is 27 less than 7 times the number. What is the number?
5. Eighteen is 5% of what number?
6. Ten is added to a number, and the result is doubled. If the answer is 34, what is the number?

*See Example 26*

7. The sum of two consecutive integers is 31. What are the numbers?
8. The sum of two consecutive integers is 255. What are the numbers?
9. The sum of two consecutive even integers is 46. What are the numbers?
10. The sum of two consecutive odd integers is 104. What are the numbers?
11. The sum of three consecutive odd integers is 57. What are the numbers?
12. The sum of three consecutive even integers is 144. What are the numbers?

*See Example 27*

13. Ms. LaRue invests \$15,000, part at 7% and the rest at 6%. In 1 year, she earns \$960 in interest. How much did she invest at each rate?
14. Mr. Jackson invests \$8000, part at 10% and the rest at 5%. In 1 year, he earns \$425 in interest. How much did he invest at each rate?
15. The Ulanowskis have some money invested at 8% and \$2500 more than that at 6%. If their total interest is \$570, how much did they invest at each rate?
16. The Garcias have some money invested at 5% and \$1500 less than that at 6%. If they earn a total of \$350 interest, how much was invested at each rate?
17. The Goldbergs invest \$5000 at interest rate $r$ and \$2000 at interest rate $(r + 1\%)$. They earn a total of \$580 interest. What are the interest rates?
18. Beaver Plains College invests \$200,000 at an interest rate $r$, \$100,000 at rate $(r + 1\%)$, and \$50,000 at rate $(r + 2\%)$. They earn a total of \$21,250 interest. What are the interest rates?

19. Frank has \$10,000 to invest, some at 5% and the rest at 8%. How much must he invest at each rate so that the interest is the same as 7% of the \$10,000?

20. Kathy has \$3000 to invest, some at 7% and the rest at 6%. How much must she invest at each rate so that the interest is the same as 6.3% of the \$3000?

*See Example 28*

21. A pharmacist has a 15% and a 25% solution of a certain drug. How much of each must he mix to get 20 ounces of a 19% solution?

22. A nurse is mixing two iodine solutions, one 30% and the other 60%. How much of each must she mix to get 100 milliliters of a 50% solution?

23. A butcher has some 70% lean hamburger and some 90% lean ground round. How much of each should she mix to get 100 pounds of 75% lean ground beef?

24. A goldsmith has a 90% gold alloy and a 50% gold alloy. How much of each must he mix to obtain 10 ounces of a 80% gold alloy?

25. A chemist has 15 liters of a 40% acid solution. How many liters of a 70% solution must be added to obtain a 50% solution?

26. How many ounces of a 10% alcohol solution must be added to 80 ounces of a 30% alcohol solution to obtain a 15% alcohol solution?

*See Example 29*

27. Ed's scores are 50, 62, 71, and 38. What must he score on the fifth test so that he will have a five-test average of at least 60?

28. An administration has a rule that the average salary in a department must be less than \$16,000. The present salaries are \$13,200, \$18,100, \$17,700, \$12,500, and \$14,400. If they hire a sixth person, what is the range that they can pay him or her?

29. The relation between Fahrenheit and Celsius is

$$F = \tfrac{9}{5}C + 32$$

A certain plant should have temperatures of 50 to 77°F. What range is this in degrees Celsius?

30. The state tax in Illinois for a family of four is given by

$$T = 0.025(I - 4000)$$

What range of incomes $I$ will produce a range of taxes $T$ between \$200 and \$800?

# CHAPTER 2 SUMMARY

## Important Words and Phrases

absolute-value equation (2.4)
absolute-value inequality (2.4)
addition property of equality (2.1)
addition property of inequality (2.3)
combining terms (2.1)
inequality (2.3)
literal equation (2.2)
multiplication property of equality (2.1)
multiplication property of inequality (2.3)
open equation (introduction)
open inequality (2.3)
solution (introduction)
solution set (introduction)

## Important Theorems

Let $a$, $b$, and $c$ be real numbers.

If $a = b$, then

1. $a + c = b + c$
2. $ac = bc$

If $a < b$, then

1. $a + c < b + c$
2. $ac < bc$ (if $c$ is positive)
3. $ac > bc$ (if $c$ is negative)

These relations are also true for $\leq$, $>$, and $\geq$.

## Important Procedures

The *goal* in solving equations and inequalities is to isolate the unknown.

*To solve equations for $x$:*
1. Remove all parentheses.
2. Combine terms on each side of the equation.
3. Move all $x$ terms to one side.
4. Move all number terms to the other side.
5. Multiply or divide to isolate $x$.

*To solve literal equations:*
1. Identify the unknown from the other letters.
2. Follow the procedure described above for solving.

*To solve inequalities:*

1. Isolate the unknown using the same procedures as for equations, except that multiplying or dividing by a negative number reverses the inequality symbol.

*To solve an absolute-value equation* such as $|P| = c$:

1. Split this into equations, $P = c$ or $P = -c$.
2. Solve each separately.

*To solve an absolute-value inequality* such as $|P| < c$:

1. Write this as $-c < P < c$.
2. Solve.

*To solve an absolute-value inequality* such as $|P| > c$:

1. Split into $P > c$ or $P < -c$.
2. Solve each separately.

## CHAPTER 2 REVIEW EXERCISES

Solve the following equations and inequalities for the unknown. Graph the solution sets for the inequalities.

1. $x + 7 = 23$
2. $7r - 13 = 15$
3. $5p - 3 = 8p + 12$
4. $6a - 2 - 3a = 4a - 3 + 9$
5. $6(b - 2) + 7 = 5(2b + 1) - b$
6. $ax - b = c$, for $x$
7. $3(2x - a) = 5(b - x) + x$, for $x$
8. $pV = nRT$, for $T$
9. $k - 3 \leq -10$
10. $-2q > -10$
11. $-3t + 5 \geq 14$
12. $4y - 3 + 2y < y - 28$
13. $2(u - 9) < 4(u + 3) + 1$
14. $2 < 4m - 3 < 7$
15. $-2 \leq \dfrac{n - 3}{4} \leq 2$
16. $x + 1 < 7$ and $3x + 1 > 7$
17. $|h| = 10$
18. $|z - 3| = 8$
19. $|4s - 3| = 7$
20. $|f - 3| < 2$
21. $|5w - 2| \leq 7$
22. $|r + 3| > 5$
23. $|7v + 5| \geq 1$
24. Six times a number is 18 more than 4 times the number. Find the number.
25. The sum of three consecutive odd integers is 81. Find the integers.

26. The Andersons have \$13,000 invested, part at 6% and the rest at 9%. If their total interest is \$840, how much did they invest at each interest rate?

27. A chemist is mixing a 20% acid solution with a 40% acid solution. How much of each must he mix to get 200 milliliters of a 25% acid solution?

28. Amy has scores of 48, 62, 53, 44, and 47. What grade must she earn on the sixth test to have a six-test average of at least 50?

# 3
# POLYNOMIALS

## 3.1 BASIC TERMS

In Chapter 2, we saw letters such as $x$ used in equations to be solved. In this chapter, we deal with **variables,** which are letters that may be replaced by numbers. For example, the area of a rectangle is given by $A = l \cdot w$. Here, $l$ (length) and $w$ (width) are variables, since we may replace $l$ and $w$ by different numbers.

Before we start with the algebraic operations, we must define a few concepts. A **term** is any product of a number, variables, and powers of variables. The numeral in the product is called the **coefficient.**

**EXAMPLE 1** The following are algebraic terms.

| TERM | COEFFICIENT |
|---|---|
| $3xy^2$ | 3 |
| $abc$ | 1 |
| $-7xm^2p^4$ | $-7$ |
| $-cd^2$ | $-1$ |

Notice in the second and fourth cases that the coefficients are 1 and $-1$ when no numeral appears in the term.

A **polynomial** is a sum of terms, such as $3xy - 5x^2 + xyz^4$. We have special names for polynomials with small numbers of terms. A **monomial** is a polynomial with one term, such as $7ab^2$. A **binomial** is a polynomial with two terms, such as $x^2 + 7a$. A **trinomial** is a polynomial with three terms, such as $x^2 + 5x - 3$. The **degree of a polynomial** is the highest power to which the variable is raised. The degree of $7a^3 - 4a^2 + a$ is 3, since the highest power of $a$ is 3.

**EXAMPLE 2** The following are various polynomials, along with their degrees and special names.

| POLYNOMIAL | DEGREE | SPECIAL NAME |
|---|---|---|
| $x^2 + 5$ | 2 | Binomial |
| $8t^4$ | 4 | Monomial |
| $9u^5 - 3u^2 + 12$ | 5 | Trinomial |
| $16y^7 - 3y^3 + 4y^2 - 2y + 6$ | 7 | Polynomial |

Polynomials are often represented by a letter; for example,

$$P(x) = 8x^3 - 5x^2 + 2x - 1$$

is a polynomial where $P$ acts like a name and $x$ is the variable. We can **evaluate** a polynomial by substituting a number for the variable.

**EXAMPLE 3** If $P(x) = 2x^2 - 3x + 5$, evaluate $P(2)$ and $P(-3)$.

**SOLUTION** $P(x)$ tells what happens to $x$: we take twice the square of $x$, subtract 3 times $x$, and then add 5. To find what $P$ does to a number, we simply substitute it into the polynomial.

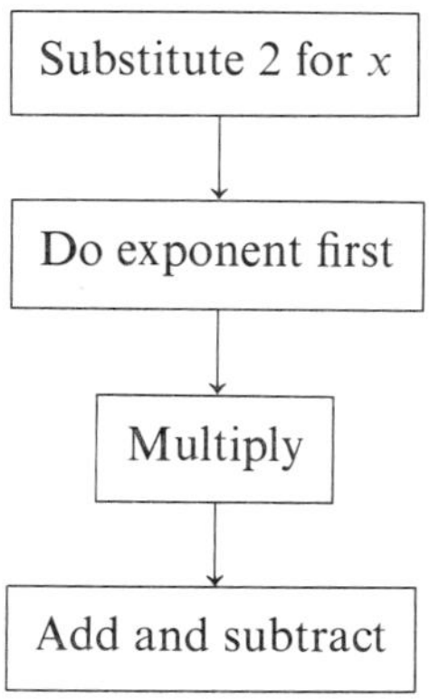


$$\begin{aligned} P(2) &= 2(2)^2 - 3(2) + 5 \\ &= 2(4) - 3(2) + 5 \\ &= 8 - 6 + 5 \\ &= 7 \end{aligned}$$

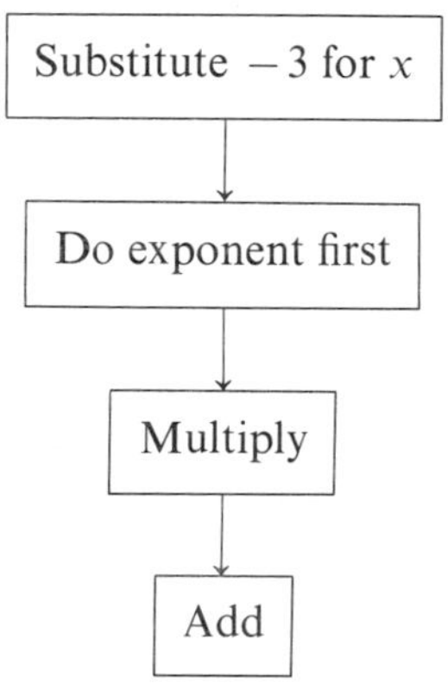


$$\begin{aligned} P(-3) &= 2(-3)^2 - 3(-3) + 5 \\ &= 2(9) - 3(-3) + 5 \\ &= 18 + 9 + 5 \\ &= 32 \end{aligned}$$

> To evaluate a polynomial:
>
> 1. Substitute the number wherever the variable appears in the polynomial.
> 2. Simplify according to the priority rules for real numbers: first parentheses, then exponents, then multiplication/division, and finally addition/subtraction.
> 3. Be careful with negative signs.

**EXAMPLE 4** $P(x) = 7x^4 - 3x^3 + 6x^2 - 2x + 1$. Use a hand calculator to evaluate $P(5)$.

**SOLUTION** We calculate each term separately and then use [M+] or [M−] to add it to and subtract it from the memory. We are calculating $P(5) = 7(5)^4 - 3(5)^3 + 6(5)^2 - 2(5) + 1$.

| PUNCH | DISPLAY | MEANING |
|---|---|---|
| [C] [MC] | 0. | Clear; clear memory |
| [7] [×] [5] [×] [5] [×] [5] [×] [5] [=] | 4375. | $7 \cdot 5^4$ |
| [M+] | 4375. | Add to memory |
| [3] [×] [5] [×] [5] [×] [5] [=] | 375. | $3 \cdot 5^3$ |
| [M−] | 375. | Subtract from memory |
| [6] [×] [5] [×] [5] [=] | 150. | $6 \cdot 5^2$ |
| [M+] | 150. | Add to memory |
| [2] [×] [5] [=] | 10. | $2 \cdot 5$ |
| [M−] | 10. | Subtract from memory |
| [1] [M+] | 1. | Add 1 to memory |
| [MR] | 4141. | Recall total (answer) |

Notice that we use [M+] and [M−] to enter numbers into the memory, and we use [MR] to recall the total from the memory.

The most general combination of the mathematical symbols with the operations of addition, subtraction, multiplication, division, and exponents is called an **expression.** For instance, $6a^2 - \dfrac{7b^3}{c + d} + \dfrac{(a - b)^2}{d}$ is an expression.

## PROBLEM SET 3.1

For each of the following terms, identify the coefficient.

*See Example 1*

1. $7x^2$
2. $9a^3$
3. $-8xy^3$
4. $13a^2b^3$
5. $x^3$
6. $-a^4$
7. $-m^2n^3p^4$
8. $r^4s^5t^6u^7$

For each of the following polynomials, give the degree and give its special name (monomial, binomial, or trinomial) if it has one.

*See Example 2*

9. $8x^5 - 3x^2$
10. $-18a^4$
11. $y^2 - 5y + 4$
12. $a^4 - 3a^6$
13. $4t^7 - 6t^2 + t$
14. $4u^7 - 8u^5 - 6u^4 + 10u + 7$
15. $x^{2n} + 3x^n + 2$
16. $a^{3n} - 3a^{2n} + 3a^n + 1$

Evaluate the following polynomials.

*See Example 3*

17. If $P(x) = x^2 + 4$, find $P(3)$, $P(0)$, and $P(-2)$.
18. If $Q(x) = 3x^2 - x + 2$, find $Q(5)$, $Q(0)$, and $Q(-3)$.
19. If $R(x) = -2x^2 - 8x + 5$, find $R(4)$, $R(0)$, and $R(-1)$.
20. If $P(x) = -3x^2 + 4x - 7$, find $P(7)$, $P(0)$, and $P(-4)$.
21. If $D(x) = (2x + 1)^2 - x^2$, find $D(2)$, $D(0)$, and $D(-4)$.
22. If $S(x) = (3x - 2)^2 + 4x + 7$, find $S(1)$, $S(0)$, and $S(-2)$.
23. If $T(x) = x^6 - x^3 + 1$, find $T(2)$, $T(0)$, and $T(-1)$.
24. If $P(x) = x^{10} + x^5 - 1$, find $P(1)$, $P(0)$, and $P(-2)$.

Use a hand calculator to evaluate the following polynomials.

*See Example 4*

25. If $P(x) = 4x^7 - 12x^4 + 15x^2 + 25x - 175$, find $P(11)$.
26. If $Q(x) = 15x^4 + 25x^3 - 35x^2 + 40x - 2500$, find $Q(45)$.
27. If $A(x) = 38x^3 - 217x^2 + 528x + 1728$, find $A(125)$.
28. If $R(x) = 255x^2 + 1758x + 23{,}725$, find $R(592)$.

*Life science application*

After organic waste is deposited in a lake, the oxygen content is given by

$$C(t) = t^3 - 45t^2 + 20{,}000$$

where $t$ is the time in days.

29. Find $C(0)$.
30. Find $C(20)$.
31. Find $C(30)$.
32. Find $C(40)$.

*Business application*

The total cost to produce a certain machine is given by

$$\text{TC}(Q) = 0.001Q^3 - 1.1Q^2 + 1180Q + 132{,}000$$

where $Q$ is the quantity produced.

33. Find TC(0).
34. Find TC(100).
35. Find TC(300).
36. Find TC(500).

*Health application*

The amount of sugar in the blood stream after an ingestion is given by

$$A(t) = 3.8 + 0.2t - 0.1t^2$$

where $t$ is the time in hours.

37. Find $A(0)$.
38. Find $A(1)$.
39. Find $A(2)$.
40. Find $A(3)$.

## 3.2 ADDING AND SUBTRACTING POLYNOMIALS

We saw in Chapter 2 how to use the distributive law to help simplify polynomials. For example:

**EXAMPLE 5** **(a)** $2x^2 + 7x^2 = (2 + 7)x^2 = 9x^2$

**(b)** $4a^3b + 6a^3b - 3a^3b = (4 + 6 - 3)a^3b = 7a^3b$

Terms that are exactly the same except for the coefficient are called **like terms.** In the foregoing examples, we have simplified the polynomials by **collecting** and **combining like terms.**

> To add polynomials:
>
> 1. Group (or line up vertically) all like terms.
> 2. Add the coefficients.

**EXAMPLE 6** Find $S = (3x^2 - 2x + 5) + (5x^3 - x - 1) + (6x^3 - x^2 - 2)$.

**SOLUTION** We line up all like terms, leaving space for "missing" powers of $x$.

| | |
|---|---|
| Line up like terms | $\begin{array}{rrrr} & 3x^2 & -\ 2x & +\ 5 \\ 5x^3 & & -\ x & -\ 1 \\ 6x^3 & -\ x^2 & & -\ 2 \\ \hline 11x^3 & +\ 2x^2 & -\ 3x & +\ 2 \end{array}$ |
| ↓ Add coefficients | |

This type of problem is often worked horizontally by grouping like terms using the commutative and associative laws.

| | |
|---|---|
| Given | $S = (3x^2 - 2x + 5) + (5x^3 - x - 1) + (6x^3 - x^2 - 2)$ |
| ↓ Group like terms | $= (5x^3 + 6x^3) + (3x^2 - x^2) + (-2x - x) + (5 - 1 - 2)$ |
| ↓ Simplify | $= 11x^3 + 2x^2 - 3x + 2$ |

Recall from Chapter 1 that the definition for subtracting real numbers was $a - b = a + (-b)$. In words, *we add the opposite.* We do the same thing when we subtract polynomials. We obtain the **opposite of a polynomial** by changing *all* its signs. For example, the opposite of $7x^3 - 4x^2 + 5x - 3$ is $-7x^3 + 4x^2 - 5x + 3$.

> To subtract polynomials:
>
> 1. Replace the polynomial to be subtracted by its opposite; that is, change the signs (all the signs!) in the polynomial to be subtracted.
> 2. Add the resulting polynomials.

**EXAMPLE 7** Find $D = (5a^3 - 6a - 7) - (8a^2 - 4a + 1)$.

**SOLUTION** We change all the signs of the second polynomial, and then add.

| | |
|---|---|
| Given | $D = (5a^3 - 6a - 7) - (8a^2 - 4a + 1)$ |
| Subtraction means addition of opposite | $= (5a^3 - 6a - 7) + (-8a^2 + 4a - 1)$ |
| Group like terms | $= 5a^3 - 8a^2 + (-6a + 4a) + (-7 - 1)$ |
| Simplify | $= 5a^3 - 8a^2 - 2a - 8$ |

Often, we must remove a series of parentheses to simplify an expression completely. We give the following rule. (We use "parentheses" to mean all symbols of inclusion including brackets and braces.)

> To simplify an expression with parentheses, brackets, or braces:
>
> 1. If parentheses are nested (parentheses within parentheses), work from inside to outside.
>
> 2a. If the sign before the parentheses is +, remove the parentheses and leave the signs inside the parentheses unchanged.
>
> 2b. If the sign before the parentheses is −, remove the parentheses and change all the signs within the parentheses. (Add the opposite.)
>
> 3. Combine like terms.

**EXAMPLE 8** Simplify $P = 6x - (3x - [2 + (x - 5)] - [5x - 1])$.

**SOLUTION** We simplify this from inside to outside. We also do it in small simple steps rather than one large step.

| | |
|---|---|
| Given | $P = 6x - (3x - [2 + (x - 5)] - [5x - 1])$ |
| + means no sign change | $= 6x - (3x - [2 + x - 5] - [5x - 1])$ |
| Combine like terms | $= 6x - (3x - [-3 + x] - [5x - 1])$ |
| − means change signs | $= 6x - (3x + 3 - x - 5x + 1)$ |
| Combine like terms | $= 6x - (-3x + 4)$ |
| − means change signs | $= 6x + 3x - 4$ |
| Combine like terms | $= 9x - 4$ |

| YES | NO |
|---|---|
| $6x - (2x + 1) = 6x - 2x - 1$ | ~~$6x - (2x + 1) = 6x - 2x + 1$~~ |

We must change *all* signs within the parentheses!

## PROBLEM SET 3.2

Simplify the following.

*See Example 5*

1. $8x - 6x + 5x - x$
2. $4a - 7a - 10a + 2a$
3. $3pq^2 - 8pq^2 - 11pq^2$
4. $a^2bc + 4a^2bc - 12a^2bc$

Add the following polynomials.

*See Example 6*

5. $$\begin{array}{r} 2x^2 - 6x + 1 \\ + \underline{6x^2 + 4x - 8} \end{array}$$

6. $$\begin{array}{r} 7a^2 - 2a + 3 \\ + \underline{3a^2 + 5a - 5} \end{array}$$

7. $$\begin{array}{rrrrr} & 8p^3 & - 4p^2 & & + 1 \\ + & \underline{3p^3} & & \underline{- 8p} & \underline{- 7} \end{array}$$

8. $$\begin{array}{rrrrrr} & 5b^4 & & + 6b^2 & & + 2 \\ + & & \underline{7b^3} & \underline{- 4b^2} & \underline{- 3b} & \end{array}$$

9. $$\begin{array}{rrrr} & 4m^2 & - 2m & + 1 \\ & 8m^2 & - 5m & - 7 \\ + & \underline{m^2} & \underline{+ 3m} & \underline{+ 2} \end{array}$$

10. $$\begin{array}{rrr} 5c^2 & - 3c & - 2 \\ 4c^2 & + 7c & - 8 \\ \underline{c^2} & \underline{- 6c} & \underline{+ 4} \end{array}$$

11. $$\begin{array}{rrrrr} & x^3 & & - 2x & + 4 \\ & 4x^3 & - 6x^2 & & - 6 \\ + & & \underline{4x^2} & \underline{+ 9x} & \underline{- 1} \end{array}$$

12. $$\begin{array}{rrrrrr} & 4r^4 & - 6r^3 & & - r & - 3 \\ & & 8r^3 & - r^2 & - 2r & + 5 \\ + & \underline{8r^4} & & & \underline{- 7r} & \underline{- 10} \end{array}$$

13. $(12z^2 - 8z - 1) + (8z^2 + 7z + 1) + (4z^2 - z + 2)$

14. $(5h^2 - 2h - 6) + (6h^2 - h + 7) + (9h^2 + 8h - 1)$

15. $(7x^4 - 3x^2 + x - 6) + (4x^4 + 5x^3 + 6x^2 + 11)$

16. $(10p^5 - 8p^3 + 2p^2 + 2) + (4p^4 + 6p^3 + 8p^2 + p - 11)$

Subtract the following polynomials.

*See Example 7*

17. $$\begin{array}{r} 8x^2 - 4x - 1 \\ - \underline{(6x^2 + 5x - 2)} \end{array}$$

18. $$\begin{array}{r} 5b^2 - 5b + 2 \\ - \underline{(2b^2 - 3b - 6)} \end{array}$$

19. $$\begin{array}{rrrrr} & 9y^3 & - 8y^2 & & + 1 \\ - & \underline{(10y^3} & \underline{+ 5y^2} & \underline{- 4y} & \underline{\quad )} \end{array}$$

20. $$\begin{array}{rrrrrr} & 4s^4 & - 8s^3 & & - s & \\ - & \underline{(3s^4} & & \underline{+ 2s^2} & \underline{+ 5s} & \underline{+ 3)} \end{array}$$

21. $$\begin{array}{rrrrrr} & 2m^4 & - 3m^3 & - 7m^2 & & + 12 \\ - & \underline{(\ m^4} & & \underline{- 8m^2} & \underline{- 11m} & \underline{- \ 8)} \end{array}$$

22. $$\begin{array}{rrrrrrr} & 5k^5 & - 9k^4 & & - k^2 & & + 3 \\ - & \underline{(\ k^5} & \underline{- 2k^4} & \underline{- 3k^3} & & \underline{+ k} & \underline{+ 1)} \end{array}$$

23. $(8t^2 - 6t - 2) - (6t^2 + 5t + 1)$

24. $(4u^3 - 3u^2 - u - 1) - (8u^3 - 9u + 4)$

25. $(6w^4 - 8w^2 + 4) - (5w^4 - 4w^3 + 6w^2 - 7w - 8)$

26. $(3n^3 - 8n^2 - 10) - (7n^4 + 2n^3 - 3n + 1)$

27. $(8a^4 - 3a^2 - a + 2) + (7a^3 - 6a^2 - 2a - 1) - (a^5 - 3a^2 + a)$
28. $(4r^3 - 5r^2 - 7r + 8) - (r^5 - r - 2) - (6r^4 - 8r^2 + 13)$

Simplify the following.

*See Example 8*

29. $6a - [2 - (3a - 5)]$
30. $8x - [5x - (2 + 7x)]$
31. $2 - (3x - [4 + (5 + 7x)])$
32. $4m + (3n - [6m - (7n - 3m)] + m)$
33. $4a - (2b + [3a - b] - [2a - (6b - a)])$
34. $(7p + 8q) - (3r - [2p - 3q] - [4p + 5r] + [6q - 5r])$
35. $5t - ([2 - (6 - 4t)] - [9t - (8 + 2t) - (3t - 7)])$
36. $9x - (3y - [2x - (y + 4x)] - [7y + (8x - 2y) - (3y - 4x)])$

*Business application*

In business, we have the relation that

$$\text{TP} = \text{TR} - \text{TC}$$

where TP is the total profit, TR the total revenue, and TC the total cost. In the following situations, find the unknown quantity. ($Q$ is the quantity produced.)

| | TP | TR | TC |
|---|---|---|---|
| 37. | ? | $6000Q$ | $0.001Q^3 + 1180Q + 131,000$ |
| 38. | ? | $1120Q$ | $0.01Q^2 + 950Q + 52,000$ |
| 39. | $-0.02Q^2 + 5Q - 1100$ | $65Q$ | ? |
| 40. | $-0.005Q^2 + 15Q - 35,000$ | ? | $0.004Q^2 + 95Q + 37,000$ |

## 3.3 MULTIPLYING AND DIVIDING MONOMIALS

Let us look at the product of $a^m \cdot a^n$. We have

$$a^m \cdot a^n = \underbrace{(a \cdot a \cdots a)}_{m \text{ factors}} \cdot \underbrace{(a \cdot a \cdots a)}_{n \text{ factors}} = \underbrace{a \cdot a \cdot a \cdots a}_{m + n \text{ factor}} = a^{m+n}$$

This becomes an important theorem.

> ***THEOREM 3.1*** For any real number $a$ and positive integers $m$ and $n$,
>
> $$a^m \cdot a^n = a^{m+n}$$

**EXAMPLE 9** **(a)** $x^5 \cdot x^7 = x^{5+7} = x^{12}$

**(b)** $10^6 \cdot 10^1 = 10^{6+1} = 10^7$

**(c)** $a^4 \cdot a^2 \cdot a^8 = a^{4+2+8} = a^{14}$

**(d)** $x^{n+1} \cdot x^{2n+3} = x^{3n+4}$

| YES | NO |
|---|---|
| $t^6 \cdot t^3 = t^9$ | ~~$t^6 \cdot t^3 = t^{18}$~~ |

We must *add* exponents when we multiply monomials.

| YES | NO |
|---|---|
| $2^5 \cdot 3^7 = 2^5 \cdot 3^7$ | ~~$2^5 \cdot 3^7 = 6^{12}$~~ |

The bases must be the *same* to use Theorem 3.1.

| YES | NO |
|---|---|
| $3^5 \cdot 3^7 = 3^{12}$ | ~~$3^5 \cdot 3^7 = 9^{12}$~~ |

The base stays the same in Theorem 3.1.

Now let us look at division. For example,

$$\frac{a^7}{a^2} = \frac{a \cdot a \cdot a \cdot a \cdot a \cdot \not{a} \cdot \not{a}}{\not{a} \cdot \not{a}} = a^5$$

Here, the exponents were subtracted. Also,

$$\frac{a^4}{a^6} = \frac{\not{a} \cdot \not{a} \cdot \not{a} \cdot \not{a}}{\not{a} \cdot \not{a} \cdot \not{a} \cdot \not{a} \cdot a \cdot a} = \frac{1}{a^2}$$

Again, we subtracted exponents, but since the denominator had the larger exponent, the resulting exponent $(6 - 4 = 2)$ is in the denominator. These examples suggest the following theorem.

> ***THEOREM 3.2*** For any real number $a$ $(\neq 0)$ and positive integers $m$ and $n$,
>
> $$\frac{a^m}{a^n} = \begin{cases} a^{m-n} & \text{if } m > n \\ 1 & \text{if } m = n \\ \dfrac{1}{a^{n-m}} & \text{if } m < n \end{cases}$$

**EXAMPLE 10** (a) $\dfrac{10^8}{10^3} = 10^{8-3} = 10^5$

(b) $\dfrac{w^{10}}{w} = w^{10-1} = w^9$

(c) $\dfrac{5^8}{5^8} = 1$

(d) $\dfrac{x^4}{x^9} = \dfrac{1}{x^{9-4}} = \dfrac{1}{x^5}$

(e) $\dfrac{a^{5n+1}}{a^{3n}} = a^{2n+1}$

> To multiply (or divide) monomials:
>
> 1. Multiply (or divide) the coefficients.
> 2. Add (or subtract) the exponents with the *same* base, using Theorems 3.1 and 3.2.

**EXAMPLE 11** Multiply $(-6a^2b^4c)(2ab^2c^7)(3b^8d)$.

**SOLUTION** We multiply the coefficients and letters separately.

| Multiply coefficients, add exponents |
|---|

$$\begin{aligned}(-6a^2b^4c)(2ab^2c^7)(3b^8d) &= (-6 \cdot 2 \cdot 3)(a^2a)(b^4b^2b^8)(cc^7)(d) \\ &= -36a^3b^{14}c^8d\end{aligned}$$

**EXAMPLE 12** Divide $\dfrac{16x^7y^2z^4}{-2x^4y^4z^4}$.

**SOLUTION** We divide each coefficient and letters separately.

| Divide coefficients, subtract exponents |
|---|

$$\frac{16x^7y^2z^4}{-2x^4y^4z^4} = \frac{-16}{2} \cdot \frac{x^7}{x^4} \cdot \frac{y^2}{y^4} \cdot \frac{z^4}{z^4}$$

$$= \frac{-8x^3}{y^2}$$

Notice that the $z^4/z^4$ term canceled completely since it is equal to 1.

## PROBLEM SET 3.3

Multiply the following monomials.

*See Example 9*

1. $x^7 \cdot x^4$
2. $a^3a^{10}$
3. $r^2r^8r^7$
4. $t^6t^7t$
5. $(5x^2)(2x^3)$
6. $(6y^4)(-3y^5)$

*See Example 11*

7. $(10ab^2)(-3a^2b^4)$
8. $(4m^2n^5)(8mn^9)$
9. $(ab^2c^3)(-9a^3b^3c^4)$
10. $(12p^2q^4r^6)(3p^2q)$
11. $(5xy^2)(3y^2z^4t)$
12. $(-4a^2b^3c)(10b^4c^2d^7)$
13. $(a^2bc)(-6a^4b^2)(2a^2b^3c^4)$
14. $(x^2y^2)(-4xy^7z^3)(-3x^2yz^9)$
15. $(-2rs^4tu)(-3r^2st^2u^4)(4r^2su^2)$
16. $(-m^4n^3p)(m^3p^2q^4)(n^8p^2r^7)$

Divide the following monomials.

*See Example 10*

17. $\dfrac{x^{12}}{x^5}$
18. $\dfrac{a^7}{a^3}$
19. $\dfrac{t^5}{t^5}$
20. $\dfrac{m^{10}}{m^{10}}$
21. $\dfrac{h^5}{h^{11}}$
22. $\dfrac{r^6}{r^7}$

*See Example 12*

23. $\dfrac{6x^{10}}{3x^5}$
24. $\dfrac{14t^9}{7t^7}$
25. $\dfrac{10x^2y^4}{-2x^5y^3}$
26. $\dfrac{20a^4b^3}{10ab^7}$
27. $\dfrac{22m^7n^4}{-11mn^4}$
28. $\dfrac{36p^7q^8}{9p^4q^{10}}$
29. $\dfrac{-49x^2y^3z^4}{7x^5y^3z}$
30. $\dfrac{-18a^8b^9c}{-9ab^5c^5}$
31. $\dfrac{17rst^2}{-17rst^2}$

32. $\dfrac{64xyz^9}{32xyz^7}$

33. $\dfrac{(2x + 1)^7}{(2x + 1)^5}$

34. $\dfrac{(6a - 7)^{10}}{(6a - 7)^2}$

35. $\dfrac{(5t + 2)^5}{(5t + 2)^5}$

36. $\dfrac{(7s - 2)^4}{(7s - 2)^5}$

Simplify the following.

*See Examples 9b and 10e*

37. $x^n \cdot x^{2n}$

38. $a^{n+5} \cdot a^{2n-1}$

39. $\dfrac{b^{5n+1}}{b^{3n}}$

40. $\dfrac{m^{2n+2}}{m^{n-3}}$

## 3.4 MULTIPLYING POLYNOMIALS

Before we multiply two polynomials, let us first multiply a monomial by a a polynomial using the distributive law.

**EXAMPLE 13** Multiply $(2x^2y)(x + 2x^2y^2 - 5x^3y)$.

**SOLUTION** By the distributive law, we multiply the $2x^2y$ by all the terms inside the parentheses, using the rules for multiplying monomials.

| Distributive law | $2x^2y(x + 2x^2y^2 - 5x^3y) = 2x^3y + 4x^4y^3 - 10x^5y^2$ |
|---|---|

To multiply polynomials in general, we use the distributive law twice.

$$(a + b)(c + d) = (a + b)c + (a + b)d = ac + bc + ad + bd$$

This suggests the following rule.

> To multiply two polynomials:
>
> 1. Multiply all terms of one by all the terms of the other.
> 2. Combine like terms.

Actually, this procedure is just like multiplying whole numbers. Notice the similarity below.

$$
\begin{array}{r}
21 \\
\times\ \underline{13} \\
63 \\
\underline{21\phantom{0}} \\
273
\end{array}
\qquad
\begin{array}{rrr}
 & 2a & +\ \ b \\
 & \times & \underline{a + 3b} \\
 & 6ab & +\ 3b^2 \\
\underline{2a^2 + ab} & & \\
2a^2 & +\ 7ab & +\ 3b^2
\end{array}
$$

We multiply all the terms in the bottom polynomial by all those in the top. Notice that we line up the like terms and add.

**EXAMPLE 14** Multiply $(2x^2 - 5x + 1)(3x^2 + 2x - 4)$.

**SOLUTION** We multiply this out vertically as we would two whole numbers.

Multiply each bottom term by each top term

↓

Combine like terms

$$
\begin{array}{rrrrr}
 & & 2x^2 & -\ 5x & +\ 1 \\
 & & \times\ 3x^2 & +\ 2x & -\ 4 \\
 & & -8x^2 & +\ 20x & -\ 4 \\
 & 4x^3 & -\ 10x^2 & +\ 2x & \\
6x^4 & -\ 15x^3 & +\ 3x^2 & & \\
\hline
6x^4 & -\ 11x^3 & -\ 15x^2 & +\ 22x & -\ 4
\end{array}
$$

We have a special rule to multiply two binomials quickly. It is called **FOIL.**

**FOIL = Firsts + Outers + Inners + Lasts**

**EXAMPLE 15** Multiply $(2x + 3)(3x - 4)$ by FOIL.

**SOLUTION**

FOIL

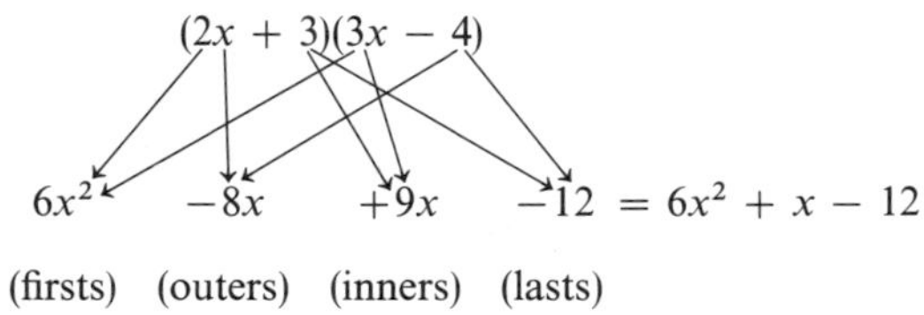

We can usually do this in our head while combining the like middle terms to get $6x^2 + x - 12$.

| YES | NO |
| --- | --- |
| $(2x + 3)(3x - 4) = 6x + x - 12$ | $(2x + 3)(3x - 4) = 6x^2 - 12$ |

We must multiply all the terms of the first by all the terms of the second.

Since $(a + b)(a - b) = a^2 - ab + ab - b^2 = a^2 - b^2$, we get the following theorem.

***THEOREM 3.3*** For any real numbers $a$ and $b$,

$$(a + b)(a - b) = a^2 - b^2$$

**EXAMPLE 16** **(a)** $(x + 5)(x - 5) = x^2 - 25$

**(b)** $(p - 7q)(p + 7q) = p^2 - 49q^2$

Since $(a + b)(a + b) = a^2 + ab + ab + b^2$, we get

***THEOREM 3.4*** For any real numbers $a$ and $b$,

$$(a + b)^2 = a^2 + 2ab + b^2$$

***THEOREM 3.5*** For any real numbers $a$ and $b$,

$$(a - b)^2 = a^2 - 2ab + b^2$$

**EXAMPLE 17** **(a)** $(t + 4)^2 = t^2 + 8t + 16$

**(b)** $(2x - 5)^2 = 4x^2 - 20x + 25$

| YES | NO |
| --- | --- |
| $(r + 7)^2 = r^2 + 14r + 49$ | $(r + 7)^2 = r^2 + 49$ |

We must also add twice the product of the terms.

**EXAMPLE 18** Multiply $(2a + 1)(3a - 2)(4a + 5)$.

**SOLUTION** Just as we multiply three numbers, we multiply the first two polynomials and then multiply that product by the third polynomial.

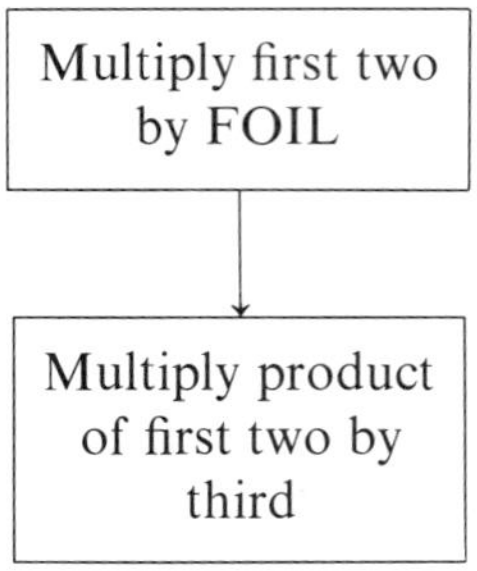


$$(2a + 1)(3a - 2) = 6a^2 - a - 2$$

$$\begin{array}{rrrr} & 6a^2 & - \ a & - \ 2 \\ \times & & 4a & + \ 5 \\ \hline & 30a^2 & - \ 5a & - \ 10 \\ 24a^3 & - \ 4a^2 & - \ 8a & \\ \hline 24a^3 & + \ 26a^2 & - \ 13a & - \ 10 \end{array}$$

The answer is $24a^3 + 26a^2 - 13a - 10$.

## PROBLEM SET 3.4

Multiply the following polynomials. Wherever possible, work the problem out in your head.

*See Example 13*

1. $4x(8x - 3y)$
2. $2a(6a - b)$
3. $3pq^2(5p - 2pq - 3q)$
4. $5x^2y^2(x + 2x^2y - 3xy^3)$
5. $-7m^2n^3(2mn - 5m^4 + 8n^3)$
6. $10rs^2t^3(2rt^4 - 3r^2s + 4s^3t^2)$

*See Examples 14 and 15*

7. $(x + 5)(x - 2)$
8. $(a + 6)(a + 5)$
9. $(2x - 3)(3x - 5)$
10. $(3x - 1)(4x + 3)$
11. $(5y - 3)(y + 7)$
12. $(8z - 1)(5z - 1)$
13. $(a + b)(a + 3b)$
14. $(p - 4q)(2p + q)$
15. $(3x - 2y)(5x + y)$
16. $(6m + n)(5m - 3n)$

*See Example 16*

17. $(r - 8)(r + 8)$
18. $(2a + 5)(2a - 5)$
19. $(4x - 3y)(4x + 3y)$
20. $(2c + 7d)(2c - 7d)$
21. $(10x + 3a)(10x - 3a)$
22. $(3a - 5b)(3a + 5b)$

*See Example 17*

23. $(t - 2)^2$
24. $(g + 6)^2$

25. $(x + 3)^2$

26. $(a - 5)^2$

27. $(p + 2q)^2$

28. $(3h - 2k)^2$

*See Example 14*

29. $(x + 5)(x^2 - 2x + 3)$

30. $(a + 7)(2a^2 - 5a + 3)$

31. $(2r - 3)(3r^2 - r + 4)$

32. $(3s + 6)(8s^2 - 5s - 2)$

33. $(m^2 + 5)(m^3 - 4m + 3)$

34. $(p^3 - 3p)(p^4 + 6p^3 - 2)$

35. $(x^2 + 2x + 3)(x^2 - 5x + 7)$

36. $(b^2 - b - 7)(2b^2 + 3b + 5)$

37. $(2a^3 - 4a - 1)(5a^3 - 2a^2 + 6)$

38. $(8t^4 - 3t^2 + 1)(5t^3 - 2t^2 - t)$

Multiply the following polynomials. To multiply three polynomials, multiply any two and then multiply the product by the third.

*See Example 18*

39. $(x + 1)(x + 2)(x + 3)$

40. $(a + 2)(a - 4)(a + 5)$

41. $(2r + 5)(r - 3)(3r - 1)$

42. $(6z + 1)(z + 2)(z - 2)$

43. $(x + 1)(x - 1)(x^2 - x + 3)$

44. $(m + 5)(m - 5)(m^2 - 5m - 6)$

45. $(x + 1)^2(x - 2)$

46. $(a - 2)^2(a + 3)$

47. $(p + 2)^2(p^2 - 4p + 3)$

48. $(k - 3)^2(2k^2 - 4k + 7)$

*Business application*

The total revenue (TR) and the total cost (TC) of producing a certain stereo system in quantity $Q$ is given by the following formulas. Write these as polynomials.

49. $\text{TR} = 200{,}000 - 25(Q - 20)^2$

50. $\text{TC} = (Q - 50)^3 + 250{,}000$

*Life science applications*

51. The total weight $W$ of an insect population is given by $W = n \cdot w$, where $n$ is the total number and $w$ the average weight. Find $W$ if $n = 3t^2 + 50$ and $w = 2t^2 - t + 3$ and $t$ is time.

52. The population of a bacteria is given by $P = 3000(2 + t)^2$. Write this as a polynomial.

*Engineering application*

53. The deflection $y$ in a beam held in a wall is given by

$$y = \frac{-P}{96EI}(L - x)^2(11x - 2L)$$

where $x$ is the distance from the wall, $P$ the weight on the beam, and $E$, $I$, and $L$ are constants of the beam's size (Fig. 3.1). Write this expression as a polynomial.

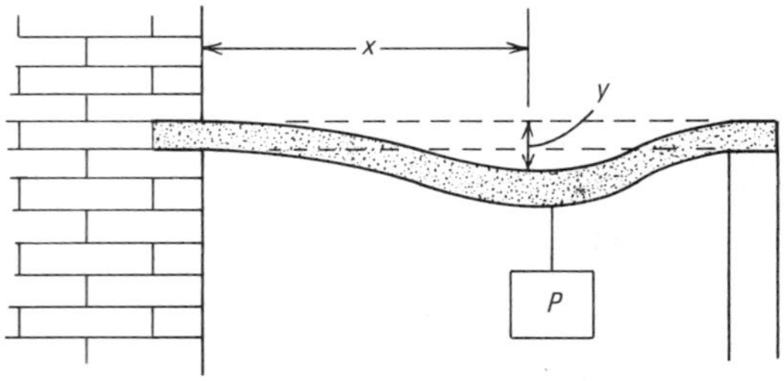


Figure 3.1

## 3.5 DIVIDING POLYNOMIALS

Let us first divide a polynomial by a monomial. Consider $\dfrac{A + B + C + \cdots}{M}$ where $A$, $B$, $C$, and $M$ are monomials. We have

| | |
|---|---|
| Division by $M$ means multiplication by $\dfrac{1}{M}$ | $\dfrac{A + B + C + \cdots}{M} = \dfrac{1}{M}[A + B + C + \cdots]$ |
| Distributive law | $= \dfrac{A}{M} + \dfrac{B}{M} + \dfrac{C}{M} + \cdots$ |

This gives us the following theorem.

***THEOREM 3.6*** For monomials $A$, $B$, $C$, and $M$,

$$\frac{A + B + C + \cdots}{M} = \frac{A}{M} + \frac{B}{M} + \frac{C}{M} + \cdots$$

**EXAMPLE 19** Divide $(15x^2y^3 + 25xy^5 - 10x^3y) \div 5xy$.

**SOLUTION** We write this division as a fraction and use Theorem 3.6 to "split the numerator."

Write as three fractions

$$\frac{15x^2y^3 + 25xy^5 - 10x^3y}{5xy} = \frac{15x^2y^3}{5xy} + \frac{25xy^5}{5xy} + \frac{-10x^3y}{5xy}$$

↓

Divide the monomials

$$= 3xy^2 + 5y^4 - 2x^2$$

| YES | NO |
|---|---|
| $\frac{6x^2 + 12x}{2x} = 3x + 6$ | ~~$\frac{2x}{6x^2 + 12x} = \frac{1}{3x} + \frac{1}{6}$~~ |

Theorem 3.6 works only when the monomial is in the denominator.

Dividing two polynomials is like dividing two whole numbers. Consider the similarity:

$$\begin{array}{r} 12 \\ 13\overline{)157} \\ 13 \\ \hline 27 \\ 26 \\ \hline 1 \end{array} \qquad \begin{array}{r} x + 2 \\ x + 3\overline{)x^2 + 5x + 7} \\ x^2 + 3x \quad \\ \hline 2x + 7 \\ 2x + 6 \\ \hline 1 \end{array}$$

In the whole-number problem, we say that the answer is $12\frac{1}{13}$. In the polynomial problem, we say that the answer is $x + 2 + \dfrac{1}{x + 3}$.

As a reminder, the key terms in a division problem are as follows:

$$\begin{array}{r} \text{Quotient} \\ \text{Divisor}\ \overline{)\text{Dividend}} \\ *** \quad\quad \\ \hline ***** \\ ***** \\ \hline \text{Remainder} \end{array}$$

We give the following rule for the long division of polynomials.

> To divide polynomials:
>
> 1. Arrange the terms of both polynomials so that all powers appear in descending order. Use zeros to fill in for missing powers.
> 2. Divide the first term (highest power) of the divisor into first term of the dividend. Put the answer in the quotient.
> 3. Multiply this quotient term by the divisor and put the product under like terms of the dividend.
> 4. Subtract (change signs and add).
> 5. Bring down next term.
> 6. Repeat steps 2 through 5 until the remainder term has degree less than that of divisor.
> 7. Write answer as Quotient $+ \dfrac{\text{Remainder}}{\text{Divisor}}$.

**EXAMPLE 20** Divide $2x^3 + 11x^2 - 25x - 28$ by $x + 7$.

**SOLUTION** Both divisor and dividend have their powers in descending order, and there are no missing powers. So we begin by dividing the highest powers, $2x^3 \div x = 2x^2$. We show the steps of the first cycle. (When we change signs to subtract, we will circle the new sign to avoid confusion.)

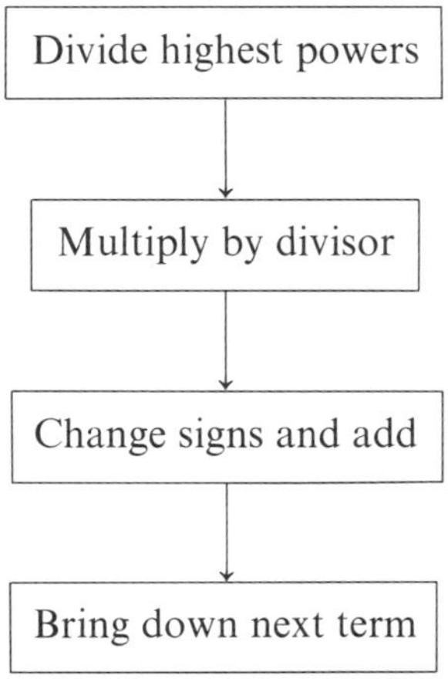


$$
\begin{array}{r}
2x^2 \phantom{{}+11x^2-25x-28} \\
x+7\overline{)\,2x^3 + 11x^2 - 25x - 28} \\
\ominus \quad\;\; \ominus \qquad\qquad\qquad\quad \\
2x^3 + 14x^2 \qquad \downarrow \quad\;\; \\
\hline
-3x^2 - 25x \qquad
\end{array}
$$

We repeat these same four steps, beginning by dividing highest powers, $-3x^2 \div x = -3x$. When completed, the division looks as follows:

$$\begin{array}{r@{\,}l}
 & 2x^2 - 3x - 4 \\ \cline{2-2}
x + 7 \big) & 2x^3 + 11x^2 - 25x - 28 \\
 & \ominus \quad \ominus \\
 & 2x^3 + 14x^2 \\ \cline{2-2}
 & -3x^2 - 25x \\
 & \oplus \quad \oplus \\
 & -3x^2 - 21x \\ \cline{2-2}
 & -4x - 28 \\
 & \oplus \quad \oplus \\
 & -4x - 28 \\ \cline{2-2}
\end{array}$$

The quotient is $2x^2 - 3x - 4$. There is no remainder. Also, we can check our answer by multiplying $2x^2 - 3x - 4$ by $x + 7$, and we get $2x^3 + 11x^2 - 25x - 28$.

**EXAMPLE 21** Divide $t^4 + 3t^3 - 5t - 6$ by $t^2 - 2t + 2$.

**SOLUTION** Notice that the dividend is missing a $t^2$-term, so we rewrite it as $t^4 + 3t^3 + 0t^2 - 5t - 6$. This acts as a placeholder just as the zero does in the number 43056. (As in Example 20, we circle the new signs when we subtract.)

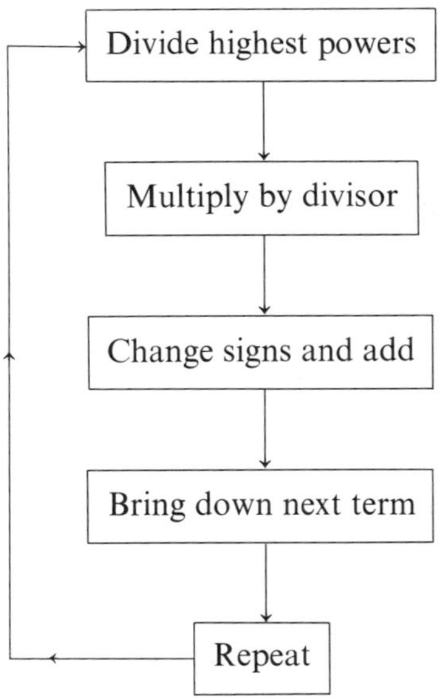


$$\begin{array}{r@{\,}l}
 & t^2 + 5t + 8 \\ \cline{2-2}
t^2 - 2t + 2 \big) & t^4 + 3t^3 + 0t^2 - 5t - 6 \\
 & \oplus \quad \ominus \\
 & t^4 - 2t^3 + 2t^2 \quad \downarrow \\ \cline{2-2}
 & 5t^3 - 2t^2 - 5t \\
 & \ominus \quad \oplus \quad \ominus \\
 & 5t^3 - 10t^2 + 10t \\ \cline{2-2}
 & 8t^2 - 15t - 6 \\
 & \ominus \quad \oplus \quad \ominus \\
 & 8t^2 - 16t + 16 \\ \cline{2-2}
 & t - 22
\end{array}$$

We say that the answer is $t^2 + 5t + 8 + \dfrac{t - 22}{t^2 - 2t + 2}$.

We study polynomial fractions such as this in more detail in Chapter 5.

## PROBLEM SET 3.5

Divide the following polynomials. Leave your answers in the form

$$\text{Quotient} + \frac{\text{Remainder}}{\text{Divisor}}$$

*See Example 19*

1. $\dfrac{8x^4 - 6x^3 + 10x^2}{2x}$

2. $\dfrac{27t^5 - 15t^3 - 6t^2}{3t^2}$

3. $\dfrac{15a^9 - 20a^5 - 25a}{5a}$

4. $\dfrac{8r^5 - 12r^4 + 16r^3}{4r^3}$

5. $\dfrac{14m^4n^3 - 7m^5n^2 + 21mn^5}{7mn^2}$

6. $\dfrac{6p^8q^3 - 5p^3q^4 + 3p^2q^{10}}{p^2q^3}$

7. $\dfrac{8s^4 - 2s^2 + 4}{2s}$

8. $\dfrac{10x^5 - 15x^3 - 5x}{5x^2}$

9. $\dfrac{4a^2b^5 - 6a^5b^3 + 8ab}{2ab^2}$

10. $\dfrac{3u^4v^5 - 6u^2v^3 + 9uv^2}{3u^2v^2}$

*See Example 20*

11. $\dfrac{x^2 - x - 6}{x + 2}$

12. $\dfrac{t^2 + 12t + 35}{t + 5}$

13. $\dfrac{a^3 - 8a^2 + 13a + 6}{a - 3}$

14. $\dfrac{2m^3 + 3m^2 - 24m - 16}{m + 4}$

15. $\dfrac{6r^3 - 11r^2 - 7r + 15}{2r - 3}$

16. $\dfrac{4y^3 - 31y + 15}{2y - 5}$

17. $\dfrac{5b^3 - 7b^2 - 3b + 2}{b - 3}$

18. $\dfrac{6k^3 - k^2 + 3k - 9}{k + 2}$

19. $\dfrac{t^4 - 6t^3 - 4t^2 - 5t - 6}{t + 2}$

20. $\dfrac{2z^4 - 3z^3 + 5z^2 - z - 1}{z - 3}$

21. $\dfrac{a^5 - 1}{a - 1}$

22. $\dfrac{b^6 + 1}{b + 1}$

23. $\dfrac{r^7}{r - 1}$

24. $\dfrac{p^6}{p + 2}$

*See Example 21*

25. $\dfrac{h^4 - 3h^3 + 4h^2 - 7h - 15}{h^2 - h + 5}$

26. $\dfrac{2s^4 + 5s^3 - 11s^2 - 10s + 8}{s^2 + 3s - 2}$

27. $\dfrac{m^4 - 3m^3 + 5m^2 - 10m + 6}{m^2 - 2m - 3}$

28. $\dfrac{x^4 - x^2 - 2x - 1}{x^2 - x + 2}$

29. $\dfrac{6k^4 - k^3 + 6k^2 - 3k + 1}{2k^2 - k + 1}$

30. $\dfrac{12r^4 - 8r^3 + 11r^2 - 15}{3r^2 - 2r + 5}$

# CHAPTER 3 SUMMARY

## Important Words and Phrases

binomial (3.1)
coefficient (3.1)
combining like terms (3.2)
degree of a monomial (3.1)
degree of a polynomial (3.1)
dividend (3.5)
divisor (3.5)
evaluate a polynomial (3.1)
expression (3.1)
FOIL (3.4)
like terms (3.2)
monomial (3.1)
opposite of a polynomial (3.2)
polynomial (3.1)
quotient (3.5)
remainder (3.5)
term (3.1)
trinomial (3.1)
variable (3.1)

## Important Theorems

For any real numbers $a$, $b$, $c$, $d$, and positive integers $m$ and $n$,

$$a^m \cdot a^n = a^{m+n}$$

$$\frac{a^m}{a^n} = \begin{cases} a^{m-n} & m > n \\ 1 & m = n \\ \dfrac{1}{a^{n-m}} & m < n \end{cases} \qquad (a \neq 0)$$

$$(a + b)(a - b) = a^2 - b^2$$

$$(a + b)^2 = a^2 + 2ab + b^2$$

$$(a - b)^2 = a^2 - 2ab + b^2$$

$$(a + b)(c + d) = ac + ad + bc + bd$$

$$\frac{A + B + C + \cdots}{M} = \frac{A}{M} + \frac{B}{M} + \frac{C}{M} + \cdots \quad \text{for monomials } A, B, C, M, \ldots$$

## Important Procedures

*To simplify an expression with parentheses:*

1. Work from inside to outside.
2. + before parentheses means leave unchanged.
3. − before parentheses means change all signs within.
4. Then combine like terms.

*To multiply two polynomials:*
1. Multiply all the terms of one by all the terms of the other.
2. Combine like terms.

*To divide two polynomials:*
1. Arrange terms in descending powers.
2. Use zeros for missing terms.
3. Divide first term of divisor into first term of dividend.
4. Multiply quotient term by divisor.
5. Subtract.
6. Bring down next term.
7. Repeat process until remainder term has degree less than divisor's.
8. Write the answer as Quotient $+ \dfrac{\text{Remainder}}{\text{Divisor}}$.

## CHAPTER 3 REVIEW EXERCISES

For the following terms, identify the coefficient.

1. $-a^5$
2. $3x^2y^3z^5$

For each of the following polynomials, give the degree and the special name.

3. $t^4 - 2t$
4. $5x^9 - 6x^4 - 10x$
5. If $P(x) = 3x^2 - 2x + 1$, find $P(2)$, $P(0)$, and $P(-1)$.

Simplify the following.

6. $a^2b - 7a^2b + 7a^2b$
7. $4r - (3s - [r - 5s - (7s - 2r)] + r - s)$

Perform the indicated operations.

8.
$$\begin{array}{rrrr} 7t^3 & -\,8t^2 & & -\,3 \\ 2t^3 & & +\,5t & +\,1 \\ + \quad & 6t^2 & -\,2t & -\,7 \\ \hline \end{array}$$

9.
$$\begin{array}{rrrrr} 10a^4 & -\,3a^3 & -\,6a^2 & & -\,10 \\ -\,(\;5a^4 & & -\,7a^2 & +\,a & -\;\;3) \\ \hline \end{array}$$

10. $(7u^3 - 4u^2 + 5u - 1) + (8u^3 - 3u^2 - 2u - 5) + (u^3 + 6u^2 + 7u + 7)$

11. $(4n^3 - 5n^2 + 6n - 1) - (7n^3 - 6n^2 - 4n - 6) - (2n^3 + 8n^2 - 5n + 3)$

12. $(6a^7) \cdot (7a^9)$

13. $(-2x^2y^4z)(6xy^7z^2)(-3x^3yz^4)$

14. $\dfrac{21t^5}{-3t^2}$

15. $\dfrac{-24p^5q^7r^2}{-3p^4q^4r^4}$

16. $\dfrac{(a + 3b)^4}{(a + 3b)^{10}}$

17. $-4xy^3(2x^2 - 3xy^5 - 4x^4y^2)$

18. $(3h + 2k)(4h - 5k)$

19. $(8u - 3)(8u + 3)$

20. $(2c - 5)^2$

21. $(m^2 + 2m + 5)(2m^2 - 3m - 4)$

22. $(t + 1)(t + 2)(t - 5)$

23. $\dfrac{10a^4 - 8a^3 + 12a}{2a}$

24. $\dfrac{6x^2y^5 - 3x^3y - 15xy^7}{3x^2y}$

25. $\dfrac{r^3 - 1}{r - 1}$

26. $\dfrac{3s^4 - 2s^3 + s^2 - 5s + 3}{s^2 + 2s - 3}$

# 4
# FACTORING

Factoring is basically the opposite of multiplying. When multiplying, we have two factors, and we wish to find the product:

$$\textit{Multiplication:}\quad (2x - 5)(x + 3) = \boxed{\quad ? \quad}$$

When factoring, we have the product, and we wish to find the factors which produced that product:

$$\textit{Factoring:}\quad 2x^2 + x - 15 = \boxed{?} \cdot \boxed{?}$$

**Factoring** is the process we use to split a number or polynomial into factors. Recall from arithmetic that a natural number, 2 or more, is either *prime* or can be written as a product of primes. For example,

$$28 = 2 \cdot 2 \cdot 7$$

$$39 = 3 \cdot 13$$

$$59 = \text{prime}$$

$$60 = 2 \cdot 2 \cdot 3 \cdot 5$$

Similarly, we can often split polynomials into factors. In this chapter, we look at some of the techniques for factoring polynomials.

## 4.1 GREATEST COMMON FACTOR

The first type of factoring uses the distributive law to factor out the **greatest common factor (GCF).** This is the largest expression that will divide evenly into all the terms of the polynomial. For example, $2x$ is the GCF of $2x^3$, $4x^2$, $8x$, since $2x$ is the largest monomial that will evenly divide those three terms.

> A polynomial has as factors:
>
> 1. The GCF.
> 2. The polynomial divided by the GCF.

**EXAMPLE 1** Factor $7x + 14y$.

**SOLUTION** Here, the GCF is 7, since 7 is the largest number that divides evenly into both terms.

| Divide out GCF, 7 | $7x + 14y = 7(x + 2y)$ |
|---|---|

**EXAMPLE 2** Factor $3x^2 + 6x$

**SOLUTION** The GCF is $3x$, since both terms have 3 and $x$ as factors.

| Divide out GCF, $3x$ | $3x^2 + 6x = 3x(x + 2)$ |
|---|---|

**EXAMPLE 3** Factor $10x^2y - 15xy^2 + 25xy$.

**SOLUTION** The GCF is $5xy$. Notice that we cannot divide $x^2$ or $y^2$ out of all the terms, so neither can be in the GCF.

| Divide out GCF, $5xy$ | $10x^2y - 15xy^2 + 25xy = 5xy(2x - 3y + 5)$ |
|---|---|

| YES | NO |
|---|---|
| $8x^2 + 4x = 4x(2x + 1)$ | ~~$8x^2 + 4x = 4x(2x)$~~ |

We must divide the GCF out of *all* the terms.

**EXAMPLE 4** Factor $7(a + b) - x(a + b)$.

**SOLUTION** Here the GCF is $(a + b)$, since it is common to both terms.

| Divide out GCF, $(a + b)$ | $7(a + b) - x(a + b) = (7 - x)(a + b)$ |
|---|---|

**EXAMPLE 5** Find the missing factor: $x^2 - 9x = -x(?)$.

**SOLUTION** Here we are factoring out $-x$ as the GCF. We still follow the same rule and divide each term by $-x$, using the rules for signed numbers.

| Divide out GCF, $-x$ | $x^2 - 9x = -x(-x + 9)$ |
|---|---|

## PROBLEM SET 4.1

Factor the following polynomials by factoring out the greatest common factor (GCF).

*See Example 1*

1. $8a + 8b$
2. $6x - 6y$
3. $4m - 8n$
4. $10p - 15q$
5. $12r - 18s + 24t$
6. $30x - 40y - 45z$

*See Example 2*

7. $2a + 3a^2$
8. $9m^4 - 8m^3$
9. $16k^2 - 12k$
10. $5a^3 - 15a^2$
11. $2x^4 - 3x^3 - 11x^2$
12. $6t - 4t^3 + 10t^5$
13. $8b^8 - 6b^6 - 4b^4 + 2b^2$
14. $12c^{12} + 9c^9 - 6c^6 - 3c^3$
15. $81ab - 9b^2$
16. $36x^2y - 24zy^2$

*See Example 3*

17. $12rs^3 + 18r^2s^2$
18. $20m^4n^3 - 15m^2n^2$
19. $25x^5y^2 + 10x^2y^3 - 30xy^7$
20. $10p^7q^2 - 12pq^3 - 14p^3q^4$
21. $21ab^2c^3 - 14a^3b^2c$
22. $15x^2y^4z^6 + 25x^3y^3z^3$
23. $16r^3p^4q^5 - 18r^4p^4$
24. $20t^4uv^2 - 22tu^5$
25. $30x^4y^2 - 25x^2z^2 + 40xyz$
26. $15a^2bc^4 - 18a^2c^2 - 24a^4b^2c$
27. $32rs^2t^3u^4 - 24r^2s^3t^3u^2$
28. $28a^2bc^4d^2 + 49a^4b^3c^2d$
29. $4w^2x^2y^2 + 6x^3yz^2 + 8w^4y^4z^3$
30. $6a^4bc^2 - 12a^2cd^2 - 15b^3c^4d^3$

*See Example 4*

31. $x(a + b) + 5(a + b)$
32. $3a(x + y) - 8(x + y)$
33. $6(p - q) - m(p - q)$
34. $r(m + n) - s(m + n)$
35. $8(2x - 1) - r(2x - 1)$
36. $7a(4y + 1) - 6b(4y + 1)$
37. $a(x + y)^2 - b(x + y)$
38. $4(2a - b) - x(2a - b)^2$

Find the missing factors.

*See Example 5*

39. $a^2 - 7a = -a(?)$
40. $8x - 10y = -2(?)$
41. $x^2 - x^3y - x^4 = -x(?)$
42. $-p^2 - p^3 = -p(?)$

43. $s^2t^2 - s^3t = -st(?)$ 44. $m^4n^5 - m^3n^7 = m^2n^4(?)$

*Business applications*

45. An item sells for a price of $p$. If there is a discount factor of $d$, the final price is reduced by $pd$. Factor the final price, $p - pd$.

46. A man invests $P$ dollars at an interest rate $r$. After 1 year this amount is $P(1 + r)$. In the second year, he earns $P(1 + r)r$ in interest. Factor and simplify this new balance, $P(1 + r) + P(1 + r)r$.

*Engineering application*

47. The maximum deflection $y$ in a beam with a weight $P$ on it is given by

$$y = \frac{PL^3}{3EI} + \frac{wL^4}{8EI}$$

where $w$, $L$, $E$, and $I$ are constants for the beam. Factor the expression on the right side of the equation.

## 4.2 FACTORING TRINOMIALS

Since factoring is the reverse of multiplying, it is useful to look again at how the FOIL method worked to multiply binomials.

$$(x + 5)(x + 2) = x^2 + 7x + 10$$
$$(x - 3)(x + 7) = x^2 + 4x - 21$$
$$(2x + 1)(3x + 2) = 6x^2 + 7x + 2$$
$$(4x - 3)(2x + 5) = 8x^2 + 14x - 15$$
$$(4x + 3)(2x - 5) = 8x^2 - 14x - 15$$

$(F)$ product of first terms ↑ (first term)

↑ (middle term) sum of outer product and inner product $(O + I)$

↑ (last term) $(L)$ product of last terms

To work backwards and factor the trinomials, we will factor the first and last terms of the trinomial. Then by trial and error, we will try to get the middle term. Let us work two simple examples before we state a procedure.

**EXAMPLE 6** Factor $x^2 + 7x + 10$.

**SOLUTION** We factor the first and last terms, and then try to get the middle term. The first term, $x^2$, factors as $x \cdot x$, while the last term, 10, can be factored as $1 \cdot 10$ and as $2 \cdot 5$.

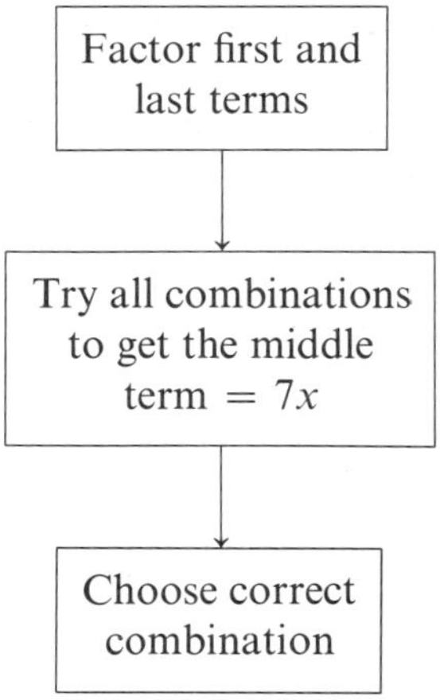


$x^2$ → $(x \quad)(x \quad)$ ; $+10$ → $(\quad +1)(\quad +10)$ or $(\quad +2)(\quad +5)$

| Combinations | Middle Term |
|---|---|
| $(x + 1)(x + 10)$ | $10x + x = 11x$ |
| $(x + 2)(x + 5)$ | $5x + 2x = 7x$ ✓ |

$$x^2 + 7x + 10 = (x + 2)(x + 5)$$

Notice that we try all the combinations of factors of the first with factors of the last. We then choose the one that gives the middle term.

**EXAMPLE 7** Factor $2m^2 - 9m + 10$.

**SOLUTION** We factor the first and last terms and try all the combinations. Notice that we will factor 10 as $(-1)(-10)$ and $(-2)(-5)$, since the middle term is negative.

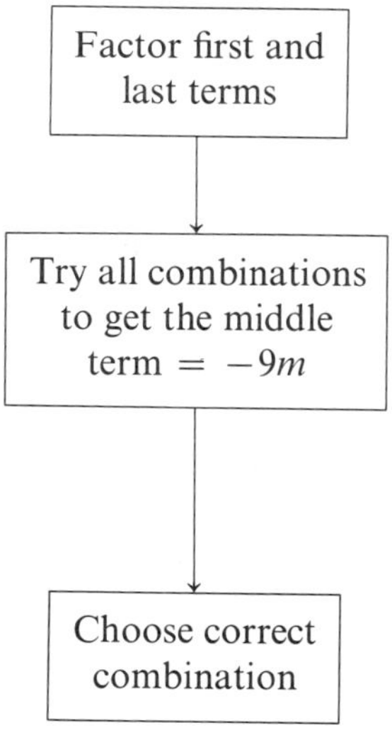


$2m^2$ → $(2m \quad)(m \quad)$ ; $+10$ → $(\quad -1)(\quad -10)$ or $(\quad -2)(\quad -5)$

| Combinations | Middle Term |
|---|---|
| $(2m - 1)(m - 10)$ | $-20m - m = -21m$ |
| $(2m - 10)(m - 1)$ | $-2m - 10m = -12m$ |
| $(2m - 2)(m - 5)$ | $-10m - 2m = -12m$ |
| $(2m - 5)(m - 2)$ | $-4m - 5m = -9m$ ✓ |

$$2m^2 - 9m + 10 = (2m - 5)(m - 2)$$

Let us now write out the general procedure.

> To factor the general trinomial, $ax^2 + bx + c$:
>
> 1. Find all the factorings of the first term, $ax^2$.
> 2. Find all the factorings of the last term, $c$.
> 3. Look at all the binomial combinations of factors of $ax^2$ with factors of $c$.
> 4. Find the combination that gives the middle term $= bx$.
> 5. If you get the middle term to be $-bx$ (right number, but wrong sign), reverse *both* signs of the factors of $c$.

As we work these out in the book, we show all the combinations together with the correct one. When you work a problem, you will no doubt stop as soon as you hit the correct combination.

**EXAMPLE 8** Factor $2x^2 + 5x - 12$.

**SOLUTION** We factor the first and last terms and try all the possibilities.

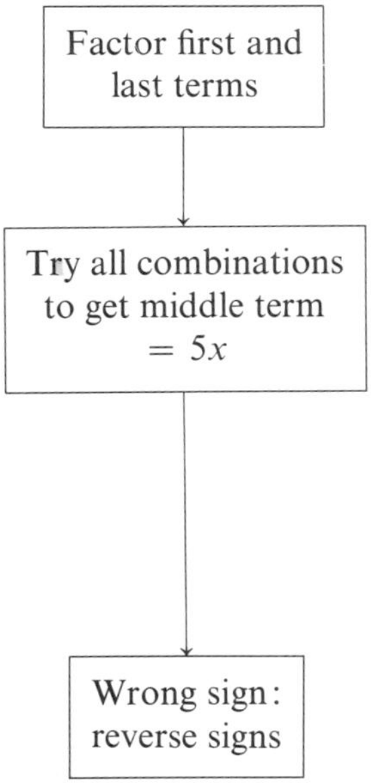


$$\underbrace{(2x \quad)(x \quad)}_{2x^2} \longleftrightarrow \underbrace{\begin{array}{l}(\quad +1)(\quad -12)\\(\quad -2)(\quad -6)\\(\quad +3)(\quad -4)\end{array}}_{-12}$$

| COMBINATIONS | MIDDLE TERM | |
|---|---|---|
| $(2x + 1)(x - 12)$ | $-24x + x = -23x$ | |
| $(2x - 12)(x + 1)$ | $2x - 12x = -10x$ | |
| $(2x + 2)(x - 6)$ | $-12x + 2x = -10x$ | |
| $(2x - 6)(x + 2)$ | $4x - 6x = -2x$ | |
| $(2x + 3)(x - 4)$ | $-8x + 3x = -5x$ | (almost) |
| $(2x - 4)(x + 3)$ | $6x - 4x = 2x$ | |

$$2x^2 + 5x - 12 = (2x - 3)(x + 4)$$

Notice that we found the middle term $= -5x$. Since only the sign was wrong, we changed $+3$ and $-4$ into $-3$ and $+4$ to get $(2x - 3)(x + 4)$. Finally, notice that there were two possibilities for each of the combinations of factors.

**EXAMPLE 9** Factor $6a^2 + 11ab - 10b^2$.

**SOLUTION** We first factor $6a^2$ and $-10b^2$ and try all the combinations, looking for $11ab$.

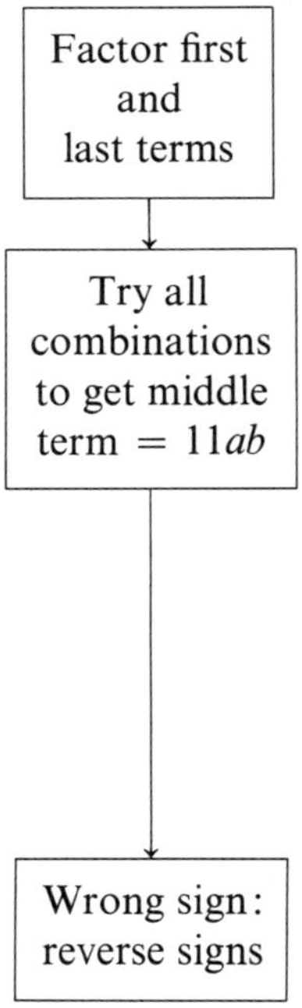


$$\overbrace{\begin{matrix}(6a \quad)(a \quad) \\ (3a \quad)(2a \quad)\end{matrix}}^{6a^2} \longleftrightarrow \overbrace{\begin{matrix}(\quad +b)(\quad -10b) \\ (\quad +2b)(\quad -5b)\end{matrix}}^{-10b^2}$$

| Combinations | Middle Term |
|---|---|
| $(6a + b)(a - 10b)$ | $-60ab + ab = -59ab$ |
| $(6a - 10b)(a + b)$ | $6ab - 10ab = -4ab$ |
| $(6a + 2b)(a - 5b)$ | $-30ab + 2ab = -28ab$ |
| $(6a - 5b)(a + 2b)$ | $12ab - 5ab = 7ab$ |
| $(3a + b)(2a - 10b)$ | $-30ab + 2ab = -28ab$ |
| $(3a - 10b)(2a + b)$ | $3ab - 20ab = -17ab$ |
| $(3a + 2b)(2a - 5b)$ | $-15ab + 4ab = -11ab$ (almost) |
| $(3a - 5b)(2a + 2b)$ | $6ab - 10ab = -4ab$ |

$$6a^2 + 11ab - 10b^2 = (3a - 2b)(2a + 5b)$$

**EXAMPLE 10** Factor $3x^2 + x + 2$.

**SOLUTION** We start by factoring both the first and last terms.

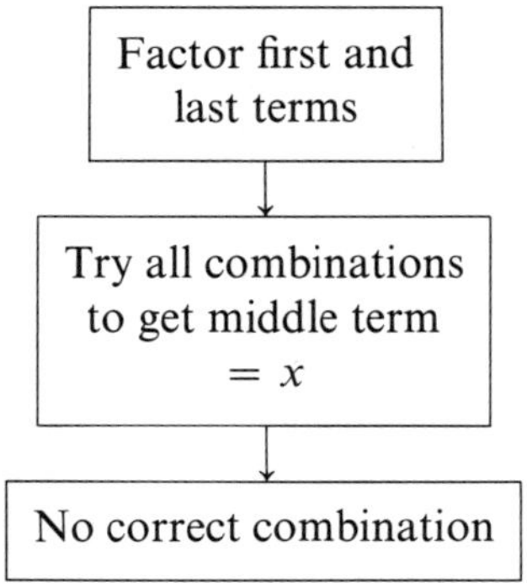


$$\overbrace{(3x \quad)(x \quad)}^{3x^2} \longleftrightarrow \overbrace{(\quad +1)(\quad +2)}^{2}$$

| Combinations | Middle Term |
|---|---|
| $(3x + 1)(x + 2)$ | $6x + x = 7x$ |
| $(3x + 2)(x + 1)$ | $3x + 2x = 5x$ |

$3x^2 + x + 2$ *cannot* be factored

Just as 7 is a prime number, so are there polynomials, such as $3x^2 + x + 2$, that cannot be factored with integer coefficients.

**EXAMPLE 11** Factor $a^4 - a^2 - 6$.

**SOLUTION** We can work this problem like the ones before: factor the first and last terms and try the combinations to get the middle term.

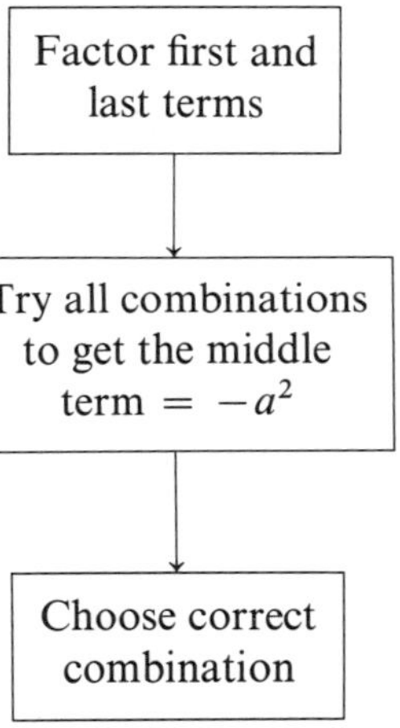


$$\overbrace{(a^2 \quad)(a^2 \quad)}^{a^4} \longleftrightarrow \overbrace{(\quad +2)(\quad -3)}^{-6}$$
$$(\quad +1)(\quad -6)$$

| COMBINATIONS | MIDDLE TERM |
|---|---|
| $(a^2 + 2)(a^2 - 3)$ | $-3a^2 + 2a^2 = -a^2$ ✓ |
| $(a^2 + 1)(a^2 - 6)$ | $-6a^2 + a^2 = -5a^2$ |

$$a^4 - a^2 - 6 = (a^2 + 2)(a^2 - 3)$$

## PROBLEM SET 4.2

Factor the following trinomials. (*Be careful:* Some cannot be factored with integers—see Example 10.)

*See Examples 6 and 7*

1. $x^2 + 6x + 5$
2. $a^2 + 8a + 12$
3. $w^2 - 7w + 12$
4. $c^2 + 3c - 10$
5. $y^2 + 8y - 6$
6. $n^2 - 7n + 10$
7. $x^2 - 15x + 56$
8. $m^2 - m - 20$
9. $p^2 + 13p + 30$
10. $x^2 + 4x - 32$
11. $a^2 + 6ab - 16b^2$
12. $r^2 - 11rs - 60s^2$
13. $u^2 + 13uv - 30v^2$
14. $h^2 - 2hk - 48k^2$
15. $x^2 - 12xy + 35y^2$
16. $p^2 + pq - 72q^2$
17. $u^2 + uv - 156v^2$
18. $a^2 - 11ab + 28b^2$
19. $p^2 - 5pq - 21q^2$
20. $s^2 - 3st - 54t^2$

*See Examples 7 and 8*

21. $6x^2 + 5x + 1$
22. $6a^2 + 23a + 20$
23. $12t^2 + 5t - 2$
24. $6r^2 + 29r + 20$
25. $9m^2 - 3m - 2$
26. $8k^2 - 25k + 15$
27. $10y^2 + 9y - 9$
28. $7z^2 + 37z + 10$
29. $20w^2 - 17w + 3$
30. $16s^2 - 2s - 3$

*See Example 9*

31. $15x^2 + 29xy - 14y^2$
32. $10a^2 - 17ab + 3b^2$
33. $6p^2 - 13pq - 63q^2$
34. $10u^2 + 23uv - 5v^2$
35. $12r^2 + 29rs - 8s^2$
36. $32h^2 + 60hk + 25k^2$
37. $5c^2 - 37cd + 42d^2$
38. $8x^2 + 22xy - 21y^2$
39. $2a^2 + 15ab + 36b^2$
40. $18p^2 - 21pq - 4q^2$

*See Example 11*

41. $a^4 + 7a^2 + 10$
42. $b^4 + 7b^2 + 6$
43. $x^4 + 5x^2 - 14$
44. $y^4 - 13y^2z^2 + 40z^4$
45. $t^4 + 11t^2s^2 + 18s^4$
46. $8u^4 + 18u^2 + 9$
47. $20r^4 + 7r^2 - 3$
48. $10p^4 - 31p^2q^2 - 63q^4$
49. $6a^4 - 19a^2b^2 + 3b^4$
50. $8x^4 + 17x^2y^2 + 7y^2$

## 4.3 SPECIAL FACTORING

There are several special factorings that come up quite a bit. The first three are the reverses of the special multiplication Theorems 3.3 to 3.5.

> ***THEOREM 4.1*** For any real numbers $a$ and $b$,
>
> $$a^2 - b^2 = (a + b)(a - b)$$

**EXAMPLE 12** These examples are all the difference of two squares, so we can use Theorem 4.1.

**(a)** $x^2 - 49 = (x + 7)(x - 7)$

**(b)** $4a^2 - 9b^2 = (2a + 3b)(2a - 3b)$

**(c)** $9 - (u + v)^2 = (3 + u + v)(3 - u - v)$

**(d)** $p^4 - q^4 = (p^2 + q^2)(p^2 - q^2)$
$\quad = (p^2 + q^2)(p + q)(p - q)$

Notice that we used Theorem 4.1 twice in the fourth example.

***THEOREM 4.2*** For any real numbers $a$ and $b$,

$$a^2 + 2ab + b^2 = (a + b)^2$$

**EXAMPLE 13** **(a)** $x^2 + 6x + 9 = (x + 3)^2$

**(b)** $4a^2 + 12ab + 9b^2 = (2a + 3b)^2$

**(c)** $1 + 14p^2 + 49p^4 = (1 + 7p^2)^2$

***THEOREM 4.3*** For any real numbers $a$ and $b$,

$$a^2 - 2ab + b^2 = (a - b)^2$$

**EXAMPLE 14** **(a)** $p^2 - 10p + 25 = (p - 5)^2$

**(b)** $16u^2 - 72uv + 81v^2 = (4u - 9v)^2$

| YES | NO |
|---|---|
| $a^2 + b^2$ *cannot* be factored | ~~$a^2 + b^2 = (a + b)^2$~~ |
| $x^2 - 25 = (x + 5)(x - 5)$ | ~~$x^2 - 25 = (x - 5)^2$~~ |

We also have two theorems for the sums and differences of cubes.

***THEOREM 4.4*** For any real numbers $a$ and $b$,

$$a^3 - b^3 = (a - b)(a^2 + ab + b^2)$$

***THEOREM 4.5*** For any real numbers $a$ and $b$,

$$a^3 + b^3 = (a + b)(a^2 - ab + b^2)$$

To help remember these theorems, consider the following display.

product

square ↓ ↓ ↓ square

$$a^3 - b^3 = (a - b)(a^2 + ab + b^2)$$
$$a^3 + b^3 = (a + b)(a^2 - ab + b^2)$$

same sign

opposite sign

**EXAMPLE 15** **(a)** $x^3 + 8 = (x + 2)(x^2 - 2x + 4)$

**(b)** $t^3 - 125 = (t - 5)(t^2 + 5t + 25)$

**(c)** $27u^3 + 64v^3 = (3u + 4v)(9u^2 - 12uv + 16v^2)$

**(d)** $1 - k^6 = (1 - k^3)(1 + k^3)$
$= (1 - k)(1 + k + k^2)(1 + k)(1 - k + k^2)$

| YES | NO |
|---|---|
| $p^3 + 27 = (p + 3)(p^2 - 3p + 9)$ | ~~$p^3 + 27 = (p + 3)^3$~~ |
| $8a^3 - 1 = (2a - 1)(4a^2 + 2a + 1)$ | ~~$8a^3 - 1 = (2a - 1)(4a^2 + 1)$~~ |

We can check $p^3 + 27 = (p + 3)(p^2 - 3p + 9)$ by multiplying.

$$\begin{array}{rrrr} & p^2 & - 3p & + 9 \\ \times & & p & + 3 \\ \hline & 3p^2 & - 9p & + 27 \\ p^3 & - 3p^2 & + 9p & \\ \hline p^3 & & & + 27 \end{array}$$

So we see that it checks.

## PROBLEM SET 4.3

Factor the following binomials. (*Be careful:* Some may factor more than once, and some may not factor at all.)

*See Example 12*

1. $x^2 - 25$
2. $a^2 - 36$
3. $1 - p^2$
4. $9 - u^2$
5. $16m^2 - 49n^2$
6. $25x^2 - 64y^2$
7. $81 - a^2b^2$
8. $w^2 - \frac{1}{4}u^2v^2$
9. $p^2 + 9q^2$
10. $16h^2 - 25m^2n^4$
11. $\frac{1}{9}u^2v^2 - a^2b^2$
12. $9r^2s^2t^2 - 16w^2t^4$
13. $(a + b)^2 - 1$
14. $(x + 2y)^2 - 9$
15. $(m + n)^2 - (a + b)^2$
16. $(2p - q)^2 - (3u - v)^2$

*See Example 12d*

17. $a^4 - b^4$
18. $y^4 - 16$
19. $x^4 - 81y^4$
20. $1 + x^4$

Factor each of the following perfect-square trinomials.

*See Examples 13 and 14*

21. $x^2 + 2x + 1$
22. $a^2 + 4a + 4$
23. $t^2 - 10t + 25$
24. $b^2 + 14b + 49$
25. $r^2 - 18r + 81$
26. $s^2 - 12s + 36$
27. $a^2 - 6ab + 9b^2$
28. $p^2 + 8pq + 16q^2$
29. $4x^2 - \frac{4}{3}x + \frac{1}{9}$
30. $t^2 + t + \frac{1}{4}$
31. $4r^2 + 36rs + 81s^2$
32. $9u^2 - 24uv + 16v^2$
33. $25a^2 - 60abc + 36b^2c^2$
34. $49x^2y^4 + 56xy^2z + 16z^2$

*See Example 13c*

35. $a^4 + 18a^2b^2 + 81b^4$
36. $4p^8 + 20p^4q^4 + 25q^8$

Factor each of the following sums and differences of cubes.

*See Example 15*

37. $x^3 + y^3$
38. $p^3 - q^3$
39. $a^3 + 1$
40. $t^3 - 8$
41. $m^3 - 125n^3$
42. $27u^3 - 8v^3$
43. $64w^3 + 1$
44. $125 - 8x^3y^3$
45. $a^3 - \frac{1}{27}$
46. $1 + \dfrac{x^3}{8}$
47. $27a^3b^3 + 1000c^3d^3$
48. $8x^3y^3z^6 - u^3v^3w^6$

*See Example 15d*

49. $x^6 - y^6$
50. $a^6 + b^6$

51. $u^9 - v^9$

52. $p^9 + q^9$

53. $m^{12} + n^{12}$

54. $r^{15} - s^{15}$

## 4.4 FACTORING BY GROUPING

Sometimes, we cannot factor a polynomial by dividing out a common factor, as a trinomial, or a special product. One last method that we will study is factoring by **grouping.**

**EXAMPLE 16** Factor $a^2 - 6a + 2ab - 12b$.

**SOLUTION** There is no common factor; it is not a trinomial; and it is not a special product. We factor the first two terms and then the last two.

Factor first and last pair separately

$$a^2 - 6a + 2ab - 12b = a(a - 6) + 2b(a - 6)$$

Divide out GCF, $(a - 6)$

$$= (a - 6)(a + 2b)$$

Notice that the first and last pair both had the same factor, $a - 6$. This is how factoring by grouping works: if possible, we factor twice.

> To factor four terms by grouping:
>
> 1. Arrange the four terms into two pairs.
> 2. Factor the first and last pairs separately.
> 3. If these pairs have a GCF, factor that out.

**EXAMPLE 17** Factor $xy + xz - 2y - 2z$.

**SOLUTION** We factor in pairs, and factor the results.

| | |
|---|---|
| Factor first and last pair separately | $xy + xz - 2y - 2z = x(y + z) - 2(y + z)$ |
| ↓ Divide out GCF, $(y + z)$ | $= (x - 2)(y + z)$ |

**EXAMPLE 18** Factor $p^2 - 6q - 3p + 2pq$.

**SOLUTION** We cannot factor this, so we rearrange the terms first.

| | |
|---|---|
| Rearrange terms | $p^2 - 6q - 3p + 2pq = p^2 + 2pq - 3p - 6q$ |
| ↓ Factor first and last pair separately | $= p(p + 2q) - 3(p + 2q)$ |
| ↓ Divide out GCF, $(p + 2q)$ | $= (p - 3)(p + 2q)$ |

**EXAMPLE 19** Factor $a + b - c + 5$.

**SOLUTION** There are no ways to group these so that we get a GCF, so it just cannot be factored.

## PROBLEM SET 4.4

Factor the following polynomials by grouping. (*Be careful:* Some cannot be factored—see Example 19; some must be rearranged first—see Example 18.)

*See Examples 16 and 17*

1. $ab + 3a + 2b + 6$
2. $xy - 7x + 5y - 35$
3. $ac + bc + ad + bd$
4. $a^2 - ab + as - sb$
5. $x^2 - 8x - xy + 8y$
6. $pq - 6p - 2q^2 + 12q$
7. $uv + 8u - 7uv - 56v$
8. $ac - 5a + 2bc - 10b$
9. $2m^2 + mn - 10m - 5n$
10. $12r^2 + 15r - 8rs - 10s$

11. $12k^2 - 21k - 20hk + 35h$
12. $24ab + 54a - 4b - 9$
13. $x - y + z - 7$
14. $18ab - 63a - 4b + 14$
15. $2ax - 4cx - 5a + 10c$
16. $2rs - 3ra - 4ab - 5bc$

## 4.5 GENERAL FACTORING

Let us combine all these rules and techniques into one general rule for factoring.

> When factoring any polynomial:
>
> 1. Factor out the greatest common factor (GCF).
>
> 2a. If there are two terms, try a special factoring.
>
> 2b. If there are three terms, try factoring as a trinomial.
>
> 2c. If there are four terms, try grouping.

**EXAMPLE 20** Factor $x^5y^3 - xy^7$.

**SOLUTION** We start with the GCF; then we use a special factoring.

| | |
|---|---|
| Divide out GCF, $xy^3$ | $x^5y^3 - xy^7 = xy^3(x^4 - y^4)$ |
| ↓ Factor difference of squares | $= xy^3(x^2 - y^2)(x^2 + y^2)$ |
| ↓ Factor difference of squares | $= xy^3(x + y)(x - y)(x^2 + y^2)$ |

**EXAMPLE 21** Factor $2a^6 + 4a^5 - 70a^4$.

**SOLUTION** We first factor out the GCF, and then factor the trinomial.

| | |
|---|---|
| Divide out GCF, $2a^4$ | $2a^6 + 4a^5 - 70a^4 = 2a^4(a^2 + 2a - 35)$ |
| ↓ Factor trinomial | $= 2a^4(a + 7)(a - 5)$ |

## PROBLEM SET 4.5

Factor the following polynomials completely.

*See Example 20*

1. $x^5 - x^3$
2. $8a^3 - 18a$
3. $5x^4 - 80x^2$
4. $27r^2s^2 - 75s^2$
5. $t^{10} - t^2$
6. $32u^5 - 2uv^4w^8$
7. $b^4 - b$
8. $z^5 + z^2$
9. $24x^4y + 81xy^4$
10. $7n^4 - 56m^3n^4$
11. $a - 64a^7b^6$
12. $3x^7y^2 - 3xy^8$

*See Example 21*

13. $4a^3 + 4a^2 - 24a$
14. $3t^4 + 6t^3 - 45t^2$
15. $5x^5 - 35x^4 + 30x^3$
16. $2wz^3 + 14z^2w^2 - 36w^3z$
17. $18a^4 - 21a^3 + 6a^2$
18. $20r^6 + 22r^5 - 12r^4$
19. $18s^3t - 69s^2t^3$
20. $16x^2z^2 + 28xyz^2 - 30y^2z^2$
21. $abc - 2ac + 7bc - 14c$
22. $u^2w + u^2t - uvw - uvt$
23. $4p^3q - 20p^2q + 8p^2q^2 - 40pq^2$
24. $40x^3y^3 - 16x^3y^2 + 30x^2y^3 - 12x^2y^2$

*Business applications*

25. A businessman discovers that the relation between the price $p$ of an item and the demand $q$ is given by

$$p = 0.0005q^2 - q + 500$$

Factor the right side of this equation. (*Hint:* Factor out 0.0005 first.)

If a manufacturer produces $x$ items, his total revenue TR $= 1000x$. His total cost TC $= x^2 + 600x + 30{,}000$.

26. Find the total profit, TP = TR − TC.
27. Factor this expression for TP.

*Health application*

28. An infection is spreading through a community at a rate given by $r = 10y - 0.0001y^2$. Factor this expression completely. (*Hint:* Factor out 0.0001 first.)

*Life science application*

29. A bacteria-culture population is given by

$$P = 12{,}000 + 12{,}000t + 3000t^2$$

where $t$ is the time in hours. Factor this completely.

## CHAPTER 4 SUMMARY

### Important Words and Phrases

factoring (introduction)
FOIL (4.2)
greatest common factor (4.1)
grouping (4.4)
prime (introduction)
special factoring (4.3)

### Important Theorems

For all real numbers $a$ and $b$,

$a^2 - b^2 = (a - b)(a + b)$

$a^2 + b^2$ does not factor

$a^2 - 2ab + b^2 = (a - b)^2$

$a^2 + 2ab + b^2 = (a + b)^2$

$a^3 - b^3 = (a - b)(a^2 + ab + b^2)$

$a^3 + b^3 = (a + b)(a^2 - ab + b^2)$

### Important Procedures

*To factor trinomials by trial and error:*
1. Factor the first and third terms.
2. Try all the binomial combinations to get the middle term of the trinomial. (If we get the middle term with the wrong sign, we reverse the signs of the factors of the third term.)

*A polynomial has as factors:*
1. The GCF.
2. The polynomial divided by the GCF.

*To factor four terms by grouping:*
1. Arrange into two pairs.
2. Factor the first and last pair separately.
3. If possible, factor out the GCF from these pairs.

*To factor any polynomial:*
1. Factor out the GCF.
2. Try a special factoring.
3. Try factoring as a trinomial.
4. Try grouping.

## CHAPTER 4 REVIEW EXERCISES

Factor the following polynomials completely.

1. $2x - 10x^7$
2. $20y^8 - 15y^6 + 10y^4 - 5y^2$
3. $3ab^2c^3 - 6a^2b^2c^3 + 9a^3b^2c$
4. $8r(a - 2b) - 7s(a - 2b)$
5. $m^2 + 3m - 40$
6. $p^2 - pq - 12q^2$
7. $24t^2 + 7t - 5$
8. $15a^2 - 26ab + 8b^2$
9. $h^4 + 6h^2 - 27$
10. $6r^4 - 7r^2 + 2$
11. $1 - 49n^2$
12. $25a^2b^2c^4 - 36u^2v^4$
13. $(x + 2y)^2 - z^2$
14. $4x^2 - 28xy + 49y^2$
15. $a^3 + 125$
16. $8u^3 - 1000v^3$
17. $a^2 + 3a - 2ab - 6b$
18. $rs - rt - s^2 + st$
19. $x^6 - x^2$
20. $x^4y^2 + xy^5$
21. $2a^7b - 128ab^7$
22. $4r^5 - 4r^4 - 80r^3$
23. $24s^2t - 42st + 9t$
24. $8p^3 - 24p^2 - 8p^2q + 24pq$

# 5
# FRACTIONS

In arithmetic we know that a **fraction** is the quotient of two integers, usually written $\frac{A}{B}$. $A$ is called the **numerator,** and $B$ is called the **denominator.** The denominator $B$ can *never* be zero.

In algebra, a fraction is still a quotient, $\frac{A}{B}$, except that both $A$ and $B$ may now be any expressions. Just as with number fractions, the denominator $B$ can *never* be zero. An algebraic fraction is also called a **rational expression.**

The following are examples of algebraic fractions:

(a) $\dfrac{x^2 + 2x + 7}{3x^3 - 5x - 8}$ (b) $\dfrac{a^2b - a^7b^3 + 3}{6a - 4b^3 + 5ab}$

(c) $\dfrac{17}{x + y}$ (d) $\dfrac{a + 2b + c}{5x}$

Throughout this chapter, it will be clear how similar algebraic fractions are to number fractions. We will constantly be using the similarity to help explain the new concepts.

## 5.1 REDUCTION OF FRACTIONS

We recall from arithmetic that

$$\frac{10}{15} = \frac{2 \cdot 5}{3 \cdot 5} = \frac{2}{3} \cdot \left(\frac{5}{5}\right) = \frac{2}{3}, \text{ since } \frac{5}{5} = 1$$

> We can multiply or divide both the numerator and denominator of a fraction by the same number without changing its value, since this is really multiplication by 1.

We can also write this as a theorem.

> ***THEOREM 5.1*** For any expressions $A$, $B$, and $C$ $(B \neq 0, C \neq 0)$:
>
> $$\frac{AB}{CB} = \frac{A}{C} \cdot \frac{B}{B} = \frac{A}{C}$$

We use Theorem 5.1 to reduce fractions. A fraction is **reduced to lowest terms** when the numerator and denominator have no common factors.

**EXAMPLE 1** Reduce $\frac{30}{84}$ to lowest terms.

**SOLUTION** We will use a method slightly different from that which you may have learned in arithmetic.

| | |
|---|---|
| Factor numerator and denominator into primes | $\frac{30}{84} = \frac{2 \cdot 3 \cdot 5}{2 \cdot 2 \cdot 3 \cdot 7}$ |
| ↓ Cancel common factors | $= \frac{5}{2 \cdot 7} \cdot \frac{\cancel{2 \cdot 3}}{\cancel{2 \cdot 3}}$ |
| ↓ Simplify remaining terms | $= \frac{5}{14}$ |

This procedure is easily seen to work with numbers. It works just as well with algebraic fractions.

> To reduce a fraction to its lowest terms:
>
> 1. Factor the numerator and denominator completely.
> 2. Cancel all common factors.
> 3. Simplify the remaining terms.

**EXAMPLE 2** Reduce $\frac{14x^3y^5z^3}{21x^7y^2z^3}$ to lowest terms.

**SOLUTION** We can use Theorem 3.2 to help us divide the monomials.

Factor completely → Cancel common factors

$$\frac{14x^3y^5z^3}{21x^7y^2z^3} = \frac{2 \cdot 7x^3y^5z^3}{3 \cdot 7x^7y^2z^3}$$

$$= \frac{2y^3}{3x^4} \cdot \frac{\cancel{7x^3y^2z^3}}{\cancel{7x^3y^2z^3}} = \frac{2y^3}{3x^4}$$

**EXAMPLE 3** Reduce $\dfrac{6x^3y^2 - 3x^2y}{3x^3y^2 + 5x^2y^3}$ to lowest terms.

**SOLUTION** We must first factor before we cancel.

Factor completely → Cancel common factors → Simplify

$$\frac{6x^3y^2 - 3x^2y}{3x^3y^2 + 5x^2y^3} = \frac{3x^2y(2x - 1)}{x^2y^2(3x + 5y)}$$

$$= \frac{3(2x - 1)}{y(3x + 5y)} \cdot \frac{\cancel{x^2y}}{\cancel{x^2y}}$$

$$= \frac{3(2x - 1)}{y(3x + 5y)}$$

| YES | NO |
|---|---|
| $\dfrac{6x^3y^2 - 3x^2y}{3x^3y^2 + 5x^2y^3} = \dfrac{3(2x - 1)}{y(3x + 5y)} \cdot \dfrac{\cancel{x^2y}}{\cancel{x^2y}}$ | ~~$\dfrac{\cancel{6x^3y^2} - 3x^2y}{\cancel{3x^3y^2} + 5x^2y^3} = 2 - \dfrac{3}{5y}$~~ |
| We must factor first! | |

**EXAMPLE 4** Reduce $\dfrac{9y^2 - 4x^2}{8x^2 - 10xy - 3y^2}$ to lowest terms.

**SOLUTION** We first factor both numerator and denominator.

| | |
|---|---|
| Factor completely | $$\frac{9y^2 - 4x^2}{8x^2 - 10xy - 3y^2} = \frac{(3y - 2x)(3y + 2x)}{(2x - 3y)(4x + y)}$$ |
| ↓ | |
| Rewrite $(3y - 2x) = (-1)(2x - 3y)$ and cancel common factor | $$= \frac{(-1)\cancel{(2x - 3y)}(3y + 2x)}{\cancel{(2x - 3y)}(4x + y)}$$ |
| ↓ | |
| Simplify | $$= \frac{-3y - 2x}{4x + y}$$ or $$-\frac{3y + 2x}{4x + y}$$ |

Notice that we wrote $3y - 2x$ as $(-1)(2x - 3y)$ so that we could cancel with the $(2x - 3y)$ term in the denominator.

## PROBLEM SET 5.1

Reduce each of the following to lowest terms.

See Example 1

1. $\dfrac{15}{21}$ 2. $\dfrac{20}{65}$ 3. $\dfrac{16}{60}$

See Example 2

4. $\dfrac{33}{121}$ 5. $\dfrac{6x^4}{8x^5}$ 6. $\dfrac{10a^6}{15a^2}$

7. $\dfrac{18a^2b^2c^2}{21ab^2c^3}$ 8. $\dfrac{20m^4n^5r^6}{28m^7n^4r^3}$ 9. $\dfrac{35x^7y^9z^{10}}{21x^{10}y^4}$

10. $\dfrac{-34ab^2c^3}{51a^6c}$ 11. $\dfrac{45a^2b^3c^7}{-40b^7d^{10}}$ 12. $\dfrac{105mnp^4}{70m^4pq}$

See Example 3

13. $\dfrac{6x + 6y}{x^2 + xy}$ 14. $\dfrac{7a - 7b}{ab - b^2}$ 15. $\dfrac{7x + x^2}{xy + x^3}$

16. $\dfrac{8a^2 - a}{ab + a^3}$ 17. $\dfrac{a - b}{b - a}$ 18. $\dfrac{7 - x}{x - 7}$

19. $\dfrac{ab + ax}{ab - ax}$ 20. $\dfrac{xy + y^2}{xy - y^2}$ 21. $\dfrac{a^3 + a^2x + ax^2}{ab}$

See Example 4

22. $\dfrac{x^2y}{x^4y^2 - x^3y + xy^5}$ 23. $\dfrac{x^2 - y^2}{xy + y^2}$ 24. $\dfrac{7a - ab}{b^2 - 49}$

25. $\dfrac{x^2 + 4x - 5}{8x + 40}$ 26. $\dfrac{5x - 35}{x^2 - 9x + 14}$ 27. $\dfrac{x^2 + 2x - 35}{x^2 - 25}$

28. $\dfrac{x^2 - 36}{x^2 + 5x - 6}$ 29. $\dfrac{x^2 + 11x + 30}{x^2 + 4x - 12}$ 30. $\dfrac{x^2 + x - 12}{x^2 - 8x + 15}$

31. $\dfrac{x^2 + 15x + 56}{x^2 + 14x + 48}$ 32. $\dfrac{x^2 - 10x + 24}{x^2 - 9x + 20}$ 33. $\dfrac{a^2 - b^2}{(a - b)^3}$

34. $\dfrac{(m - 5)^2}{m^2 - 25}$ 35. $\dfrac{3x^2 + 16x + 5}{2x^2 + 11x + 5}$ 36. $\dfrac{3a^2 + 19a - 14}{3a^2 - 17a + 10}$

37. $\dfrac{6x^2 + x - 12}{9x^2 - 6x - 8}$ 38. $\dfrac{12a^2 - 11a - 15}{15a^2 - 19a - 10}$

39. $\dfrac{x^2 + 5x + xy + 5y}{2x^2 - xy + 10x - 5y}$ 40. $\dfrac{5a - a^2 - 5b + ab}{10a - 2a^2 + 5b - ab}$

41. $\dfrac{x^2 - 5x + xy - 5y}{x^2 - 11x + 30}$ 42. $\dfrac{p^2 - p - 42}{2p^2 - pq + 12p - 6q}$

43. $\dfrac{a^3 - 8}{a^2 - 4}$ 44. $\dfrac{x^2 + 3x - 10}{x^3 + 125}$

## 5.2 MULTIPLICATION OF FRACTIONS

The rule for multiplying algebraic fractions is the same as for number fractions. We have the following definition.

> **DEFINITION 5.1** For any expressions $A$, $B$, $C$, and $D$ $(B \neq 0, D \neq 0)$,
>
> $$\frac{A}{B} \cdot \frac{C}{D} = \frac{A \cdot C}{B \cdot D}$$

**EXAMPLE 5** Multiply $\dfrac{9}{12} \cdot \dfrac{4}{15}$.

**SOLUTION** Let us work this by factoring first, and then multiplying and canceling.

**Factor**

$$\frac{9}{16}\cdot\frac{4}{15}=\frac{3\cdot 3}{2\cdot 2\cdot 2\cdot 2}\cdot\frac{2\cdot 2}{3\cdot 5}$$

**Multiply across**

$$=\frac{3\cdot 3\cdot 2\cdot 2}{2\cdot 2\cdot 2\cdot 2\cdot 3\cdot 5}$$

**Cancel common factors**

$$=\frac{3}{2\cdot 2\cdot 5}\cdot\frac{\cancel{3\cdot 2\cdot 2}}{\cancel{3\cdot 2\cdot 2}}$$

**Simplify**

$$=\frac{3}{20}$$

This is the technique that we will use when we multiply algebraic fractions.

> To multiply algebraic fractions:
>
> 1. Factor all polynomials.
> 2. Multiply across: numerator times numerator, denominator times denominator.
> 3. Cancel any common factors.
> 4. Reduce, if necessary.

**EXAMPLE 6** Multiply $\dfrac{4a^2b}{9xy^5}\cdot\dfrac{3x^2y^3}{8a^4b^3}$

**SOLUTION** Since there are no polynomials, we can start by multiplying across.

**Multiply across**

$$\frac{4a^2b}{9xy^5}\cdot\frac{3x^2y^3}{8a^4b^3}=\frac{12a^2bx^2y^3}{72xy^5a^4b^3}$$

**Cancel common factors**

$$=\frac{x}{6y^2a^2b^2}\cdot\frac{\cancel{12xy^3a^2b}}{\cancel{12xy^3a^2b}}$$

**Simplify**

$$=\frac{x}{6y^2a^2b^2}$$

**EXAMPLE 7** Multiply $\dfrac{x^2 + x}{x^2 - 1} \cdot \dfrac{x^2 + 6x - 7}{x^3 - 5x^2}$.

**SOLUTION** We factor first.

**Factor**

$$\frac{x^2 + x}{x^2 - 1} \cdot \frac{x^2 + 6x - 7}{x^3 - 5x^2} = \frac{x(x + 1)}{(x + 1)(x - 1)} \cdot \frac{(x + 7)(x - 1)}{x^2(x - 5)}$$

**Multiply across**

$$= \frac{x(x + 1)(x + 7)(x - 1)}{(x + 1)(x - 1)x^2(x - 5)}$$

**Cancel common factors**

$$= \frac{x + 7}{x(x - 5)} \cdot \frac{\cancel{x(x + 1)(x - 1)}}{\cancel{x(x + 1)(x - 1)}}$$

**Simplify**

$$= \frac{x + 7}{x(x - 5)}$$

**EXAMPLE 8** Simplify $F = \dfrac{x^2 - 4y^2}{x^2 + 7xy + 10y^2} \cdot \dfrac{x^2 + 5xy}{x^2 - 4xy - 21y^2} \cdot \dfrac{x^2 + 4xy + 3y^2}{x^2 + xy}$.

**SOLUTION** We multiply three fractions the same as two.

**Factor**

$$F = \frac{(x + 2y)(x - 2y)}{(x + 5y)(x + 2y)} \cdot \frac{x(x + 5y)}{(x - 7y)(x + 3y)} \cdot \frac{(x + 3y)(x + y)}{x(x + y)}$$

**Multiply across**

$$= \frac{(x + 2y)(x - 2y)(x)(x + 5y)(x + 3y)(x + y)}{(x + 5y)(x + 2y)(x - 7y)(x + 3y)(x)(x + y)}$$

**Cancel common factors**

$$= \frac{x - 2y}{x - 7y} \cdot \frac{\cancel{(x + 2y)(x)(x + 5y)(x + 3y)(x + y)}}{\cancel{(x + 2y)(x)(x + 5y)(x + 3y)(x + y)}}$$

**Simplify**

$$= \frac{x - 2y}{x - 7y}$$

## PROBLEM SET 5.2

Multiply the following fractions, and reduce all products to lowest terms.

*See Example 5*

1. $\dfrac{5}{12} \cdot \dfrac{4}{15}$

2. $\dfrac{-51}{10} \cdot \dfrac{25}{34}$

3. $\dfrac{-3}{10} \cdot \dfrac{15}{77} \cdot \dfrac{-14}{39}$

4. $\dfrac{18}{35} \cdot \dfrac{21}{50} \cdot \dfrac{45}{72}$

*See Example 6*

5. $\dfrac{10ab}{9c^2} \cdot \dfrac{3c}{5a}$

6. $\dfrac{14xy^2}{5a^3} \cdot \dfrac{15ab^2c^3}{28x^5yz}$

7. $\dfrac{-33m^2n^3}{18ab^2c^3} \cdot \dfrac{21a^2b^2c}{44m^8}$

8. $\dfrac{16r^8s^9t^{10}}{28s^2tu^5} \cdot \dfrac{7t^3u^4v^5}{25rstuv^2}$

9. $\dfrac{-16ab^2}{9bc^5} \cdot \dfrac{15c^3d^2}{14bx^2y}$

10. $\dfrac{12uvw^2}{-11vw^3x^3} \cdot \dfrac{-22x^2y^3z^4}{-15u^2vwz}$

11. $\dfrac{-ab^2}{5a^2x^2} \cdot \dfrac{15abx}{14bx^2y} \cdot \dfrac{7ay^5}{-10x^2y^4}$

12. $\dfrac{-xyz}{16ax^4} \cdot \dfrac{8aby^3}{-5byz^4} \cdot \dfrac{15ab^3x^2}{-8a^4xz^5}$

*See Examples 7 and 8*

13. $\dfrac{x^2 + 5x}{x^4} \cdot \dfrac{x^2}{x^2 - 25}$

14. $\dfrac{a^5}{a^3 + 7a^2} \cdot \dfrac{a^2 - 49}{a}$

15. $\dfrac{(x - 2)^3}{x + 3} \cdot \dfrac{x^2 - 9}{x^2 - 4}$

16. $\dfrac{x^2 - 36}{x + 4} \cdot \dfrac{x^2 - 16}{(x - 6)^2}$

17. $\dfrac{x^5 + x^4}{x^2 - 1} \cdot \dfrac{x^2 + 4x - 5}{x^4 + 5x^3}$

18. $\dfrac{a^6 - 6a^5}{a^2 - a - 30} \cdot \dfrac{a^2 - 25}{a^4 - 36a^2}$

19. $\dfrac{x^2 - x - 42}{x^2 + 11x + 30} \cdot \dfrac{x^2 + 13x + 40}{x^2 + 4x - 32}$

20. $\dfrac{x^2 + 8x - 20}{x^2 + 6x - 16} \cdot \dfrac{x^2 + 3x - 40}{x^2 + 6x - 40}$

21. $\dfrac{a^2 - 25}{a^2 + 12a + 35} \cdot \dfrac{a^2 - 49}{a^2 - 16a + 63}$

22. $\dfrac{b^2 - 16}{b^2 - b - 20} \cdot \dfrac{b^2 + b - 30}{b^2 - 36}$

23. $\dfrac{3x^2 + 31x + 10}{x^2 + 11x + 10} \cdot \dfrac{x^2 - 1}{x^2 + 4x - 5}$

24. $\dfrac{4x^2 + 23x + 15}{4x^2 - 21x - 18} \cdot \dfrac{x^2 - 7x + 6}{x^2 - 25}$

25. $\dfrac{6x^2 - 2xy + 15x - 5y}{2x^2 + 2xy + 5x + 5y} \cdot \dfrac{3x^2 + 3xy + x + y}{3x^2 - 6x - xy + 2y}$

26. $\dfrac{2a^2 + ac + 2ab + bc}{a^2 + ab + 7a + 7b} \cdot \dfrac{a^2 + 3ac + 7a + 21c}{ac + ab + 3c^2 + 3cb}$

*See Example 8*

27. $\dfrac{x^2 + 5x}{x^2 + 6x - 7} \cdot \dfrac{x^2 + 8x - 9}{x^2 - 9} \cdot \dfrac{x^2 + 10x + 21}{x^4 + 9x^3}$

28. $\dfrac{x^2 - 3x - 28}{x^2 - 9x + 8} \cdot \dfrac{x^5 - x^4}{x^2 + 7x + 12} \cdot \dfrac{x^2 - 12x + 32}{x^4 - 7x^3}$

29. $\dfrac{x^2 - 3x}{x^4 + 5x^3} \cdot \dfrac{x^3 - 8}{x^2 + x - 12} \cdot \dfrac{x^2 - 16}{x^4 + 2x^3 + 4x^2}$

30. $\dfrac{x^2 - 25}{x^3 - 5x^2} \cdot \dfrac{x^2 - 9x + 18}{x^3 + 1} \cdot \dfrac{x^2 - 3x - 4}{x^2 - 10x + 24}$

## 5.3 DIVISION OF FRACTIONS

In division, $10 \div 2 = x$ means $2 \cdot x = 10$. Similarly,

| | |
|---|---|
| Definition of division | $x = \dfrac{A}{B} \div \dfrac{C}{D}$ means $x \cdot \dfrac{C}{D} = \dfrac{A}{B}$ |
| ↓ Multiply both sides by $\dfrac{D}{C}$ | $x \cdot \dfrac{C}{D} \cdot \dfrac{D}{C} = \dfrac{A}{B} \cdot \dfrac{D}{C}$ |
| ↓ Since $\dfrac{C}{D} \cdot \dfrac{D}{C} = 1$ | $x = \dfrac{A}{B} \cdot \dfrac{D}{C}$ |

Thus, we have the following theorem.

> ***THEOREM 5.2*** For expressions $A, B, C, D$ ($B, C,$ and $D \neq 0$),
>
> $$\frac{A}{B} \div \frac{C}{D} = \frac{A}{B} \cdot \frac{D}{C} = \frac{AD}{BC}$$

This theorem tells us to invert the divisor (the second fraction) and then work as a multiplication problem. We can also say that we are replacing the divisor by its reciprocal and multiplying.

**EXAMPLE 9** Divide $\dfrac{5}{16} \div \dfrac{3}{8}$.

**SOLUTION** We invert and multiply.

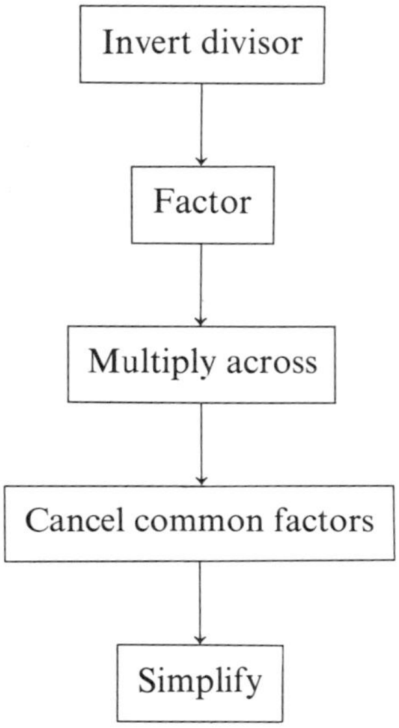


$$\frac{5}{16} \div \frac{3}{8} = \frac{5}{16} \cdot \frac{8}{3}$$

$$= \frac{5}{2 \cdot 2 \cdot 2 \cdot 2} \cdot \frac{2 \cdot 2 \cdot 2}{3}$$

$$= \frac{5 \cdot 2 \cdot 2 \cdot 2}{2 \cdot 2 \cdot 2 \cdot 2 \cdot 3}$$

$$= \frac{5}{2 \cdot 3} \cdot \frac{\cancel{2 \cdot 2 \cdot 2}}{\cancel{2 \cdot 2 \cdot 2}}$$

$$= \frac{5}{6}$$

> To divide algebraic fractions:
>
> 1. Invert the divisor (second fraction).
> 2. Factor all polynomials.
> 3. Multiply across: numerator times numerator, denominator times denominator.
> 4. Cancel any common factors.
> 5. Reduce, if necessary.

Notice that steps 2 to 5 are exactly the same as for multiplying algebraic fractions.

**EXAMPLE 10** Divide $\dfrac{6a^4b^3}{77xy^5} \div \dfrac{15ab^7}{44x^2y^2}$.

**SOLUTION** We invert and multiply.

**Invert divisor**

$$\frac{6a^4b^3}{77xy^5} \div \frac{15ab^7}{44x^2y^2} = \frac{6a^4b^3}{77xy^5} \cdot \frac{44x^2y^2}{15ab^7}$$

**Multiply across**

$$= \frac{6 \cdot 44a^4b^3x^2y^2}{77 \cdot 15xy^5ab^7}$$

**Factor and cancel common factors**

$$= \frac{2 \cdot 2 \cdot 2a^3x}{7 \cdot 5b^4y^3} \cdot \frac{\cancel{3 \cdot 11ab^3xy^2}}{\cancel{3 \cdot 11ab^3xy^2}}$$

**Simplify**

$$= \frac{8a^3x}{35b^4y^3}$$

**EXAMPLE 11** Divide $\dfrac{2x^2 + 9x + 4}{x^2 - 16} \div \dfrac{2x^2 + 11x + 5}{x^2 + 2x - 15}$.

**SOLUTION** We first invert the divisor and then follow the rules for multiplying (factor, cancel, and multiply).

**Invert divisor**

$$\frac{2x^2 + 9x + 4}{x^2 - 16} \div \frac{2x^2 + 11x + 5}{x^2 + 2x - 15} = \frac{2x^2 + 9x + 4}{x^2 - 16} \cdot \frac{x^2 + 2x - 15}{2x^2 + 11x + 5}$$

**Factor**

$$= \frac{(2x + 1)(x + 4)}{(x + 4)(x - 4)} \cdot \frac{(x + 5)(x - 3)}{(2x + 1)(x + 5)}$$

**Multiply across**

$$= \frac{(2x + 1)(x + 4)(x + 5)(x - 3)}{(x + 4)(x - 4)(2x + 1)(x + 5)}$$

**Cancel common factors**

$$= \frac{x - 3}{x - 4} \cdot \frac{\cancel{(2x + 1)(x + 4)(x + 5)}}{\cancel{(2x + 1)(x + 4)(x + 5)}}$$

**Simplify**

$$= \frac{x - 3}{x - 4}$$

**EXAMPLE 12** Simplify $F = \frac{x+4}{x^2+5x} \cdot \frac{x^2-25}{x^2-x-2} \div \frac{x^6+4x^5}{x^3+3x^2}$.

**SOLUTION** As before, we invert the last fraction and change the division to multiplication. Then we multiply as we would any three fractions.

**Invert divisor**

$$F = \frac{x+4}{x^2+5x} \cdot \frac{x^2-25}{x^2-x-2} \cdot \frac{x^3+3x^2}{x^6+4x^5}$$

**Factor**

$$= \frac{x+4}{x(x+5)} \cdot \frac{(x+5)(x-5)}{(x-2)(x+1)} \cdot \frac{x^2(x+3)}{x^5(x+4)}$$

**Multiply across**

$$= \frac{(x+4)(x+5)(x-5)x^2(x+3)}{x^6(x+5)(x-2)(x+1)(x+4)}$$

**Cancel common factors**

$$= \frac{(x-5)(x+3)}{x^4(x-2)(x+1)} \cdot \frac{(x+4)(x+5)x^2}{(x+4)(x+5)x^2}$$

**Simplify**

$$= \frac{x^2-2x-15}{x^6-x^5-2x^4}$$

## PROBLEM SET 5.3

Divide the following fractions, and reduce all quotients to lowest terms.

*See Example 9*

1. $\frac{4}{9} \div \frac{16}{27}$
2. $\frac{18}{35} \div \frac{40}{77}$
3. $\frac{8}{15} \div \frac{36}{55}$
4. $\frac{51}{52} \div \frac{17}{18}$

*See Example 10*

5. $\frac{6a^2b^5}{7xy^3} \div \frac{12ab^7}{21x^6y^6}$
6. $\frac{14m^5n^3}{9a^7} \div \frac{21m^4n^4}{4a^3b^3}$

7. $\dfrac{-25ab^2c^3}{18x^3y^3z^3} \div \dfrac{15a^2b^2c^2}{27x^5y^4z^3}$

8. $\dfrac{16p^2q^7}{15r^5s^7t} \div \dfrac{8p^3q^8r}{35s^5r^4u^2}$

9. $\dfrac{18a^2b^3c}{-45c^4d^5e} \div \dfrac{9a^2b^5}{-15b^5c^4d^3}$

10. $\dfrac{4xy^2z^7}{5u^6v^8w^3} \div \dfrac{-8w^4x^6y^3}{15u^3v^4z^5}$

11. $\dfrac{2a^4bcd^3}{3wx^2y^3z^4} \div \dfrac{8abc^2}{9x^4y^5z^9}$

12. $\dfrac{r^3st^4}{3tu^3v^4} \div \dfrac{7q^8rs^9}{12s^2t^2u^2v^8}$

See Example 11

13. $\dfrac{x^2+4x}{x^4} \div \dfrac{x^2-16}{x^7}$

14. $\dfrac{b^2}{b^6-36b^4} \div \dfrac{b^8}{b^3+6b^2}$

15. $\dfrac{x-3}{x+5} \div \dfrac{x^2-9}{x^2-25}$

16. $\dfrac{x-2}{(x+6)^3} \div \dfrac{x^2-x-2}{x^2-36}$

17. $\dfrac{x^4+2x^3}{x^2-4} \div \dfrac{x^3+7x^2}{x^2+x-6}$

18. $\dfrac{a^5+5a^4}{a^2-4a+3} \div \dfrac{a^2+8a}{a^2+5a-24}$

19. $\dfrac{p^2+9p+20}{p^2+4p-21} \div \dfrac{p^2+7p+10}{p^2+9p+14}$

20. $\dfrac{r^2-4r-12}{r^2-3r-18} \div \dfrac{r^2-6r-16}{r^2-4r-21}$

21. $\dfrac{x^2-49}{x^2-2x-35} \div \dfrac{x^2+2x-8}{x^2+7x+12}$

22. $\dfrac{a^2+19a+90}{a^2-100} \div \dfrac{a^2-a-42}{a^2-3a-28}$

23. $\dfrac{3x^2+16+5}{x^2+3x-10} \div \dfrac{3x^2+25x+8}{x^2+x-56}$

24. $\dfrac{4x^2+25x-21}{5x^2+32x-21} \div \dfrac{x^2-3x+2}{5x^2-8x+3}$

25. $\dfrac{3ax^2+5ax+6x+10}{3bx^2+5bx+15x+25} \div \dfrac{ax^2+7ax+2x+14}{4bx^2-3bx+20x-15}$

26. $\dfrac{30x^2+6ax-25x-5a}{24x^2-6ax-20x+5a} \div \dfrac{10x^2-35x+2ax-7a}{12x^2-8x-3ax+2a}$

Simplify the following.

See Example 12

27. $\dfrac{x^3-4x^2}{x^2-x-6} \cdot \dfrac{x^4-3x^3}{x^2-16} \div \dfrac{x^4}{x+2}$

28. $\dfrac{x^2-4x-45}{x^3+5x^2} \cdot \dfrac{x^6+6x^5}{x^2-9x} \div \dfrac{x^2-x-42}{x+4}$

29. $\dfrac{x^3 - 8}{x^2 - 3x - 40} \cdot \dfrac{x^2 - 6x - 16}{x^2 - 4} \div \dfrac{x^2 + 2x - 63}{x^2 - 2x - 35}$

30. $\dfrac{x^2 - 9}{x^2 + 13x + 40} \cdot \dfrac{x^2 + 4x - 32}{x^3 + 27} \div \dfrac{x^3 - 4x^2}{x^5 - 3x^4 + 9x^3}$

## 5.4 ADDITION AND SUBTRACTION OF FRACTIONS

Adding or subtracting fractions is easy if they have the same denominator. We have the following definition.

**DEFINITION 5.2** For expressions $A$, $B$, and $C$ $(C \neq 0)$,

$$\frac{A}{C} + \frac{B}{C} = \frac{A + B}{C}$$

For example,

$$\frac{4}{13} + \frac{3}{13} = \frac{7}{13}$$

What if the fractions do not have the same denominator? Then we must find the **lowest common denominator (LCD),** which is the smallest number or expression that all the denominators will divide into evenly. Sometimes this is easy to find; sometimes it is difficult to find. Once we find the LCD of the denominators, we must rewrite all of the old fractions as new fractions with this LCD.

Since it is often difficult (especially with algebraic fractions) to see the LCD immediately, we give a foolproof method for adding or subtracting fractions of any type. The method involves factoring the denominators. The first example involves number fractions.

**EXAMPLE 13** Add $\dfrac{3}{10} + \dfrac{7}{12} + \dfrac{4}{15}$.

**SOLUTION** We begin by factoring to find the LCD.

**Factor denominators; line up common factors**

$$\begin{aligned} 10 &= 2 \cdot 5 \\ 12 &= 2 \cdot \phantom{5 \cdot} 2 \cdot 3 \\ 15 &= \phantom{2 \cdot} 5 \phantom{\cdot 2} \cdot 3 \end{aligned}$$

↓

**Fill in missing factors for each column**

$$\begin{aligned} 10 &\to 2 \cdot 5 \cdot \textcircled{2} \cdot \textcircled{3} \\ 12 &\to 2 \cdot \textcircled{5} \cdot 2 \cdot 3 \\ 15 &\to \textcircled{2} \cdot 5 \cdot \textcircled{2} \cdot 3 \end{aligned}$$

↓

**Multiply factors from each column to get LCD**

$$\text{LCD} = 2 \cdot 5 \cdot 2 \cdot 3 = 60$$

↓

**Multiply each fraction by $1 = \frac{\text{missing factors}}{\text{missing factors}}$ to get fractions with LCD = 60**

$$\frac{3}{10} + \frac{7}{12} + \frac{4}{15}$$

$$= \frac{3}{10} \cdot \frac{\textcircled{6}}{\textcircled{6}} + \frac{7}{12} \cdot \frac{\textcircled{5}}{\textcircled{5}} + \frac{4}{15} \cdot \frac{\textcircled{4}}{\textcircled{4}}$$

↓

**Add numerators**

$$= \frac{18}{60} + \frac{35}{60} + \frac{16}{60} = \frac{69}{60}$$

↓

**Reduce**

$$= \frac{23}{20}$$

Notice that we factor the denominators and line up common factors. This process gives us the LCD. Also, by filling in the missing factors, we know what to multiply the numerators and denominators by to get the new fractions, all with the same denominator, the LCD.

Let us summarize this procedure, since we will use it with algebraic fractions.

To find the LCD and add fractions:

1. Factor all denominators completely.
2. Line up the common factors in columns, with unlike factors in different columns.
3. Fill in the missing factors in each column (circle these for later use).
4. Bring down a factor from each column. The product is the LCD.

5. Multiply each fraction by $1 = \dfrac{\text{missing factor}}{\text{missing factor}}$ to get the fractions to have the same denominator, the LCD.
6. Add the numerators.
7. Reduce, if necessary.

**EXAMPLE 14** Add $\dfrac{6}{a^2b} + \dfrac{5}{ab^3c} + \dfrac{8x}{a^3b^2c^2}$.

**SOLUTION** We will add these fractions using the procedure given above.

| | |
|---|---|
| Factor denominators; line up common factors | $a^2b = a \cdot a \cdot \quad \cdot b$<br>$ab^3c = a \cdot \quad \quad \cdot b \cdot b \cdot b \cdot c$<br>$a^3b^2c^2 = a \cdot a \cdot a \cdot b \cdot b \cdot \quad \cdot c \cdot c$ |
| ↓ Fill in missing factors in each column | $a^2b \to a \cdot a \cdot \textcircled{a} \cdot b \cdot \textcircled{b} \cdot \textcircled{b} \cdot \textcircled{c} \cdot \textcircled{c}$<br>$ab^3c \to a \cdot \textcircled{a} \cdot \textcircled{a} \cdot b \cdot b \cdot b \cdot c \cdot \textcircled{c}$<br>$a^3b^2c^2 \to a \cdot a \cdot a \cdot b \cdot b \cdot \textcircled{b} \cdot c \cdot c$ |
| ↓ Multiply factors from each column to get LCD | $\text{LCD} = a \cdot a \cdot a \cdot b \cdot b \cdot b \cdot c \cdot c = a^3b^3c^2$ |
| ↓ Multiply each fraction by $1 = \dfrac{\text{missing factor}}{\text{missing factor}}$ to get fractions with LCD $= a^3b^3c^2$ | $\dfrac{6}{a^2b} + \dfrac{5}{ab^3c} + \dfrac{8x}{a^3b^2c^2}$<br>$= \dfrac{6}{a^2b} \cdot \dfrac{(ab^2c^2)}{(ab^2c^2)} + \dfrac{5}{ab^3c} \cdot \dfrac{(a^2c)}{(a^2c)} + \dfrac{8x}{a^3b^2c^2} \cdot \dfrac{(b)}{(b)}$ |
| ↓ Simplify and add numerators | $= \dfrac{6ab^2c^2 + 5a^2c + 8xb}{a^3b^3c^2}$ |

Notice that the LCD is the highest power of each of the bases.

**EXAMPLE 15** Add $\dfrac{3}{x+5} + \dfrac{x}{x+7}$.

**SOLUTION** Here there is nothing to factor. Thus, the LCD is simply the product of the denominators. To help demonstrate the method, we continue to use the procedure given above.

| | |
|---|---|
| Fill in missing factors in each column | $x+5 \rightarrow (x+5)(x+7)$<br>$x+7 \rightarrow (x+5)(x+7)$ |
| Multiply factors from each column to get LCD | $\text{LCD} = (x+5)(x+7)$ |
| Multiply each fraction by $1 = \dfrac{\text{missing factor}}{\text{missing factor}}$ to get fractions with $\text{LCD} = (x+5)(x+7)$ | $\dfrac{3}{x+5} + \dfrac{x}{x+7}$<br>$\dfrac{3}{x+5} \cdot \dfrac{(x+7)}{(x+7)} + \dfrac{x}{x+7} \cdot \dfrac{(x+5)}{(x+5)}$ |
| Simplify | $= \dfrac{3x+21}{(x+5)(x+7)} + \dfrac{x^2+5x}{(x+5)(x+7)}$ |
| Add numerators | $= \dfrac{x^2+8x+21}{(x+5)(x+7)}$ or $\dfrac{x^2+8x+21}{x^2+12x+35}$ |

| YES | NO |
|---|---|
| $\dfrac{3}{x+5} + \dfrac{x}{x+7} = \dfrac{x^2+8x+21}{x^2+12x+35}$ | ~~$\dfrac{3}{x+5} + \dfrac{x}{x+7} = \dfrac{3+x}{2x+12}$~~ |

We must find the common denominator.

| YES | NO |
|---|---|
| $\dfrac{3}{x+5} + \dfrac{x}{x+7} = \dfrac{x^2+8x+71}{x^2+12x+35}$ | ~~$\dfrac{3}{x+5} + \dfrac{x}{x+7} = \dfrac{3+x}{x^2+12+35}$~~ |

We must find the new numerators.

Subtracting fractions is like subtracting numbers or polynomials: *we add the opposite of the fraction to be subtracted.* We find the **opposite of a fraction** by changing *all* the signs of the terms in the numerator.

**EXAMPLE 16** Find $D = \dfrac{2x}{x^2 + x - 6} - \dfrac{5}{x^2 + 3x - 10}$.

**SOLUTION** The LCD is not at all obvious, so we will have to factor the denominators. But first, to make things easier, we change the subtraction to the addition of the opposite by changing the sign of the second numerator.

| | |
|---|---|
| Change sign and add | $D = \dfrac{2x}{x^2 + x - 6} - \dfrac{5}{x^2 + 3x - 10}$ <br> $= \dfrac{2x}{x^2 + x - 6} + \dfrac{-5}{x^2 + 3x - 10}$ |
| Factor denominators, line up common factors | $x^2+x-6=(x-2)(x+3)$ <br> $x^2+3x-10=(x-2)\qquad(x+5)$ |
| Fill in missing factors in each column | $x^2+x-6\to(x-2)(x+3)\boxed{(x+5)}$ <br> $x^2+3x-10\to(x-2)\boxed{(x+3)}(x+5)$ |
| Multiply factors from each column to get LCD | $\text{LCD}=(x-2)(x+3)(x+5)$ |
| Multiply each fraction by $1 = \dfrac{\text{missing factor}}{\text{missing factor}}$ to get fraction with LCD | $D=\dfrac{2x}{(x-2)(x+3)}+\dfrac{-5}{(x-2)(x+5)}$ <br> $=\dfrac{2x}{(x-2)(x+3)}\cdot\dfrac{(x+5)}{(x+5)}+\dfrac{-5}{(x-2)(x+5)}\cdot\dfrac{(x+3)}{(x+3)}$ |
| Simplify | $=\dfrac{2x^2+10x}{(x-2)(x+3)(x+5)}+\dfrac{-5x-15}{(x-2)(x+5)(x+3)}$ |
| Add numerators | $=\dfrac{2x^2+5x-15}{(x-2)(x+3)(x+5)}$ |

**EXAMPLE 17** Simplify $T = \dfrac{x}{x^2 - 9x + 20} + \dfrac{6}{x^2 + x - 20} - \dfrac{3x + 2}{x^2 - 25}$.

**SOLUTION** Before factoring, we change the last subtraction to addition of the opposite by changing the signs (all the signs!) in the last numerator.

**Change signs and add**

$$T = \frac{x}{x^2-9x+20} + \frac{6}{x^2+x-20} + \frac{-3x-2}{x^2-25}$$

↓

**Factor denominators, line up common factors**

$$\begin{aligned} x^2-9x+20 &= (x-4)(x-5) \\ x^2+x-20 &= (x-4)\phantom{(x-5)}(x+5) \\ x^2-25 &= \phantom{(x-4)}(x-5)(x+5) \end{aligned}$$

↓

**Fill in missing factors in each column**

$$\begin{aligned} x^2-9x+20 &\to (x-4)(x-5)\boxed{(x+5)} \\ x^2+x-20 &\to (x-4)\boxed{(x-5)}(x+5) \\ x^2-25 &\to \boxed{(x-4)}(x-5)(x+5) \end{aligned}$$

↓

**Multiply factors from each column to get LCD**

$$\text{LCD} = (x-4)(x-5)(x+5)$$

↓

**Multiply each fraction by $1 = \dfrac{\text{missing factor}}{\text{missing factor}}$ to get fractions with LCD**

$$T = \frac{x}{(x-4)(x-5)} + \frac{6}{(x-4)(x+5)} + \frac{-3x-2}{(x+5)(x-5)}$$

$$= \frac{x}{(x-4)(x-5)} \cdot \frac{\boxed{x+5}}{\boxed{x+5}} + \frac{6}{(x-4)(x+5)} \cdot \frac{\boxed{x-5}}{\boxed{x-5}}$$

$$+ \frac{(-3x-2)}{(x+5)(x-5)} \cdot \frac{\boxed{x-4}}{\boxed{x-4}}$$

↓

**Simplify**

$$= \frac{x^2+5x}{(x-4)(x-5)(x+5)} + \frac{6x-30}{(x-4)(x-5)(x+5)}$$

$$+ \frac{-3x^2+10x+8}{(x-4)(x-5)(x+5)}$$

↓

**Add numerators**

$$= \frac{-2x^2+21x-22}{(x-4)(x-5)(x+5)}$$

In arithmetic, a number such as $4\frac{2}{3}$ is called a mixed number. Similarly, in algebra, the sum of a polynomial and an algebraic fraction is called a **mixed expression.** As examples,

$$x + \frac{x+1}{x+2} \qquad \text{and} \qquad x^2 + 5 - \frac{x-1}{x-3}$$

are mixed expressions. In arithmetic, we write

$$4\frac{2}{3} = 4 + \frac{2}{3} = \frac{12}{3} + \frac{2}{3} = \frac{14}{3}$$

We use a similar procedure in algebra.

> To rewrite a mixed expression as a single fraction:
>
> 1. Find the LCD for all denominators.
> 2. Multiply the nonfraction polynomial by $1 = \dfrac{\text{LCD}}{\text{LCD}}$.
> 3. Simplify as in any addition or subtraction of fractions.

**EXAMPLE 18** Write $x + \dfrac{x+1}{x+5}$ as a single fraction.

**SOLUTION** The LCD is $x + 5$, since that is the only denominator.

| | |
|---|---|
| Multiply $x$ by $1 = \dfrac{x+5}{x+5}$ | $x + \dfrac{x+1}{x+5} = \dfrac{x(x+5)}{x+5} + \dfrac{x+1}{x+5}$ |
| ↓ Simplify | $= \dfrac{x^2+5x}{x+5} + \dfrac{x+1}{x+5}$ |
| ↓ Add numerators | $= \dfrac{x^2+6x+1}{x+5}$ |

## PROBLEM SET 5.4

Simplify the following by addition or subtraction. Be sure to indicate the LCD.

*See Example 13*

1. $\frac{4}{11} + \frac{5}{11} - \frac{2}{11}$

2. $\frac{15}{19} - \frac{18}{19} - \frac{7}{19}$

3. $\frac{1}{10} + \frac{4}{15} - \frac{7}{18}$

4. $\frac{5}{28} - \frac{5}{12} - \frac{8}{21}$

5. $\frac{x+1}{10} + \frac{3x-1}{15} + \frac{2x+1}{20}$

6. $\frac{a}{6} + \frac{2a+1}{8} + \frac{3a+2}{10}$

*See Examples 14 and 15*

7. $\frac{7}{ax} - \frac{3}{ax}$

8. $\frac{12}{b^2c} + \frac{-7}{b^2c}$

9. $\frac{x+1}{6a} + \frac{3x-4}{6a} - \frac{4x-1}{6a}$

10. $\frac{r-5}{4s} - \frac{2r+1}{4s} + \frac{r}{4s}$

11. $\frac{1}{p^3} + \frac{1}{p^5} + \frac{1}{p^7}$

12. $\frac{3}{b} - \frac{4}{b^4} + \frac{7}{b^6}$

13. $\frac{3}{mn^2} - \frac{5}{m^2n}$

14. $\frac{17}{x^2y^2} + \frac{12}{xy^5}$

15. $\frac{4}{abc^2} + \frac{3}{a^2b^2c} - \frac{5}{a^3c^4}$

16. $\frac{9}{x^3yz^2} - \frac{5}{x^4y^6} + \frac{4}{xz^3}$

17. $\frac{x+y}{x^2y} + \frac{x^2-y^2}{xy^3}$

18. $\frac{a-b^2}{a^3b} - \frac{a^2-b}{ab^3}$

*See Examples 15 and 16*

19. $\frac{1}{x+4} + \frac{7}{x-3}$

20. $\frac{3}{a+2} - \frac{4}{a+6}$

21. $\frac{r}{r+3} - \frac{5}{r-2}$

22. $\frac{7}{2t+1} + \frac{t}{t-5}$

23. $\frac{k}{2k-1} + \frac{2k-3}{k}$

24. $\frac{z+6}{z} - \frac{z}{2z+7}$

25. $\frac{5}{x+h} - \frac{5}{x}$

26. $\frac{1}{(x+h)^2} - \frac{1}{x^2}$

27. $\frac{x+1}{x+5} + \frac{3x-7}{x-4}$

28. $\frac{x-5}{2x+1} - \frac{3x-7}{2x-3}$

*See Example 16*

29. $\dfrac{6}{x^2 - 1} + \dfrac{7}{x^2 - x - 2}$

30. $\dfrac{8}{x^2 - 12x + 35} - \dfrac{3}{x^2 - 25}$

31. $\dfrac{2}{x^2 + 12x + 36} - \dfrac{4}{x^2 + 5x - 6}$

32. $\dfrac{8}{x^2 - 4x + 4} + \dfrac{5}{x^2 - 7x + 10}$

33. $\dfrac{x}{x^2 - 3x - 28} + \dfrac{x - 5}{x^2 - 7x}$

34. $\dfrac{a + 7}{a^2 + 13a + 40} - \dfrac{a - 4}{a^2 + 5a - 24}$

35. $\dfrac{x}{x^3 - 8} + \dfrac{4}{x^2 - 4}$

36. $\dfrac{t - 1}{t^3 + 125} - \dfrac{8}{t^2 - 25}$

*See Example 17*

37. $\dfrac{5}{x^2 + 3x + 2} + \dfrac{6}{x^2 + 4x + 3} - \dfrac{7}{x^2 + 5x + 6}$

38. $\dfrac{2}{x^2 - 9x + 20} - \dfrac{4}{x^2 - 11x + 30} + \dfrac{6}{x^2 - 10x + 24}$

39. $\dfrac{x}{x^2 + 8x + 7} + \dfrac{x + 1}{2x^2 + 3x + 1} - \dfrac{x - 2}{2x^2 + 15x + 7}$

40. $\dfrac{a - 1}{a^2 - 10a + 16} - \dfrac{a + 1}{3a^2 - 26a + 16} - \dfrac{a + 3}{3a^2 - 8a + 4}$

*See Example 18*

41. $6 + \dfrac{1}{x}$

42. $12 - \dfrac{3}{z}$

43. $x + \dfrac{2}{x}$

44. $2x - \dfrac{7}{x}$

45. $4x - \dfrac{3}{x + 5}$

46. $6x + \dfrac{4}{x - 7}$

47. $x + 5 + \dfrac{x}{2x + 1}$

48. $y - 3 - \dfrac{2y}{5 - y}$

49. $1 + \dfrac{2}{x} + \dfrac{4}{x^2}$

50. $2 - \dfrac{3}{4t} + \dfrac{5}{2t^3}$

51. $7 + \dfrac{2}{x + 5} - \dfrac{3}{x + 7}$

52. $8 - \dfrac{4}{z - 1} + \dfrac{5}{z + 6}$

53. $x - \dfrac{6}{x} + \dfrac{5}{x + 4}$

54. $a - \dfrac{7}{a} - \dfrac{a}{a + 9}$

*Life science applications*

55. The number of bacteria in a culture, $N$, is given by

$$N = 10{,}000 + \frac{20{,}000t}{100 + t^2}$$

where $t$ is the time in hours. Write this as a single fraction.

56. The probability $p$ of a certain plant dying within $x$ hours is given by

$$p = 1 - \frac{1}{1 + x^2}$$

Write this as a single fraction.

*Optical application*

57. The *Lensmaker's Equation* is

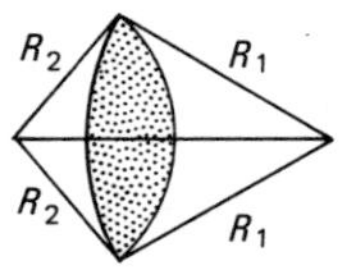


$$\frac{1}{f} = (n - 1)\left(\frac{1}{R_1} - \frac{1}{R_2}\right)$$

where $f$ is the focal length of lens, $n$ the index of refraction of glass, and $R_1$ and $R_2$ are the radii of curvature of the lens, as shown. Write the right side as a single fraction.

*Physical application*

In atomic physics, the frequency $f$ of the light emitted by an excited hydrogen atom is given by the four relations below. Write each one as a single fraction. Here $Rc = 3.3 \times 10^{15}$ hertz and $n$ is as indicated.

58. Lyman series: $f = Rc\left(1 - \frac{1}{n^2}\right), \quad n = 2, 3, 4, \ldots$

59. Balmer series: $f = Rc\left(\frac{1}{2^2} - \frac{1}{n^2}\right), \quad n = 3, 4, 5, \ldots$

60. Paschen series: $f = Rc\left(\frac{1}{3^2} - \frac{1}{n^2}\right), \quad n = 4, 5, 6, \ldots$

61. Brackett series: $f = Rc\left(\frac{1}{4^2} - \frac{1}{n^2}\right), \quad n = 5, 6, 7, \ldots$

*Engineering application*

62. The maximum deflection $y$ in a beam is given by

$$y = \frac{PL^3}{3EI} + \frac{wL^4}{8EI}$$

where $L$ is the length of the beam, $P$ the load on the beam, and $E$, $w$, and $I$ are constants of the shape. Write this as a single fraction.

*Electricity application*

63. In a series of capacitors $C_1$, $C_2$, and $C_3$, an equivalent capacitance $C$ is given by

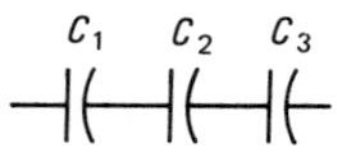


$$\frac{1}{C} = \frac{1}{C_1} + \frac{1}{C_2} + \frac{1}{C_3}$$

Write the right side as a single fraction.

*Social application*

64. One of Pareto's laws is that $P_n = P_1/n$, where $P_n$ is the population of the $n$th largest city and $P_1$ is the population of the largest city. If this is true,

the total population of the top six cities is given by

$$P_1 + \frac{P_1}{2} + \frac{P_1}{3} + \frac{P_1}{4} + \frac{P_1}{5} + \frac{P_1}{6}$$

Simplify this expression.

*Business applications*

65. The sales $y$ of an item are related to the amount of advertising $x$ by

$$y = A + \frac{B}{1 + kx}$$

where $A$, $B$, and $k$ are constants. Write this as a simple fraction.

66. The inventory costs $C$ for a company are given by

$$C = \frac{C_1 Q}{2} + \frac{C_3 R}{Q}$$

where $C_1$ is the storage charge, $Q$ the quantity ordered, $R$ the demand, and $C_3$ the shipping cost. Write this as a simple fraction.

## 5.5 COMPLEX FRACTIONS

A **complex fraction** is a fraction that has other fractions in its numerator or denominator, or both. As examples,

$$\frac{1 + \frac{2}{5}}{2 - \frac{3}{10}}, \quad \frac{8}{\frac{x^2}{a} - \frac{x}{a^2}}, \quad \frac{x + \frac{1}{x-5}}{x - \frac{1}{x-7}}, \quad \frac{\frac{4}{a}}{\frac{6}{xy}}$$

are typical complex fractions.

Usually, we want to rewrite the complex fraction as a simple fraction. One way to do this is to simplify the numerator and denominator separately, and then divide them. But this is usually a more difficult procedure. We will show an easier method.

**EXAMPLE 19** Write $\dfrac{\frac{2}{3}}{\frac{5}{x}}$ as a simple fraction.

**SOLUTION** Although we could rewrite this as $\frac{2}{3} \div \frac{5}{x}$ and divide, we find the LCD, $3x$, for both denominators. We then multiply top and bottom by this LCD, $3x$.

| | |
|---|---|
| Multiply by $1 = \dfrac{\text{LCD}}{\text{LCD}} = \dfrac{3x}{3x}$ | $\dfrac{\frac{2}{3}}{\frac{5}{x}} = \dfrac{3x\left(\frac{2}{3}\right)}{3x\left(\frac{5}{x}\right)}$ |
| ↓ | |
| Cancel and simplify | $= \dfrac{2x}{15}$ |

Notice that the denominators all cancel out as we multiply by the LCD.

> To simplify a complex fraction:
>
> 1. Determine the LCD for all the denominators of the fractions in the numerator and denominator.
> 2. Multiply the fraction by $1 = \dfrac{\text{LCD}}{\text{LCD}}$. Use the distributive law. The LCD will cancel all the denominators of the fractions.
> 3. Simplify.
> 4. Reduce, if necessary.

**EXAMPLE 20** Write $\dfrac{\frac{1}{4} + \frac{1}{3a}}{\frac{1}{12a} - \frac{1}{6a^2}}$ as a simple fraction.

**SOLUTION** We look at all the fractions and we can see that the LCD is $12a^2$. So we multiply everything by the fraction $1 = \dfrac{12a^2}{12a^2}$.

| | |
|---|---|
| Multiply by $1 = \dfrac{12a^2}{12a^2}$ | $\dfrac{\frac{1}{4} + \frac{1}{3a}}{\frac{1}{12a} - \frac{1}{6a^2}} = \dfrac{12a^2\left(\frac{1}{4} + \frac{1}{3a}\right)}{12a^2\left(\frac{1}{12a} - \frac{1}{6a^2}\right)}$ |
| ↓ | |
| Use distributive law; cancel | $= \dfrac{3a^2 + 4a}{a - 2}$ |

**EXAMPLE 21** Write $\dfrac{5 - \dfrac{1}{x-2}}{\dfrac{3}{x^2-4} + \dfrac{5}{x+2}}$ as a simple fraction.

**SOLUTION** Again, we first find the LCD of all the fractions. Perhaps it is clear to the student that it is $x^2 - 4$. If this is not obvious, use the procedure given in Section 5.4.

| | |
|---|---|
| Factor denominators, line up common factors | $x - 2 = (x-2)$<br>$x + 2 = \quad\quad (x+2)$<br>$x^2 - 4 = (x-2)(x+2)$ |
| Multiply factors from each column to get LCD | $\text{LCD} = (x-2)(x+2)$ |
| Multiply by $1 = \dfrac{(x-2)(x+2)}{(x-2)(x+2)}$ | $\dfrac{5 - \dfrac{1}{x-2}}{\dfrac{3}{x^2-4} + \dfrac{5}{x+2}} = \dfrac{(x-2)(x+2)\left(5 - \dfrac{1}{x-2}\right)}{(x-2)(x+2)\left(\dfrac{3}{x^2-4} + \dfrac{5}{x+2}\right)}$ |
| Use distributive law; cancel | $= \dfrac{5(x-2)(x+2) - (x+2)}{3 + 5(x-2)}$ |
| Simplify | $= \dfrac{5x^2 - 20 - x - 2}{3 + 5x - 10}$ |
| Simplify | $= \dfrac{5x^2 - x - 22}{5x - 7}$ |

Notice that the LCD, $(x-2)(x+2)$, cancels out the denominators in all the fractions.

## PROBLEM SET 5.5

Write the following fractions as simple fractions.

*See Example 19*

1. $\dfrac{\frac{2}{3}}{\frac{4}{5}}$

2. $\dfrac{\frac{1}{9}}{\frac{5}{12}}$

3. $\dfrac{\frac{3}{10x}}{\frac{4}{15y}}$

4. $\dfrac{\frac{4}{3a}}{\frac{7}{6b}}$

5. $\dfrac{\frac{a}{b}}{\frac{c}{d}}$

6. $\dfrac{\frac{u}{v}}{\frac{x}{y}}$

7. $\dfrac{\frac{5}{12ab}}{\frac{7}{18bc}}$

8. $\dfrac{\frac{4}{9xy^2}}{\frac{7}{15x^2y}}$

9. $\dfrac{2 + \frac{1}{3}}{4 - \frac{1}{4}}$

*See Example 20*

10. $\dfrac{3 - \frac{1}{5}}{5 + \frac{1}{3}}$

11. $\dfrac{5 + \frac{1}{x}}{3 + \frac{1}{x}}$

12. $\dfrac{5 - \frac{1}{a^2}}{6 - \frac{2}{a}}$

13. $\dfrac{x + \frac{1}{x}}{7 + \frac{1}{x^2}}$

14. $\dfrac{z^2 - \frac{2}{z^2}}{z + \frac{1}{z}}$

15. $\dfrac{\frac{1}{a} + \frac{1}{b}}{\frac{1}{a} - \frac{1}{b}}$

16. $\dfrac{\frac{x}{y} - \frac{a}{b}}{\frac{x}{y} + \frac{a}{b}}$

17. $\dfrac{\frac{3}{10x^2} - \frac{1}{15x}}{\frac{6}{x} + \frac{1}{20}}$

18. $\dfrac{\frac{8}{9a} - \frac{2}{18a^2}}{\frac{4}{3a} - \frac{1}{6}}$

19. $\dfrac{\frac{2}{ax^2} + \frac{y}{a^3x}}{\frac{3b}{a^2x^4} - \frac{4}{x^3}}$

20. $\dfrac{\frac{1}{uv^2} - \frac{2}{u^4}}{\frac{x}{u^3v^2} - \frac{8}{v^3}}$

21. $\dfrac{\frac{x}{abc^2} - \frac{y}{a^2bc} + \frac{z}{ab^2c}}{\frac{5}{ab} - \frac{6}{ac} + \frac{7}{bc}}$

22. $\dfrac{\frac{4}{xy^2} - \frac{2}{xz^2} + \frac{3}{yz}}{\frac{a}{xyz} - \frac{b}{xyz^2} + \frac{c}{x^2yz}}$

See Example 21

23. $$\frac{2 + \frac{3}{a + 2}}{4 - \frac{1}{a - 3}}$$

24. $$\frac{7 - \frac{4}{x + 7}}{5 - \frac{3}{x + 5}}$$

25. $$\frac{\frac{8}{p} + \frac{2}{p + 1}}{\frac{6}{p + 1} - \frac{5}{p^2}}$$

26. $$\frac{\frac{9}{y + 2} - \frac{10}{y^2}}{\frac{6}{y + 2} + \frac{3}{y}}$$

27. $$\frac{3 + \frac{2}{x + 3}}{\frac{1}{x - 3} + \frac{4}{x^2 - 9}}$$

28. $$\frac{4 - \frac{1}{x - 2}}{\frac{7}{x^2 + 5x - 14} + \frac{8}{x + 7}}$$

*Business applications*

29. The *elasticity* of an item's demand is given by

$$e = -\frac{\frac{\Delta Q}{Q}}{\frac{\Delta P}{P}}$$

where $Q$ is the demand, $\Delta Q$ the change in demand, $P$ the price, and $\Delta P$ the change in price. Simplify this fraction.

30. The monthly payment $M$ on a 36-month loan is given by

$$M = \frac{rB}{1 - \frac{1}{(1 + r)^{36}}}$$

where $B$ is the amount borrowed and $r$ the rate of interest. Write this as a simple fraction.

*Engineering application*

31. A column will support a weight if the following relation holds:

$$\frac{\frac{P}{A}}{\sigma_p} + \frac{\frac{M}{z}}{\sigma_m} \leq 1$$

Write the left-hand side as a simple fraction.

*Television application*

32. In television technology, we have the formula

$$R_k = \frac{\frac{1}{g_m} \cdot Z_0}{\frac{1}{g_m} - Z_0}$$

(see page 31). Write this as a simple fraction.

## 5.6 EQUATIONS WITH FRACTIONS

In Chapter 2, we studied linear equations that contained no fractions. In this section, we study equations with fractions, or **fractional equations.** As examples,

$$\frac{8}{x} - \frac{x-5}{3x} = 12 \qquad \text{and} \qquad \frac{x+1}{x-5} + \frac{x+3}{x-4} = 0$$

are fractional equations. The technique that we use to solve these is to multiply both sides of the equation by the LCD. This clears away the fractions, and then we can solve the equation as we did in Chapter 2.

**EXAMPLE 22** Solve $\frac{2x+1}{3} - \frac{x+1}{4} = \frac{4x-4}{6} + \frac{5-x}{2}$.

**SOLUTION** This is very difficult because of the fractions. To clear the equation of fractions, we find the LCD, which is 12. We then multiply both sides of the equation by 12. (Recall from Chapter 2 that both sides will still be equal.)

Notice that the 12 will cancel with all the denominators: for instance, $\frac{12}{3} = 4$.

Multiply both sides by LCD, 12

$$12\left[\frac{2x+1}{3} - \frac{x+1}{4}\right] = 12\left[\frac{4x-4}{6} + \frac{5-x}{2}\right]$$

↓

Cancel denominators

$$4(2x+1) - 3(x+1) = 2(4x-4) + 6(5-x)$$

↓

| | |
|---|---|
| Distributive law | $8x + 4 - 3x - 3 = 8x - 8 + 30 - 6x$ |
| Combine terms | $5x + 1 = 2x + 22$ |
| Subtract $2x$ from both sides | $3x + 1 = 22$ |
| Subtract 1 from both sides | $3x = 21$ |
| Divide both sides by 3 | $x = 7$ |

We check this by substituting $x = 7$ into the original equation:

$$\text{Left side} = \frac{15}{3} - \frac{8}{4} = 5 - 2 = 3$$

$$\text{Right side} = \frac{24}{6} + \frac{-2}{2} = 4 - 1 = 3$$

> To solve fractional equations:
>
> 1. Determine the LCD for all the denominators of the fractions.
> 2. Multiply both sides of the equation by this LCD. (All denominators should cancel out when this is done.)
> 3. Solve as a linear equation. (Review Chapter 2.)
> 4. Be sure to check and see if the answer is, in fact, a solution to the equation.

As in Chapter 2, not all of these may be needed in an equation. We simply skip an unneeded step.

**EXAMPLE 23** Solve $\frac{12}{x} + \frac{9x - 4}{x^2} = \frac{20}{x}$, for $x$.

**SOLUTION** We first find the LCD, which is $x^2$. By multiplying by $x^2$, we will clear the fractions of their denominators.

| | |
|---|---|
| Multiply both sides by LCD, $x^2$ | $x^2\left[\frac{12}{x} + \frac{9x - 4}{x^2}\right] = x^2\left[\frac{20}{x}\right]$ |
| ↓ Cancel denominators | $12x + 9x - 4 = 20x$ |
| ↓ Combine terms | $21x - 4 = 20x$ |
| ↓ Subtract $20x$ from both sides | $x - 4 = 0$ |
| ↓ Add 4 to both sides | $x = 4$ |

We can check this solution to see if it is correct. We substitute $x = 4$ into both the left- and right-hand sides.

$$\text{Left side} = \frac{12}{4} + \frac{32}{16} = 5$$

$$\text{Right side} = \frac{20}{4} = 5$$

Therefore, $x = 4$ is the solution.

**EXAMPLE 24** Solve $\dfrac{3}{x-1} - \dfrac{4}{x+1} = \dfrac{6}{x^2-1}$, for $x$.

**SOLUTION** The LCD is $x^2 - 1 = (x+1)(x-1)$. It will be easier to cancel on the left if we write the LCD as $(x+1)(x-1)$, which is just another way to write it.

| | |
|---|---|
| Multiply both sides by LCD, $x^2 - 1 = (x+1)(x-1)$ | $(x+1)(x-1)\left[\dfrac{3}{x-1} - \dfrac{4}{x+1}\right] = (x^2-1)\left[\dfrac{6}{x^2-1}\right]$ |
| ↓ Cancel denominators | $3(x+1) - 4(x-1) = 6$ |
| ↓ Distributive law | $3x + 3 - 4x + 4 = 6$ |
| ↓ Combine terms | $-x + 7 = 6$ |
| ↓ Subtract 7 from both sides | $-x = -1$ |
| ↓ Multiply both sides by $-1$ | $x = 1$ |

Let us now check this solution, $x = 1$.

$$\text{Left side} = \frac{3}{1-1} - \frac{4}{1+1} = \frac{3}{0} - \frac{4}{2} \quad ????$$

$$\text{Right side} = \frac{6}{1^2-1} = \frac{6}{0} \quad ????$$

What happened? We get meaningless expressions such as 3/0 and 6/0. Our procedure was correct. The fact is that $x = 1$ is *not* really a solution. It is called an **extraneous solution** (but it is not really a solution at all). Therefore, there is no solution.

The quickest way to spot an extraneous solution is as follows: *If the solution produces a zero in a denominator, it is extraneous and not really a solution.* Therefore, Example 24 has no solution, or we say that the solution set is $\varnothing$.

**EXAMPLE 25** The total resistance $R$ in a parallel circuit with resistances $R_1$ and $R_2$ is given by

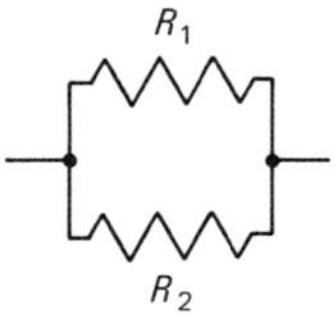


$$\frac{1}{R} = \frac{1}{R_1} + \frac{1}{R_2}$$

Solve this for $R_1$.

**SOLUTION** Recall from Chapter 2 how we solve literal equations (those with letters and numbers). We identify the unknown, perhaps with a circle, and then try to get it alone.

We solve this exactly as the one before. Here, the LCD is $RR_1R_2$. We will circle the unknown, $R_1$.

Multiply both sides by LCD, $RR_1R_2$

$$R\textcircled{R_1}R_2\left[\frac{1}{R}\right] = R\textcircled{R_1}R_2\left[\frac{1}{\textcircled{R_1}} + \frac{1}{R_2}\right]$$

↓

Cancel denominators

$$\textcircled{R_1}R_2 = RR_2 + R\textcircled{R_1}$$

↓

Subtract $RR_1$ from both sides

$$\textcircled{R_1}R_2 - R\textcircled{R_1} = RR_2$$

↓

Factor

$$\textcircled{R_1}(R_2 - R) = RR_2$$

↓

Divide both sides by $(R_2 - R)$

$$\textcircled{R_1} = \frac{RR_2}{R_2 - R}$$

Notice that we carefully move all the $R_1$ terms to one side, and eventually move the other terms, $R$ and $R_2$, to the other side.

## PROBLEM SET 5.6

Solve the following equations. (Be sure to check for extraneous solutions—see Example 24.)

*See Example 22*

1. $\frac{2x}{3} + \frac{5x}{6} = \frac{9}{2}$
2. $\frac{7z}{3} = \frac{3z}{4} + \frac{38}{6}$
3. $\frac{4a}{5} + \frac{2a - 1}{10} = \frac{a}{15}$
4. $\frac{5x}{4} - \frac{x - 6}{6} = \frac{x - 7}{12}$
5. $\frac{3x + 5}{8} - \frac{4x - 1}{4} = \frac{x}{16}$
6. $\frac{8t - 1}{9} - \frac{3t + 2}{6} = \frac{t + 5}{18}$

*See Example 23*

7. $\frac{1}{x} + \frac{1}{2} = \frac{2}{x}$
8. $\frac{2}{y} + \frac{4}{3} = \frac{6}{y}$
9. $\frac{2}{a} - \frac{1}{3} = \frac{5}{3a}$
10. $\frac{4}{3t} - \frac{7}{2t} = \frac{1}{6}$
11. $\frac{x - 1}{x} - \frac{1}{4} = \frac{x + 2}{2x}$
12. $\frac{1}{10} - \frac{r - 4}{5} = \frac{r + 3}{20}$
13. $\frac{1}{x + 4} - \frac{1}{4} = \frac{1}{8}$
14. $\frac{2}{3} - \frac{5}{a - 7} = \frac{3}{6}$
15. $\frac{x}{x - 5} + \frac{1}{2} = \frac{1}{2x - 10}$
16. $\frac{1}{y + 4} + \frac{y}{3y + 12} = \frac{1}{6}$
17. $\frac{x + 1}{3x - 9} - \frac{1}{2} = \frac{8}{6x - 18}$
18. $\frac{a}{2a - 6} - \frac{2a - 1}{3a - 9} = \frac{1}{6}$
19. $\frac{1}{x^2} + \frac{1}{x} = \frac{3}{x^2}$
20. $\frac{5}{p^2} + \frac{2}{p} = \frac{11}{p^2}$
21. $\frac{x + 1}{2x^2} - \frac{3}{4x} = \frac{6}{8x^2}$
22. $\frac{5}{3a^2} - \frac{2}{6a} = \frac{a + 3}{9a^2}$
23. $\frac{2}{x} + \frac{5}{x + 3} = \frac{4}{x}$
24. $\frac{3}{x} + \frac{1}{x + 3} = \frac{7}{x}$
25. $\frac{3}{p + 1} + \frac{3}{p} = \frac{3}{4p}$
26. $\frac{6}{a - 1} - \frac{7}{a} = \frac{6}{a - 1}$

*See Example 24*

27. $\dfrac{2}{r-5} + \dfrac{3}{r+5} = \dfrac{3}{r-5}$

28. $\dfrac{4}{t-2} - \dfrac{3}{t+1} = \dfrac{2}{t+1}$

29. $\dfrac{6}{x+1} - \dfrac{7}{x-2} = \dfrac{3}{x+1}$

30. $\dfrac{7}{m+1} - \dfrac{2}{m+3} = \dfrac{1}{m+3}$

31. $\dfrac{2}{s+5} - \dfrac{1}{s-5} = \dfrac{3}{s^2-25}$

32. $\dfrac{2}{y-2} + \dfrac{3}{y+2} = \dfrac{3}{y^2-4}$

33. $\dfrac{3}{c-3} + \dfrac{5}{c+5} = \dfrac{c-7}{c^2+2x-15}$

34. $\dfrac{1}{z+2} - \dfrac{2}{z-1} = \dfrac{z-5}{z^2+z-2}$

35. $\dfrac{7}{x-3} + \dfrac{3}{x-4} = \dfrac{x-1}{x^2-7x+12}$

36. $\dfrac{6}{x-2} - \dfrac{3}{x+4} = \dfrac{6}{x^2+2x-8}$

Solve the following equations for the indicated variable.

*See Example 25*

37. $\dfrac{1}{a} + \dfrac{1}{b} = \dfrac{1}{x}$, for $x$

38. $\dfrac{1}{c} - \dfrac{1}{d} = \dfrac{1}{e}$, for $d$

39. $A = \dfrac{m-n}{mn}$, for $m$

40. $\dfrac{a}{b} = \dfrac{c}{c+d}$, for $c$

41. $S = \dfrac{a-rl}{l-r}$, for $r$

42. $A = \dfrac{24I}{B(n+1)}$, for $n$

43. $\dfrac{1}{s^2} = \dfrac{1}{f_1} + \dfrac{1}{f_2}$, for $f_1$

44. $F = \dfrac{-GmM}{r^2}$, for $m$

45. $Y_2 = \dfrac{vY_1}{v-s} - \dfrac{Z}{v-s}$, for $v$

46. $I = \dfrac{a}{s^2} + \dfrac{b}{t^2}$, for $a$

*Health application*

47. Young's formula for the children's dosage of aspirin is

$$D_c = \frac{A}{A+12} \cdot D_a$$

Solve for the age of the child $A$ if $D_a$ = adult dosage = 5 grains and $D_c$ = child's dosage = 2 grains.

48. In the conversion of protein to amino acid, we have

$$M = \frac{28}{t+2}$$

At what time (in hours) $t$ will the mass of amino acid $M = 7$?

*Life science application*

49. The following is Kelker's formula for a buck population.

$$\frac{S_1}{P_1} = \frac{S_1 - S_2}{H}$$

where $S_1$ is the prehunt male–female ratio, $S_2$ the posthunt male–female ratio, $P_1$ the prehunt buck population, and $H$ the total harvest. Solve for $S_1$ in terms of the other variables.

*Population applications*

50. A certain population grows according to the formula

$$N = \frac{20{,}000t}{100 + t}$$

Solve for the time $t$ in terms of the population $N$.

51. The death rate $m_x$ at age $x$ is given by

$$m_x = \frac{q_x}{1 - \frac{1}{2}q_x}$$

Solve for $q_x$, the probability of dying at age $x$, in terms of $m_x$.

*Chemical application*

52. At the equilibrium of a cyanidation of ammonium, the concentration of ammonia, $x$, is given by

$$\frac{x}{1 - x} = 1.18$$

Solve for $x$.

*Photographic application*

53. For a lens we have the formula

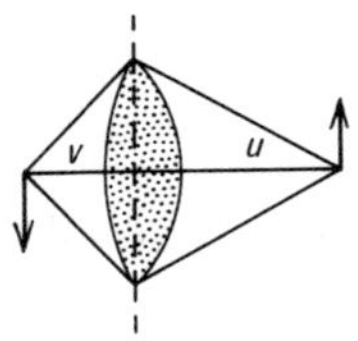


$$\frac{1}{f} = \frac{1}{u} + \frac{1}{v}$$

where $f$ is the focal length, $u$ the distance to the image, and $v$ the distance to the object. Solve for $v$ in terms of $f$ and $u$.

*Materials application*

54. The volume of mortar yield, $y$, is given by

$$y = \frac{100V_m}{100 - a}$$

where $V_m$ is the absolute solid volume of the mortar and $a$ the air content as a percent. What is the percent of air $a$ if $V_m = 2$ tons and $y = 2.5$ tons?

*Electricity applications*

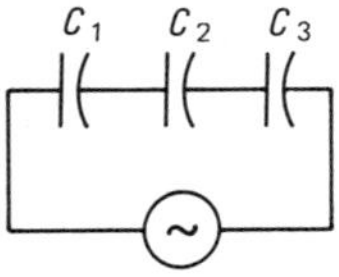


55. In a series of capacitors, $C_1$, $C_2$, and $C_3$, the net capacitance $C$ is given by

$$\frac{1}{C} = \frac{1}{C_1} + \frac{1}{C_2} + \frac{1}{C_3}$$

Solve this for $C_1$.

56. In electronics we have the formula

$$\beta = \frac{\alpha}{1 - \alpha}$$

where $\beta$ is the current gain in a transistor and $\alpha$ the forward-current-transfer ratio. Solve for $\alpha$ in terms of $\beta$.

## 5.7 WORD PROBLEMS

There are several types of applications that involve fractional equations. Among these are work problems, motion problems, and mixture problems. These can seem tricky; but they each have a pattern, and once the pattern is learned, the problems become easier.

### Translation Problems

Usually, these can be solved by translating the English information into mathematics and solving.

**EXAMPLE 26** What number must be added to both the numerator and denominator of $\frac{5}{11}$ to give the fraction $\frac{5}{7}$?

**SOLUTION** We let $x$ be the unknown number. We add this to both the top and bottom of $\frac{5}{11}$ and set the result equal to $\frac{5}{7}$.

| | |
|---|---|
| Given | Add $x$ to top and bottom of $\frac{5}{11}$ to get $\frac{5}{7}$ |
| Translate | $\frac{5+x}{11+x} = \frac{5}{7}$ |
| Multiply by LCD, $7(11 + x)$ | $7(11+x)\left[\frac{5+x}{11+x}\right] = 7(11+x)\left[\frac{5}{7}\right]$ |
| Cancel denominators | $7(5 + x) = (11 + x)5$ |
| Distributive law | $35 + 7x = 55 + 5x$ |
| Subtract $5x$ from both sides | $35 + 2x = 55$ |
| Subtract 35 from both sides | $2x = 20$ |
| Divide both sides by 2 | $x = 10$ |

We can check this: $\frac{5+10}{11+10} = \frac{15}{21}$, which reduces to $\frac{5}{7}$.

## Work Problems

These problems usually involve two people (or machines) working together at different rates. What we want is their combined rate.

The way that we will solve these is by looking at how much of a job can be done in 1 period of time. For example, if Tom can paint a room in 2 hours, then in 1 hour he can paint $\frac{1}{2}$ of a room; if it takes Pat 10 minutes to clean a room, then in 1 minute she can clean $\frac{1}{10}$ of a room. In general, we use the following formula.

> ***FORMULA 5.1*** $\dfrac{1}{T_1} + \dfrac{1}{T_2} = \dfrac{1}{x}$
>
> where $T_1$ and $T_2$ are given times and $x$ is the unknown combined time. [We subtract if the people (or whatever) are working against each other.]

**EXAMPLE 27** In a certain post office, Alice can sort a stack of mail in 30 minutes; Bob can sort the same stack in 40 minutes. If they work together, how fast can they sort the stack?

**SOLUTION** Alice can sort $\frac{1}{30}$ of a stack in 1 minute, and Bob can sort $\frac{1}{40}$. If their combined time is $x$, they can sort $\dfrac{1}{x}$ of a stack in 1 minute.

| | |
|---|---|
| Identify times | $30 =$ Alice's time to sort stack<br>$40 =$ Bob's time to sort stack<br>$x =$ combined time to sort stack |
| Determine amount they do in one period of time | $\dfrac{1}{30} =$ amount Alice does in 1 minute<br>$\dfrac{1}{40} =$ amount Bob does in 1 minute<br>$\dfrac{1}{x} =$ amount both do in 1 minute |
| Formula 5.1 | $\dfrac{1}{30} + \dfrac{1}{40} = \dfrac{1}{x}$ |
| Multiply by LCD, $120x$ | $120x\left[\dfrac{1}{30} + \dfrac{1}{40}\right] = 120x\left[\dfrac{1}{x}\right]$ |
| Cancel denominators | $4x + 3x = 120$ |
| Combine terms | $7x = 120$ |

| Divide both sides by 7 |
|---|

$$x = \frac{120}{7}$$

So, together they can do the job in $\frac{120}{7}$, or about 17.1 minutes.

## Motion Problems

These problems all involve distance, velocity, and time. These three variables are related by the following equivalent equations.

> ***FORMULA 5.2*** Let $d$ be the distance, $v$ the velocity, and $t$ the time.
>
> $$d = vt \qquad v = \frac{d}{t} \qquad t = \frac{d}{v}$$

Usually, one of the terms will be the unknown, a second will be given completely, and the third can be written in terms of the first two (using Formula 5.2). Generally, there will be some equation involving that third term.

**EXAMPLE 28** In the same time, Becky drives 600 miles, while Lisa drives 450 miles. Becky drives 10 miles per hour faster than Lisa. What are the cars' velocities?

**SOLUTION** We start by completing the $d$–$v$–$t$ table. Here $v$ is the unknown, $d$ is completely known, and we find $t$ in terms of $v$ and $d$. We let Becky's rate be $x$, and Lisa's be $x - 10$.

| | $d$ | $v$ | $t = \frac{d}{v}$ |
|---|---|---|---|
| Becky | 600 | $x$ | $\frac{600}{x}$ |
| Lisa | 450 | $x - 10$ | $\frac{450}{x - 10}$ |

We now use the fact that their times were the same.

| | |
|---|---|
| Given equal times | Becky's time = Lisa's time |
| ↓ Substitute | $\frac{600}{x} = \frac{450}{x - 10}$ |
| ↓ Multiply by LCD, $x(x - 10)$ | $x(x - 10)\left[\frac{600}{x}\right] = x(x - 10)\left[\frac{450}{x - 10}\right]$ |
| ↓ Cancel denominators | $(x - 10)600 = 450x$ |
| ↓ Distributive law | $600x - 6000 = 450x$ |
| ↓ Subtract $450x$ | $150x - 6000 = 0$ |
| ↓ Add 6000 | $150x = 6000$ |
| ↓ Divide by 150 | $x = 40$ |

Therefore, Becky was driving at 40 miles per hour, and Lisa was driving at 30 miles per hour.

## Mixture Problems

In these problems, we have a quantity that is some percent of a whole, such as a solution that is 40% alcohol or a company that is 20% women. We then wish to change the percent.

> To work mixture problems:
>
> 1. Determine the original quantity and total.
> 2. Let $x$ be the amount to be added.
> 3. To strengthen the mixture, $x$ is added to both quantity and total; to weaken the mixture, $x$ is added only to the total.
> 4. Solve the proportion
>
> $$\frac{\text{New quantity}}{\text{New total}} = \frac{\text{New percent}}{100}$$

**EXAMPLE 29** A company of 1000 is 20% women. How many women must be hired to bring the percent to 35%?

**SOLUTION** Since there are 20% women, there are 20% of 1000 = 200 women.

| | |
|---|---|
| Determine original terms | Original quantity = 200 women<br>Original total = 1000 people |
| Identify unknown | $x$ = women to be added |
| Add $x$ to both to increase percent | New quantity = $200 + x$ women<br>New total = $1000 + x$ people |
| $\dfrac{\text{New quantity}}{\text{New total}} = \dfrac{\text{New percent}}{100}$ | $\dfrac{200 + x}{1000 + x} = \dfrac{35}{100}$ |
| Multiply by LCD, $(1000 + x)100$ | $(1000 + x)100\left[\dfrac{200 + x}{1000 + x}\right] = (1000 + x)100\left[\dfrac{35}{100}\right]$ |
| Cancel denominators | $100(200 + x) = (1000 + x)35$ |
| Distributive law | $20{,}000 + 100x = 35{,}000 + 35x$ |
| Subtract $35x$ | $20{,}000 + 65x = 35{,}000$ |
| Subtract 20,000 | $65x = 15{,}000$ |
| Divide by 65 | $x = \dfrac{15{,}000}{65} \approx 231$ |

So, by hiring 231 women, the company will be 35% women. (*Check:* 431/1231 = 0.35.)

## PROBLEM SET 5.7

*See Example 26*

1. What number must be added to both the numerator and the denominator of $\frac{4}{9}$ to give the fraction $\frac{1}{2}$?
2. What number must be subtracted from both the numerator and denominator of $\frac{23}{30}$ to give the fraction $\frac{2}{3}$?
3. What number must be added to the denominator of $\frac{9}{14}$ to give the fraction $\frac{3}{7}$?
4. What number must be added to the numerator of $\frac{7}{24}$ to give the fraction $\frac{5}{8}$?
5. The sum of the reciprocals of two consecutive integers is equal to 11 divided by the product of the integers. Find the integers.
6. The sum of the reciprocals of two consecutive odd integers is equal to 16 divided by the product of the integers. Find the integers.

*See Example 27*

7. Bobby can mow a lawn in 2 hours. Billy can mow the same lawn in 3 hours. How long will it take if they work together?
8. Ann can do a job in 7 hours, while Mary can do the job in 9 hours. How long will it take them working together?
9. Allen, Bonnie, and Carol are cleaning up the litter in a park. Alone, Allen could do it in 4 hours, Bonnie in 3 hours, and Carol in 5 hours. Working together, how long will it take them?
10. Ron, Sally, and Tony are going door to door campaigning for a local candidate. Alone, Ron could do the precinct in 30 hours, Sally in 40 hours, and Terry in 50 hours. Working together, how long will it take them?
11. Pipe $P$ can fill a pool in 10 hours; pipe $Q$ can empty it in 25 hours. If both pipes are open, how long will it take to fill the pool?
12. Fran can clean a room in 10 minutes, and Becky can mess it up in 30 minutes. How long will it take Fran to clean the room if Becky is messing it up while Fran is cleaning?

*See Example 28*

13. In the same time, Bob drove 1000 miles, while his mother drove 600 miles. If Bob averaged 20 miles per hour faster than his mother, what were the cars' velocities?
14. In the same time, jet A flies 1200 miles while jet B flies 1550 miles. Jet B was averaging 150 miles per hour faster than jet A. What were their velocities?
15. Mike made a 3000-mile trip in 35 hours. He traveled part of it in a 50 mile-per-hour car and part in a 100 mile-per-hour train. What distances did he travel at each velocity?

16. Michele traveled 960 miles in 27 hours. She made part of the trip at 30 miles per hour and part at 40 miles per hour. What distances did she drive at each velocity?

17. A jetliner made a 2050-mile trip in 4.5 hours. Part of the trip was at 400 miles per hour, and part of it was at 500 miles per hour. What times were spent at each velocity?

18. Connie and Donna made a 1210-mile trip in 23 hours. Connie drove at 55 miles per hour, while Donna drove at 50 miles per hour. How much time did each of them drive?

*See Example 29*

19. Pete Lily has been at bat 148 times and has 41 hits for a batting average of .277 (he has gotten hits 27.7% of the time). How many straight hits must he get to bring his average up to .300?

20. A 3-gallon mixture in Bob's car radiator is 25% antifreeze. How much antifreeze must he add to make the mixture 35% antifreeze?

21. A 10-quart solution is 45% alcohol. How much water must be added to reduce the strength to 30% alcohol?

22. A 32-person committee is 75% conservative. How many liberals must be added so that the committee is only 60% conservative?

23. A record store has an inventory of 10,000 records, 70% of which are rock music. How many non-rock-music records must the store get to be stocked with only 55% rock?

24. A 4-pound recipe of cookies is 25% chocolate chips. How much cookie batter must be added so that the cookies are only 15% chips?

# CHAPTER 5 SUMMARY

## Important Words and Phrases

complex fractions (5.5)
denominator (introduction)
extraneous solution (5.6)
fraction (introduction)
fractional equation (5.6)
invert and multiply (5.3)
lowest common denominator (5.4)
mixed expression (5.4)
mixture problems (5.7)
motion problems (5.7)
numerator (introduction)
opposite of a fraction (5.4)
rational expression (introduction)
reduced to lowest terms (5.1)
reducing fractions (5.1)
translation problems (5.7)
work problems (5.7)

## Important Theorems and Definitions

$$\frac{AB}{CB} = \frac{A}{C} \cdot \frac{B}{B} = \frac{A}{C}$$

$$\frac{A}{B} \cdot \frac{C}{D} = \frac{AC}{BD}$$

$$\frac{A}{B} \div \frac{C}{D} = \frac{A}{B} \cdot \frac{D}{C} = \frac{AD}{BC}$$

$$\frac{A}{C} + \frac{B}{C} = \frac{A + B}{C}$$

## Important Procedures

*To reduce a fraction:*

1. Factor the numerator and denominator.
2. Cancel common factors.
3. Simplify.

*To find the LCD and add fractions:*

1. Factor the denominators and line up common factors.
2. Fill in missing factors (circled).
3. Multiply a factor from each column to get LCD.
4. Multiply each fraction by $1 = \dfrac{\text{missing factor}}{\text{missing factor}}$ to get new fractions.
5. Add numerators and reduce.

*To write a mixed expression as a fraction:*

1. Find the LCD of the denominators.
2. Multiply the nonfraction by $1 = \dfrac{\text{LCD}}{\text{LCD}}$.
3. Simplify.

*To simplify complex fractions:*

1. Find the LCD for all the denominators of the fractions.
2. Multiply by $1 = \dfrac{\text{LCD}}{\text{LCD}}$.
3. Simplify.

*To solve fractional equations:*

1. Find the LCD.
2. Multiply both sides by the LCD.
3. Remove parentheses and combine terms.
4. Isolate the unknown and solve.

## CHAPTER 5 REVIEW EXERCISES

Reduce each of the following to lowest terms.

1. $\dfrac{36x^5y^3z^6}{27x^3y^7z^3}$

2. $\dfrac{7x - x^2}{xy + x^5}$

3. $\dfrac{m^2 - 36}{m^2 + 3m - 18}$

4. $\dfrac{p^2 - 9}{p^3 + 27}$

Simplify the following (and reduce to lowest terms).

5. $\dfrac{8a^2b^3c^4}{5b^2cd^5} \cdot \dfrac{15c^5de^8}{16d^3e^2f}$

6. $\dfrac{x^4}{x^2 + x} \cdot \dfrac{x^2 - 1}{x^3}$

7. $\dfrac{x^2 + 3x + 2}{x^2 - 4} \cdot \dfrac{x^2 - 5x + 6}{x^2 + 2x - 15}$

8. $\dfrac{-15m^2n^3p}{8r^3s^4} \div \dfrac{5m^4np^3}{2rs^2t^2}$

9. $\dfrac{a^4}{a^2 - a} \div \dfrac{a^3}{a^2 - 1}$

10. $\dfrac{x^2 - 6x + 9}{x^2 + 8x + 15} \div \dfrac{x^2 - 5x + 6}{x^2 + x - 6}$

11. $\dfrac{x^2 + 6x - 7}{x^3} \cdot \dfrac{x^2 - x - 12}{x^2 + 2x - 3} \div \dfrac{x^2 - 6x + 8}{x^5}$

12. $\dfrac{x + y}{x^3y} + \dfrac{x - y}{x^2y^2}$

13. $\dfrac{6}{a} + \dfrac{7}{a^2} - \dfrac{8}{a^3}$

14. $\dfrac{6}{x^2 - 4x - 5} + \dfrac{8}{x^2 + 5x + 4}$

15. $\dfrac{a + 1}{a^3 + 8} - \dfrac{7}{a^2 - 4}$

16. $8r + \dfrac{7}{r - 3}$

17. $m + \dfrac{1}{m} - \dfrac{7}{m + 3}$

18. $\dfrac{\dfrac{6}{11a^2}}{\dfrac{5}{22a}}$

19. $\dfrac{\dfrac{1}{9r} + \dfrac{2}{3r^2}}{\dfrac{4}{27} - \dfrac{1}{r}}$

20. $\dfrac{2 + \dfrac{1}{x - 3}}{\dfrac{4}{x^2 - 9} - \dfrac{2}{x + 3}}$

Solve the following.

21. $\dfrac{1}{a - 5} - \dfrac{1}{a} = \dfrac{1}{8a}$

22. $\dfrac{x}{4x - 2} - \dfrac{x + 1}{6x - 3} = \dfrac{1}{6}$

23. $\dfrac{3}{x + 6} - \dfrac{4}{x - 6} = \dfrac{1}{x^2 - 36}$

24. $\dfrac{m}{n} = \dfrac{r}{r + s}$, for $r$

25. What number must be added to both the numerator and denominator of $\frac{2}{11}$ to give $\frac{1}{2}$?

26. Fred can paint a house in 8 hours; Larry can paint it in 10 hours. How long will it take if they work together?

27. In the same time, Sue drives 1000 miles, while Carol drives 900 miles. Sue drives 5 miles per hour faster than Carol. What are their speeds?

28. A 100-pound hamburger mixture is 65% lean meat. How much pure lean meat must be added to the mixture to make it 75% lean?

# 6
# EXPONENTS AND RADICALS

## 6.1 RULES FOR EXPONENTS

In Chapters 1 and 3, we saw what exponents were and two of the rules for dealing with exponents. We repeat these rules now.

> ***THEOREM 6.1*** For any real number $a$ and positive integers $m$ and $n$,
>
> $$a^n \cdot a^m = a^{n+m}$$

> ***THEOREM 6.2*** For any real number $a$ and positive integers $m$ and $n$,
>
> $$\frac{a^m}{a^n} = \begin{cases} a^{m-n} & \text{if } m > n \\ 1 & \text{if } m = n \\ \dfrac{1}{a^{n-m}} & \text{if } m < n \end{cases}$$
>
> for all $a \neq 0$

Let us look at some new theorems. First, consider $(a^3)^4$.

| | |
|---|---|
| Definition of fourth power | $(a^3)^4 = a^3 \cdot a^3 \cdot a^3 \cdot a^3$ |
| ↓ Theorem 6.1 | $= a^{3+3+3+3} = a^{12}$ |

Since $3 \cdot 4 = 12$, this suggests the next theorem.

> ***THEOREM 6.3*** For all real numbers $a$ and positive integers $m$ and $n$,
>
> $$(a^m)^n = a^{m \cdot n}$$

**EXAMPLE 1** **(a)** $(x^5)^3 = x^{15}$
**(b)** $(2^8)^4 = 2^{32}$
**(c)** $(t^n)^2 = t^{2n}$

Suppose that we have $(2r)^5$:

| | |
|---|---|
| Definition of fifth power | $(2r)^5 = 2r \cdot 2r \cdot 2r \cdot 2r \cdot 2r$ |
| ↓ Associative and commutative laws | $= 2 \cdot 2 \cdot 2 \cdot 2 \cdot 2 \cdot r \cdot r \cdot r \cdot r \cdot r$ |
| ↓ Definition of fifth power | $= 2^5 \cdot r^5 = 32r^5$ |

This suggests the following theorem.

> ***THEOREM 6.4*** For all real numbers $a$ and $b$ and positive integer $m$,
>
> $$(ab)^m = a^m b^m$$

**EXAMPLE 2** **(a)** $(5x)^2 = 25x^2$
**(b)** $(2abc)^3 = 8a^3b^3c^3$
**(c)** $(3xy^2)^4 = 81x^4(y^2)^4 = 81x^4y^8$
**(d)** $(4p^nq^{n+1})^2 = 16p^{2n}q^{2(n+1)} = 16p^{2n}q^{2n+2}$

Finally, let us look at the quotient $\left(\dfrac{x}{2}\right)^4$.

| | |
|---|---|
| Definition of fourth power | $\left(\dfrac{x}{2}\right)^4 = \dfrac{x}{2} \cdot \dfrac{x}{2} \cdot \dfrac{x}{2} \cdot \dfrac{x}{2}$ |
| ↓ Multiplication of fractions | $= \dfrac{x \cdot x \cdot x \cdot x}{2 \cdot 2 \cdot 2 \cdot 2}$ |
| ↓ Definition of fourth power | $= \dfrac{x^4}{2^4} = \dfrac{x^4}{16}$ |

This suggests the last theorem in this section.

> ***THEOREM 6.5*** For all real numbers $a$ and $b$ ($b \neq 0$) and positive integer $m$,
>
> $$\left(\frac{a}{b}\right)^m = \frac{a^m}{b^m}$$

**EXAMPLE 3** **(a)** $\left(\frac{3}{t}\right)^3 = \frac{27}{t^3}$

**(b)** $\left(\frac{ab}{5}\right)^2 = \frac{a^2b^2}{25}$

**(c)** $\left(\frac{p^3q^4}{r^2}\right)^3 = \frac{(p^3)^3(q^4)^3}{(r^2)^3} = \frac{p^9q^{12}}{r^6}$

**(d)** $\left(\frac{x^n}{y^{n+2}}\right)^3 = \frac{(x^n)^3}{(y^{n+2})^3} = \frac{x^{3n}}{y^{3n+6}}$

| YES | NO |
|---|---|
| $(a^5)^2 = a^{10}$ | ~~$(a^5)^2 = a^{25}$~~ |

We multiply the exponents.

**EXAMPLE 4** Simplify $(a^2b)^5 \cdot (2a^4b^2)^2$.

**SOLUTION** We first simplify each term, and then we combine them.

| | |
|---|---|
| Theorem 6.5 | $(a^2b)^5 \cdot (2a^4b^2)^2 = a^{10}b^5 \cdot 4a^8b^4$ |
| ↓ | |
| Theorem 6.1 | $= 4a^{18}b^9$ |

If we have a complicated expression, we simplify it in "slow motion."

**EXAMPLE 5** Simplify $\left[\left(\frac{a^3b}{c^3}\right)^4 \cdot \left(\frac{c^2}{a^3b^2}\right)^3\right]^2$.

**SOLUTION** Rather than trying to work this out in one big step, we work it in smaller steps. We start by simplifying the inside terms.

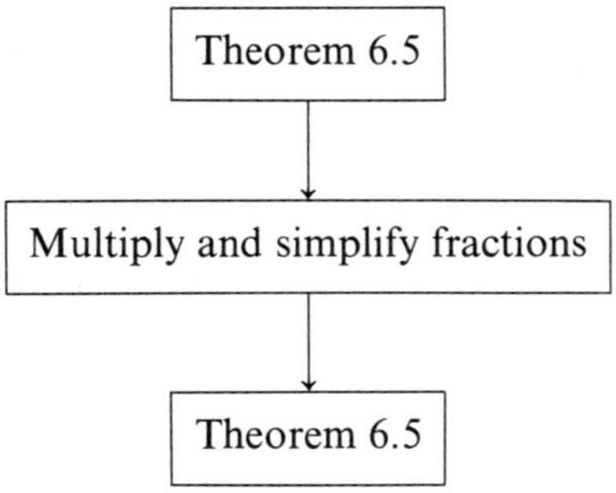


$$\left[\left(\frac{a^3b}{c^3}\right)^4 \cdot \left(\frac{c^2}{a^3b^2}\right)^3\right]^2 = \left[\frac{a^{12}b^4}{c^{12}} \cdot \frac{c^6}{a^9b^6}\right]^2$$

$$= \left[\frac{a^3}{b^2c^6}\right]^2$$

$$= \frac{a^6}{b^4c^{12}}$$

## PROBLEM SET 6.1

Simplify each of the following expressions so that each variable appears only once.

*See Example 9 (Chapter 3)*

1. $x^7 \cdot x^4$
2. $a^5 \cdot a^{10}$
3. $t^3t^4t^5$
4. $(3a^2)(4a^9)$
5. $(-4y^8z)(5x^2y^2z^3)$
6. $(-10p^3q^4r^5)(-3pq^7)$

*See Example 10 (Chapter 3)*

7. $\dfrac{r^9}{r^5}$
8. $\dfrac{h^7}{h^7}$
9. $\dfrac{m^8}{m^{11}}$
10. $\dfrac{12x^5}{-2x^7}$
11. $\dfrac{24a^2b^3c^4}{3ab^4c^7}$
12. $\dfrac{-100k^4m^3n^2}{-20k^5mp^4}$

*See Example 1*

13. $(x^2)^5$
14. $(a^4)^3$
15. $(6^9)^2$
16. $(2^6)^4$
17. $(10^n)^3$
18. $(e^x)^4$

*See Example 2*

19. $(2a)^5$
20. $(-4t)^3$
21. $(5m)^2$
22. $(3ab)^3$
23. $(-7xy^2)^2$
24. $(10r^2s^3)^4$
25. $(2m^2n^3)^6$
26. $(5p^2q^3r^4)^4$
27. $(-200a^4b^n)^2$
28. $(3x^ky^4)^3$
29. $(a^nb^{n+1}c^{n+2})^3$
30. $(p^tq^{2t}r^{3t})^2$

*See Example 3*

31. $\left(\dfrac{x}{3}\right)^4$
32. $\left(\dfrac{a}{6}\right)^2$
33. $\left(\dfrac{5}{t}\right)^3$
34. $\left(\dfrac{ab}{7}\right)^2$
35. $\left(\dfrac{-3a^2}{4b}\right)^3$
36. $\left(\dfrac{2}{r^3s^4}\right)^4$

37. $\left(\frac{4x^3y^2}{z^2w^4}\right)^3$

38. $\left(\frac{2p^4q^5}{r^6s^7}\right)^5$

39. $\left(\frac{10x^2}{-y^n}\right)^3$

40. $\left(\frac{-3t^r}{5}\right)^3$

41. $\left(\frac{2a^k}{b^{k+1}}\right)^4$

42. $\left(\frac{3m^xn^{x+2}}{k^{x+4}}\right)^2$

*See Example 4*

43. $(ab^2)^4(a^2b)^3$

44. $(-2x^2y^2)^3(3xy^3)^4$

45. $(5m^4n)^3(2m^2n^2)^2$

46. $(4p^2q)^2(3pq^2r^3)^3(pq)^4$

47. $\left(\frac{a^2b^2}{c^3}\right)^3\left(\frac{c}{ab^4}\right)^2$

48. $\left(\frac{-4u}{v^2w^3}\right)^2\left(\frac{v^3}{2uw}\right)^3$

49. $\frac{(2a^2b)^3}{(2ab^2)^2}$

50. $\frac{(3x^2y^3)^4}{(-9x^4y)^2}$

51. $\frac{(6m^2n^5)^2}{(2m^4n)^3}$

52. $\frac{(2a^2b^3c^4)^5}{(4bc^4d^7)^2}$

53. $\frac{(a^2b^2)^3(2ab^3)^4}{(4a^3bc)^2}$

54. $\frac{(3u^3v^2)^3(uv^4)^2}{(9uv^2w^3)^2}$

*See Example 5*

55. $\left[\frac{(2x^2y)^3(3x^2y^3)^2}{(6xy^4)^2}\right]^4$

56. $\left[\frac{(3a^2b)^3(2a^3b^2c)^4}{(-12ab^2c^3)^2}\right]^2$

57. $\left[\frac{(4p^2q^5)^2(p^6qr)^3}{(2p^7q^3r)^5}\right]^3$

58. $\left[\frac{(5u^3vw^2)^2(-2uv^2w)^3}{(10u^2v^2w^2)^4}\right]^2$

59. $\left[\left(\frac{p^3q^4}{r^5}\right)^2\left(\frac{r^2}{pq^5}\right)^3\right]^2$

60. $\left[\left(\frac{k^2m}{n^4}\right)^3\left(\frac{k^2n^2}{-m^4}\right)^4\right]^3$

61. $\left[\left(\frac{4a^2b^3}{c^4}\right)^3\left(\frac{c^2}{-2ab^3}\right)^5\right]^3$

62. $\left[\left(\frac{10x^3yz}{w^4}\right)^4\left(\frac{x^4y^2}{5z^2w^3}\right)^3\right]^2$

63. $\left[\left(\frac{p^2q^2}{r^3}\right)^3\left(\frac{pr^4}{q^4}\right)^2\left(\frac{r^2q}{p^3}\right)^4\right]^2$

64. $\left[\left(\frac{3k^2m}{2n^4}\right)^3\left(\frac{-4k^3n^2}{9m^3}\right)^2\left(\frac{2m^3n}{k^5}\right)^3\right]^3$

*Business applications*

65. If 6% interest is compounded quarterly, after 1 year the balance is multiplied by $(1.015)^4$. After 5 years, this is $[(1.015)^4]^5$. Rewrite this with a single exponent.

66. In a training class, there is a $\frac{7}{8}$ chance that the trainee will be successful. For 10 trainees, there is a $(\frac{7}{8})^{10}$ chance that all 10 will be successful. Write this as the ratio of two exponent terms.

*Life science application*

67. If a man and a woman each have $2^{23}$ possible combinations on the 23 chromosomes, together they have $2^{23} \cdot 2^{23}$. Write this with a single exponent.

*Health application*

68. There is a 99 in 100 chance that a patient will not experience a harmful side effect from a drug. For 10 patients, the chance that none of them will have a harmful side effect is $(\frac{99}{100})^{10}$. Write this as the ratio of two exponent terms.

*Optical applications*

Every f/stop on a camera aperature is a power of 2. Consider the table below:

| f/STOP | 1.4 | 2 | 2.8 | 4 | 5.6 | 8 | 11 | 16 |
|---|---|---|---|---|---|---|---|---|
| | ↓ | ↓ | ↓ | ↓ | ↓ | ↓ | ↓ | ↓ |
| RELATIVE AMOUNT OF LIGHT | $2^7$ | $2^6$ | $2^5$ | $2^4$ | $2^3$ | $2^2$ | $2^1$ | 1 |

To find the change in light intensity, we divide the new relative amount of light by the old. Find the change in light intensity for the following situations. (For example, f/2.8 to f/8 is $\frac{2^2}{2^5} = \frac{1}{2^3} = \frac{1}{8}$ change.)

69. f/5.6 to f/2

70. f/1.4 to f/8

71. f/4 to f/11

72. f/8 to f/2

*Social application*

73. In a recent year $10^8$ automobiles drove $10^{12}$ miles in America. Write (as an exponent) the average number of miles each car drove.

## 6.2 ZERO AND NEGATIVE EXPONENTS

So far, we have seen only positive exponents, such as $x^5$. In this section, we look at zero and negative exponents. First consider Theorem 6.2. We can work a problem two ways and get

$$\frac{a^7}{a^7} = 1 \quad \text{and} \quad \frac{a^7}{a^7} = a^{7-7} = a^0$$

Since we want these to be equal, we have the following definition.

> **DEFINITION 6.1** For any real number $a$ ($a \neq 0$),
>
> $$a^0 = 1$$

Also,

$$\frac{a^3}{a^8} = \frac{1}{a^5} \quad \text{and} \quad \frac{a^3}{a^8} = a^{3-8} = a^{-5}$$

Since we would like these two to be the same, we get the following definitions.

**DEFINITION 6.2** For all real numbers $a$ ($a \neq 0$) and positive integer $n$,

$$a^{-n} = \frac{1}{a^n}$$

**DEFINITION 6.3** For all real numbers $a$ ($a \neq 0$) and positive integer $n$,

$$\frac{1}{a^{-n}} = a^n$$

**EXAMPLE 6** **(a)** $10^0 = 1$

**(b)** $p^0 = 1$

**(c)** $x^{-4} = \frac{1}{x^4}$

**(d)** $2^{-5} = \frac{1}{2^5} = \frac{1}{32}$

**(e)** $\frac{1}{t^{-6}} = t^6$

| YES | NO |
|---|---|
| $5^{-2} = \frac{1}{5^2} = \frac{1}{25}$ | ~~$5^{-2} = -5^2 = -25$~~ |

Negative exponents mean that we must invert or take the reciprocal.

Definitions 6.2 and 6.3 can be combined.

To write a fraction with only positive exponents:

1. Invert each factor in the numerator with a negative exponent to the denominator, and write with a positive exponent.
2. Invert each factor in the denominator with a negative exponent to the numerator, and write with a positive exponent.

**EXAMPLE 7** Write $\frac{x^2y^{-2}}{z^{-3}}$ with only positive exponents.

**SOLUTION** We invert the factors with negative exponents and write them with positive exponents.

| Invert and write with positive exponents |
|---|

$$\frac{x^2y^{-2}}{z^{-3}} = \frac{x^2z^3}{y^2}$$

We can now rewrite Theorem 6.2 as follows.

> ***THEOREM 6.6*** For any real number $a$ ($\neq 0$) and any integers $m$ and $n$,
>
> $$\frac{a^m}{a^n} = a^{m-n}$$

All of the other theorems, Theorems 6.1 to 6.5, now hold for zero and negative exponents.

**EXAMPLE 8** Simplify $x^2 \cdot x^{-5}$.

**SOLUTION** By Theorem 6.1, $x^2 \cdot x^{-5} = x^{2+(-5)} = x^{-3} = \frac{1}{x^3}$.

**EXAMPLE 9** Simplify $\frac{y^{-5}}{y^{-3}}$.

**SOLUTION** By Theorem 6.6, $\frac{y^{-5}}{y^{-3}} = y^{-5-(-3)} = y^{-2} = \frac{1}{y^2}$.

**EXAMPLE 10** Simplify $\left(\frac{x^2y^3z^{-4}}{x^4y^{-2}z^{-3}}\right)^{-2}$.

**SOLUTION** As in the last section, we simplify these slowly, a step at a time.

Theorem 6.5: Multiply all exponents by $-2$

$$\left(\frac{x^2y^3z^{-4}}{x^4y^{-2}z^{-3}}\right)^{-2} = \frac{x^{-4}y^{-6}z^8}{x^{-8}y^4z^6}$$

Invert factors with negative exponents and write with positive exponents.

$$= \frac{x^{-4}y^{-6}z^8}{x^{-8}y^4z^6}$$

$$= \frac{x^8z^8}{x^4y^6y^4z^6}$$

Simplify

$$= \frac{x^4z^2}{y^{10}}$$

This is not the only way the problem could have been worked; for instance, we could have simplified the inside first and then raised everything to the $-2$ power.

> To simplify negative-exponent expressions with additions and subtractions:
>
> 1. First simplify all terms by changing the negative exponents to terms with positive exponents.
> 2. Add or subtract the remaining terms.

**EXAMPLE 11** Simplify $\left(x^{-1} + \dfrac{1}{y^{-1}}\right)^{-1}$.

**SOLUTION** We first simplify the terms with negative exponents.

Invert terms with negative exponents

$$\left(x^{-1} + \frac{1}{y^{-1}}\right)^{-1} = \left(\frac{1}{x} + y\right)^{-1}$$

Find LCD, $x$

$$= \left(\frac{1}{x} + \frac{yx}{x}\right)^{-1}$$

Add numerators

$$= \left(\frac{1 + yx}{x}\right)^{-1}$$

Invert

$$= \frac{x}{1 + yx}$$

| YES | NO |
|---|---|
| $\left(x^{-1} + \dfrac{1}{y^{-1}}\right)^{-1} = \left(\dfrac{1}{x} + y\right)^{-1}$ | ~~$\left(x^{-1} + \dfrac{1}{y^{-1}}\right)^{-1} = x + \dfrac{1}{y}$~~ |

We simplify the negative exponents within the parentheses first. Theorem 6.4 states that $(ab)^m = a^m b^m$, but $(a + b)^m \neq a^m + b^m$.

## PROBLEM SET 6.2

Simplify the following expressions so that each variable appears only once and all exponents are positive, or as a single number if possible.

*See Example 6*

1. $8^0$
2. $t^0$
3. $3^{-2}$
4. $10^{-3}$
5. $12^0$
6. $\dfrac{1}{2^{-5}}$
7. $\dfrac{1}{x^{-5}}$
8. $\dfrac{1}{(-5)^{-2}}$
9. $\dfrac{1}{m^0}$

*See Example 7*

10. $x^3y^{-2}$
11. $a^{-3}b^4$
12. $x^4y^{-5}z^{-6}$
13. $\dfrac{p^4q^{-3}}{r^{-7}}$
14. $\dfrac{5x^{-4}}{y^{-2}}$
15. $\dfrac{10a^{-2}b^3}{c^{-4}d^5}$
16. $\dfrac{2^{-5}4^3}{8^{-2}}$
17. $\dfrac{5 \cdot 3^{-3}}{2^{-4}}$
18. $\dfrac{10 \cdot 5^{-2}}{2^{-3}}$

*See Example 8*

19. $x^4 \cdot x^{-10}$
20. $t^{-4} \cdot t^{-5}$
21. $m^{-5} \cdot m^7$
22. $z^{-2} \cdot z^{-3} \cdot z^4$
23. $b^8 \cdot b^{-7} \cdot b^{-6}$
24. $a^{-1} \cdot a^{-2} \cdot a^{-3} \cdot a^{-4}$

*See Example 9*

25. $\dfrac{x^{-3}}{x^4}$
26. $\dfrac{a^{-10}}{a^{-5}}$
27. $\dfrac{r^{-4}}{r^{-7}}$
28. $\dfrac{t^9}{t^{-10}}$
29. $\dfrac{f^4f^{-7}}{f^{-11}}$
30. $\dfrac{k^{-3}k^5}{k^{-4}}$

*See Example 10*

31. $(x^2y^{-1})^{-2}$
32. $(2a^{-2}b)^{-3}$
33. $(4p^4q^{-3})^3$
34. $\left(\dfrac{u^2v^{-3}}{z^{-4}}\right)^2$
35. $\left(\dfrac{mn^{-4}}{k^5}\right)^0$
36. $\left(\dfrac{r^{-4}s^0}{t^5w^{-2}}\right)^{-2}$
37. $\dfrac{a^{-2}b^3}{(ab^{-4})^2}$
38. $\dfrac{x^{-5}y^{-3}}{(x^2y^{-2})^{-3}}$
39. $\dfrac{(u^6v^{-5}w^{-4})^3}{(u^{-7}vw^{-2})^{-1}}$

40. $\left[\left(\frac{m^4n^{-4}}{k^3}\right)^{-2}\right]^3$

41. $\left[\left(\frac{a^{-1}b^2}{c^{-5}}\right)^3\right]^{-1}$

42. $\left[\left(\frac{p^2q^{-2}}{r^3t^{-4}}\right)^{-3}\right]^{-2}$

43. $\left[\left(\frac{x^4y^{-2}z^{-3}}{x^2y^2z^{-5}}\right)^{-1}\right]^2$

44. $\left[\left(\frac{uvw^{-3}}{u^{-2}v^0w^{-1}}\right)^{-1}\right]^{-1}$

45. $\left[\left(\frac{m^2nk^{-2}}{m^{-1}n^2k^0}\right)^{-8}\right]^0$

*See Example 11*

46. $5^1 + 5^{-1}$

47. $2^3 - 2^{-3}$

48. $10^2 + 10^{-3}$

49. $k^{-1} + k^{-2}$

50. $x^{-1} - y^{-1}$

51. $a^{-2} + a^{-3}$

52. $\frac{m}{n^{-1}} + \frac{m^{-1}}{n}$

53. $uv^{-1} + u^{-1}v$

54. $(x^{-1} + y^{-1})^{-1}$

55. $(r^{-2} - s^{-2})^{-1}$

56. $\frac{x^{-1} - x^{-2}}{x^{-3} + x^{-4}}$

57. $\frac{a^{-1} + b^{-2}}{a^{-3} - b^{-2}}$

*Social application*

One of Pareto's rules for city populations is that $P_n = P_1 \cdot n^{-1}$, where $P_1$ is the population of first-ranked city and $P_n$ the population of $n$th-ranked city. Assuming this is true and that the population of New York City is about 7,500,000, compute the population of the cities ranked 2 through 10.

58. Chicago (2)

59. Los Angeles (3)

60. Philadelphia (4)

61. Detroit (5)

62. Houston (6)

63. Baltimore (7)

64. Dallas (8)

65. San Diego (9)

66. San Antonio (10)

*Business applications*

67. With 7% interest, \$100 ten years from now has a present value of $100 \cdot (1.07)^{-10}$. Write this with a positive exponent.

68. The monthly payment $M$ to amortize (or pay off) a loan is given by

$$M = \frac{rP}{1 - (1 + r)^{-n}}$$

where $r$ is the monthly interest rate, $P$ the principal borrowed, and $n$ the numbers of months of the loan. Rewrite this as a simple fraction with only positive exponents.

69. A fast-food restaurant has determined that an average of 2 customers per minute enter for service. The chance that exactly 5 customers will enter during a time $t$ is given by

$$\frac{e^{-2t}(2t)^5}{120}$$

where $e$ is a constant. Rewrite this with positive exponents and no parentheses.

*Life science application*

The weight of a certain bacteria culture is given by

$$W = \frac{20}{2 - 2^{-t}}$$

where $t$ is the time in hours. Determine the weight at the following times.

70. $t = 0$

71. $t = 1$

72. $t = 2$

73. $t = 3$

74. Rewrite the equation as a simple fraction with only positive exponents.

## 6.3 SCIENTIFIC NOTATION

**Scientific notation** is a notation used to express all numbers, from very small to very large, in a compact manner. The standard form is as follows:

> Any number = Number between 1 and 10 × Power of 10

> To express any number in scientific notation:
>
> 1. If the number is less than 1:
>    a. Move the decimal point $q$ places right until the number is between 1 and 10.
>    b. The power of 10 is $10^{-q}$.
> 2. If the number is between 1 and 10, leave it alone.
> 3. If the number is 10 or more:
>    a. Move the decimal point $p$ places left until the number is between 1 and 10.
>    b. The power of 10 is $10^p$.

**EXAMPLE 12** The following are numbers greater than 10, so we move the decimal point to the left to write them in scientific notation. Notice that the exponent is positive.

**(a)** $93{,}000{,}000 = 9_{\uparrow}3{,}000{,}000 = 9.3 \times 10^7$

**(b)** $14{,}200{,}000{,}000 = 1_{\uparrow}4{,}200{,}000{,}000 = 1.42 \times 10^{10}$

**EXAMPLE 13** The following numbers are less than 1, so we move the decimal point to the right to write them in scientific notation. Notice that the exponent is negative.

**(a)** $0.00023 = 0.0002_\wedge 3 = 2.3 \times 10^{-4}$

**(b)** $0.00000000871 = 0.000000008_\wedge 71 = 8.71 \times 10^{-9}$

Notice that we move the decimal point until the number is between 1 and 10.

> When changing a number in scientific notation into standard form:
>
> 1. Multiplying by $10^p$ means move the decimal point $p$ places right.
> 2. Multiplying by $10^{-q}$ means move the decimal point $q$ places left.

**EXAMPLE 14** The following have positive powers of 10, so we move the decimal point to the right.

**(a)** $2.91 \times 10^6 = 2_\wedge 910000_\wedge = 2{,}910{,}000$

**(b)** $8.3 \times 10^2 = 8_\wedge 30_\wedge = 830$

**EXAMPLE 15** The following have negative powers of 10, so we move the decimal point to the left.

**(a)** $5.2 \times 10^{-3} = 0_\wedge 005_\wedge 2 = 0.0052$

**(b)** $1.23 \times 10^{-10} = 0_\wedge 0000000001_\wedge 23 = 0.000000000123$

> To multiply (or divide) numbers in scientific notation:
>
> 1. Multiply (or divide) the digits part of the number.
> 2. Add (or subtract) the powers of 10.
> 3. Simplify, if necessary.

**EXAMPLE 16** Multiply $(2.3 \times 10^4) \times (1.5 \times 10^8)$.

**SOLUTION**

| Multiply digits, add powers of 10 |
|---|

$$(2.3 \times 10^4) \times (1.5 \times 10^8) = (2.3 \times 1.5) \times (10^4 \times 10^8) = 3.45 \times 10^{12}$$

**EXAMPLE 17** Divide $\dfrac{2 \times 10^{-5}}{8 \times 10^{-2}}$.

**SOLUTION**

Divide digits and powers of 10 separately

$$\frac{2 \times 10^{-5}}{8 \times 10^{-2}} = \frac{2}{8} \times \frac{10^{-5}}{10^{-2}}$$

$$= 0.25 \times 10^{-3}$$

Rewrite $0.25 = 2.5 \times 10^{-1}$ and simplify

$$= 2.5 \times 10^{-1} \times 10^{-3} = 2.5 \times 10^{-4}$$

Notice that we had to write $0.25 = 2.5 \times 10^{-1}$ in order to get the answer in proper form.

**EXAMPLE 18** Find $(1.5 \times 10^4)^3$.

**SOLUTION** We use Theorem 6.4 and raise both 1.5 and $10^4$ to the third power.

Theorem 6.4

$$(1.5 \times 10^4)^3 = 1.5^3 \times (10^4)^3$$

Simplify

$$= 3.375 \times 10^{12}$$

## PROBLEM SET 6.3

Write the following numbers in scientific notation.

See Example 12

1. 51,000
2. 820,000
3. 40,500,000
4. 25
5. 625,000,000,000
6. 8.9

See Example 13

7. 0.004
8. 0.31
9. 0.000705
10. 0.01234
11. 4.2
12. 0.00000000092

Write the following numbers in standard base-10 form.

See Example 14

13. $3 \times 10^4$
14. $2.3 \times 10^8$
15. $4.21 \times 10^{10}$
16. $3.05 \times 10^1$
17. $8.2 \times 10^0$
18. $7.777 \times 10^2$

*See Example 15*

19. $4 \times 10^{-2}$ 20. $8.3 \times 10^{-3}$ 21. $2.91 \times 10^{-1}$

22. $4.9 \times 10^{-9}$ 23. $3.33 \times 10^{-6}$ 24. $1.06 \times 10^{-4}$

Compute the following.

*See Example 16*

25. $(2 \times 10^5) \times (3 \times 10^3)$ 26. $(1.5 \times 10^4) \times (4 \times 10^{-6})$

27. $(4.2 \times 10^2) \times (2.3 \times 10^{-3})$ 28. $(1.9 \times 10^{-4}) \times (4.6 \times 10^{-1})$

29. $(5.6 \times 10^6) \times (3.1 \times 10^2)$ 30. $(4.8 \times 10^{-3}) \times (8.4 \times 10^7)$

31. $(6.5 \times 10^{-3}) \times (7.1 \times 10^5) \times (4 \times 10^{-2})$

32. $(7.9 \times 10^{-2}) \times (1.2 \times 10^{-3}) \times (8 \times 10^{-4})$

*See Example 17*

33. $\dfrac{8 \times 10^5}{2 \times 10^3}$ 34. $\dfrac{9 \times 10^{-3}}{3 \times 10^{-4}}$

35. $\dfrac{1.8 \times 10^5}{6 \times 10^{-2}}$ 36. $\dfrac{5.1 \times 10^{-6}}{1.7 \times 10^{-8}}$

37. $\dfrac{8.1 \times 10^{-3}}{9 \times 10^8}$ 38. $\dfrac{1.2 \times 10^3}{1.8 \times 10^{10}}$

39. $\dfrac{2.1 \times 10^2 \times 3.6 \times 10^{-5}}{5.4 \times 10^{-3}}$ 40. $\dfrac{6.4 \times 10^{-10}}{4.5 \times 10^8 \times 1.3 \times 10^{-3}}$

*See Example 18*

41. $(2 \times 10^5)^3$ 42. $(3 \times 10^4)^2$ 43. $(4 \times 10^{-4})^3$

44. $(5 \times 10^{-6})^4$ 45. $(1.1 \times 10^4)^5$ 46. $(2.1 \times 10^{-3})^4$

*Miscellaneous applications*

Write the numbers in the following statements in scientific notation.

47. The sun is 93,000,000 miles from Earth.
48. India has an area of 3,290,000 square kilometers.
49. The population of the world is about 4,500,000,000.
50. The weight of the earth is 6,586,000,000,000,000,000,000 tons.
51. In a recent year, Americans used 106,000,000,000 gallons of gasoline.
52. In 1970, there was 0.0000000007 death for every mile of railroad traffic.
53. The chance of winning a certain lottery is 0.000000044.
54. A modern computer can do an addition problem in about 0.0000001 second.
55. The gross national product of the United States will soon be $2,000,000,000,000.

56. An insurance company estimates the likelihood of a disaster in an area to be 0.0000053.
57. A patient is prescribed 0.0006 gram of atropine sulfate.
58. A patient is prescribed 250,000 units of penicillin.

Write the numbers in the following statements in the usual base-10 notation.

59. Avogadro's number is $6 \times 10^{23}$ molecules per mole.
60. One atomic mass unit is $1.66 \times 10^{-24}$ gram.
61. The speed of light is $3 \times 10^{10}$ centimeters per second.
62. The wavelength of green light is $5 \times 10^{-7}$ meters.
63. The core of a certain virus has a mass of $2 \times 10^{-16}$ gram.
64. Living cells have the following range in volume: smallest virus ($7.0 \times 10^{-7}$ cubic micron) to ostrich egg ($1.1 \times 10^{15}$ cubic microns).
65. There are approximately $10^{40}$ possible checker games.

## 6.4 RADICALS AND RATIONAL EXPONENTS

Just as factoring is the undoing of multiplication, roots are the undoing of exponents. With exponents, we raise a base to a power.

$$\textit{Power:} \quad 2^5 = \boxed{?}$$

With roots, we are finding what base gives the result.

$$\textit{Root:} \quad \boxed{?}^5 = 32$$

Since $2^5 = 32$, we say that 2 is the fifth root of 32. This is written $2 = \sqrt[5]{32}$. In general,

> **DEFINITION 6.4** For any positive real number $a$ and positive integer $n$,
>
> $$x = \sqrt[n]{a} \quad \text{means} \quad a = x^n$$

Here, $x$ is called the ***n*th root of *a*,** and answers the question: What number to the $n$th power gives $a$? Also, $n$ is called the **index,** $a$ is called the **radicand,** and

$\sqrt[n]{a}$ is called the **radical.** Finally, $\sqrt[2]{a}$ (the square root) is usually written $\sqrt{a}$. Although both $3^2$ and $(-3)^2$ are 9, we say that $\sqrt{9} = 3$. By convention, *radicals are always positive or zero.*

**EXAMPLE 19** **(a)** $\sqrt[3]{125} = 5$ since $5^3 = 125$

**(b)** $\sqrt{49} = 7$ since $7^2 = 49$

**(c)** $\sqrt[4]{81a^8} = 3a^2$ since $(3a^2)^4 = 81a^8$

**(d)** $\sqrt[3]{8b^3} = 2b$ since $(2b)^3 = 8b^3$

**(e)** $\sqrt{-16}$ is not a real number, since no real number squared gives $-16$

Another way to write roots is with **fractional exponents.** Consider $x = a^{1/3}$. By Theorem 6.3,

$$x^3 = (a^{1/3})^3 = a^1 = a$$

Since $x^3 = a$, then $x = \sqrt[3]{a}$. Since $x = a^{1/3}$, we get $a^{1/3} = \sqrt[3]{a}$. In general,

**DEFINITION 6.5** For any positive real number $a$ and positive integer $n$,

$$a^{1/n} = \sqrt[n]{a}$$

**EXAMPLE 20** **(a)** $16^{1/4} = \sqrt[4]{16} = 2$

**(b)** $(27)^{1/3} = \sqrt[3]{27} = 3$

| YES | NO |
|---|---|
| $16^{1/4} = 2$ | ~~$16^{1/4} = 4$~~ |

The fraction exponent means take a root, not multiply.

We now consider the more general fractional-exponent expression $a^{m/n}$. We can write the fraction $\frac{m}{n}$ as $m \cdot \frac{1}{n}$. Now we can consider $a^{m/n}$ in two ways.

***THEOREM 6.7*** For any positive real number $a$ and positive integers $m$ and $n$,

$$a^{m/n} = (a^m)^{1/n} = \sqrt[n]{a^m}$$

***THEOREM 6.8*** For any positive real number $a$ and positive integers $m$ and $n$,

$$a^{m/n} = (a^{1/n})^m = (\sqrt[n]{a})^m$$

Both theorems are correct and will give the same answer. In most cases, however, we will see that Theorem 6.8 is usually easier to work with.

With rational exponents $a^{m/n}$:

1. $m$ always represents a power.
2. $n$ always represents a root.

**EXAMPLE 21** Using Theorems 6.7 and 6.8, we can write fractional-exponent expressions as radicals. (All variables are positive.)

Denominator = root
Numerator = power

**(a)** $a^{5/7} = \sqrt[7]{a^5}$ or $(\sqrt[7]{a})^5$

**(b)** $x^{-3/2} = \sqrt{x^{-3}} = \sqrt{\dfrac{1}{x^3}}$ or $(\sqrt{x})^{-3} = \dfrac{1}{(\sqrt{x})^3}$

**EXAMPLE 22** Using Theorems 6.7 and 6.8, we can write radicals as fractional-exponent expressions. (All variables are positive.)

Denominator = root
Numerator = power

**(a)** $\sqrt{t^9} = t^{9/2}$

**(b)** $\dfrac{1}{\sqrt[3]{(a+b)^2}} = \dfrac{1}{(a+b)^{2/3}} = (a+b)^{-2/3}$

**EXAMPLE 23** Let us now simplify some numerical expressions using Theorem 6.8 (which is easier to use).

**(a)** $16^{3/4} = (16^{1/4})^3$
$= 2^3$
$= 8$

**(b)** $(27)^{4/3} = [(27)^{1/3}]^4$
$= (3)^4$
$= 81$

**(c)** $32^{-2/5} = (32^{1/5})^{-2}$
$= 2^{-2}$
$= \dfrac{1}{2^2}$
$= \dfrac{1}{4}$

| YES | NO |
|---|---|
| $-16^{1/2} = -4$ | ~~$(-16)^{1/2} = -4$~~ |
| $-\sqrt{16} = -4$ | ~~$\sqrt{-16} = -4$~~ |

$(-16)^{1/2}$ and $\sqrt{-16}$ are not real numbers. (See Section 7.3.)

Most hand calculators have a square-root key, but none have a cube-root key. However, many calculators do have an exponent key, such as $\boxed{y^x}$. To find $\sqrt[3]{4.5}$, we rewrite

$$\sqrt[3]{4.5} = 4.5^{1/3} = 4.5^{0.3333333}$$

| PUNCH | DISPLAY | MEANING |
|---|---|---|
| C | 0. | Clear |
| 4 · 5 | 4.5 | Enter base |
| $y^x$ · 3 3 3 3 3 3 3 | 0.3333333 | Enter exponent |
| = | 1.6509635 | Answer: $\sqrt[3]{4.5}$ |

To operate with fractional-exponent expressions, we use the usual rules for exponents, Theorems 6.1 to 6.6, and Definitions 6.1 to 6.3.

**EXAMPLE 24** Simplify $x^{1/2}(x^{1/3} + x^{-1/5})$.

**SOLUTION** This is just like the problem $x^3(x^2 + x^7) = x^5 + x^{10}$: use the distributive law and add exponents. Even though we have fractions for exponents, we do the same here.

Distributive law $\quad x^{1/2}(x^{1/3} + x^{-1/5}) = x^{1/2}x^{1/3} + x^{1/2}x^{-1/5}$

↓

Find common denominators $\quad = x^{3/6}x^{2/6} + x^{5/10}x^{-2/10}$

↓

Add exponents $\quad = x^{5/6} + x^{3/10}$

**EXAMPLE 25** Simplify $\left(\dfrac{a^{1/2}b^{-2/3}c^{-3/5}}{a^{1/4}b^{1/6}c^{-1/10}}\right)^{-1/2}$.

**SOLUTION** We do this in "slow motion," just as in Example 10.

Theorem 6.5: multiply all exponents by $-\frac{1}{2}$

$$\left(\frac{a^{1/2}b^{-2/3}c^{-3/5}}{a^{1/4}b^{1/6}c^{-1/10}}\right)^{-1/2} = \frac{a^{-1/4}b^{1/3}c^{3/10}}{a^{-1/8}b^{-1/12}c^{1/20}}$$

Invert factors with negative exponents and write with positive exponents

$$= \frac{a^{-1/4}\,b^{1/3}\,c^{3/10}}{a^{-1/8}\,b^{-1/12}\,c^{1/20}}$$

$$= \frac{a^{1/8}b^{1/3}b^{1/12}c^{3/10}}{a^{1/4}\qquad c^{1/20}}$$

Find common denominators

$$= \frac{a^{1/8}b^{4/12}b^{1/12}c^{6/20}}{a^{2/8}\qquad c^{1/20}}$$

Simplify

$$= \frac{b^{5/12}c^{5/20}}{a^{1/8}} = \frac{b^{5/12}c^{1/4}}{a^{1/8}}$$

## PROBLEM SET 6.4

Write the following fractional-exponent expressions as radicals.

*See Example 21*

1. $x^{1/2}$
2. $a^{1/7}$
3. $(m + n)^{1/5}$
4. $(r - 2s)^{1/9}$
5. $t^{2/3}$
6. $h^{5/2}$
7. $(x + 2y)^{8/9}$
8. $(1 - 3g)^{10/7}$
9. $p^{-1/3}$
10. $(x + t)^{-1/6}$
11. $k^{-4/5}$
12. $(1 + 5z)^{-5/3}$

Write the following radicals in fractional-exponent notation.

*See Example 22*

13. $\sqrt[3]{m}$
14. $\sqrt{a + b}$
15. $\sqrt[10]{r}$
16. $\sqrt[5]{p + q}$
17. $\sqrt{u^5}$
18. $\sqrt[3]{(x + 5y)^7}$
19. $(\sqrt[3]{k})^4$
20. $(\sqrt[5]{1 + 5c^2})^2$
21. $\sqrt[4]{\frac{1}{z}}$
22. $\sqrt{\frac{1}{1 + r}}$
23. $\frac{1}{\sqrt[5]{f}}$
24. $\frac{1}{\sqrt[6]{a + b + c}}$

Simplify each of the following as much as possible. [Be careful: some may not be real—see Example 19(e).]

*See Example 19*

25. $\sqrt{64}$
26. $\sqrt[3]{64}$
27. $\sqrt[5]{32}$
28. $\sqrt[3]{8}$
29. $\sqrt{-9}$
30. $\sqrt[4]{16}$
31. $\sqrt[3]{1000x^3}$
32. $\sqrt[5]{a^{10}}$
33. $\sqrt{\dfrac{1}{4r^6}}$
34. $\sqrt[3]{\dfrac{1}{27p^9}}$
35. $\sqrt[3]{\dfrac{8r^3}{125s^6}}$
36. $\sqrt[4]{\dfrac{16t^8}{81w^{12}}}$

*See Example 20*

37. $81^{1/2}$
38. $125^{1/3}$
39. $(8)^{1/3}$
40. $-8^{1/3}$
41. $-16^{1/4}$
42. $(-16)^{1/4}$
43. $216^{1/3}$
44. $10{,}000^{1/4}$
45. $\left(\dfrac{1}{8}\right)^{1/3}$
46. $\left(\dfrac{1}{100}\right)^{1/2}$
47. $\left(\dfrac{64}{125}\right)^{1/3}$
48. $\left(\dfrac{10{,}000}{81}\right)^{1/4}$

*See Example 23*

49. $8^{2/3}$
50. $25^{3/2}$
51. $81^{3/4}$
52. $(125)^{4/3}$
53. $\left(\dfrac{8}{27}\right)^{4/3}$
54. $\left(\dfrac{9}{4}\right)^{5/2}$
55. $64^{-4/3}$
56. $16^{-1/4}$
57. $36^{-3/2}$
58. $100{,}000^{-2/5}$
59. $\left(\dfrac{9}{16}\right)^{-3/2}$
60. $\left(\dfrac{81}{16}\right)^{-3/4}$

Simplify the following as much as possible.

*See Example 24*

61. $x^{1/2}(x^{1/2} + x^{1/3})$
62. $a^{1/3}(a^{2/3} + a^{5/3})$
63. $t^{1/5}(t^{1/3} - t^{1/4})$
64. $z^{2/3}(z^{4/5} - z^{2/7})$
65. $y^{1/3}(y^{1/4} + y^{-1/4})$
66. $k^{-4/5}(k^{2/3} - k^{-3/4})$
67. $u^{-1/2}(u^{1/3} + u^{-1/4} + u^{1/5})$
68. $b^{-2/3}(b^{-1/2} + b^{-2/3} - b^{2/3})$

*See Example 25*

69. $(x^{1/2}y^{1/3})^{1/4}$
70. $(a^{2/3}b^{1/2}c^{4/5})^{1/4}$
71. $(u^{1/3}v^{-1/4})^{2/3}$
72. $(k^{-2/3}m^{1/2}n^{-1/4})^{-3/2}$
73. $\left(\dfrac{x^{1/3}y^{-1/7}}{x^{1/6}y^{1/5}}\right)^{1/2}$
74. $\left(\dfrac{a^{2/3}b^{-1/4}}{a^{-1/9}b^{-1/6}}\right)^{-1/3}$
75. $\left(\dfrac{u^{1/3}v^{-1/4}w^{-1/5}}{u^{-1/12}v^{3/8}w^{-7/10}}\right)^{-3/2}$
76. $\left(\dfrac{p^{1/9}q^{-1/4}r^{1/6}}{p^{-1/6}q^{-5/8}r^{-5/18}}\right)^{-1/5}$

*Health applications*

The surface area of a human being is given by $S = 2W^{0.4}H^{0.7}$, where $W$ is weight in kilograms and $H$ is height in meters.

77. Rewrite this formula using fractional exponents.

78. Rewrite this formula using radicals.

79. The 24-hour heat loss in an infant is given approximately by $H = 0.02LW^{2/3}$. Compute the 24-hour heat loss if $L = 22$ inches and $W = 8$ pounds.

*Home cooling application*

80. In air-conditioning work there is a formula relating the area of the vents $A$, to the air flow $Q$, the height between vents $H$, and the difference between indoor and outdoor temperature $\Delta T$:

$$A = \frac{Q}{9.4\sqrt{H \cdot \Delta T}}$$

Rewrite this with a fractional exponent in the numerator.

*Life science applications*

81. The formula for the heat loss an animal experiences in the wind is given by

$$H = a(T - T_0)v^{1/3}$$

Find the heat loss $H$ if $a = 0.02$, $T$ the body temperature $= 37°\text{C}$, $T_0$ the air temperature $= 11°\text{C}$, and $v$ the wind velocity $= 27$ kilometers per hour.

In order for an animal's skeleton structure to support itself, it seems to be necessary that $L \leq 7h^{2/3}$, where $L$ is the length of the torso (in centimeters) and $h$ the height of the torso (in centimeters). Find the maximum torso lengths $L$ for animals with the following torso heights.

82. Elephant (125 cm)
83. Horse (64 cm)
84. Dog (27 cm)
85. Cat (8 cm)
86. Mouse (1 cm)

*Business applications*

87. The minimum cost for an inventory–production situation is given when

$$q = \sqrt{\frac{2krc}{a(k - r)}}$$

Find the lot size $q$ if $k$ is the production rate 48,000, $r$ the demand $= 3000$, $c$ the cost per setup $= 250$, and $a$ the storage costs $= \$10$ per unit.

88. The sales of a product are estimated by

$$s = (20{,}000 - 1200p)n^{1/2}a^{2/3}$$

Find the sales $s$ if $p$ is the price = \$6, $n$ the number of salespeople = 100, and $a$ the advertising budget = \$8000.

*Electronics application* The horizontal distance an antenna will carry a signal is given by

$$d = 1.22\sqrt{h}$$

where $d$ is the distance in miles and $h$ is the height of the antenna in feet. For the following heights $h$, compute the horizontal distances $d$.

89. 1296 ft
90. 900 ft
91. 729 ft
92. 1521 ft

## 6.5 SIMPLIFYING RADICALS

In this section, we learn how to simplify radicals. Table 6.1 gives the conditions for a radical to be in simplified form (together with examples of radicals that violate each condition).

**Table 6.1**

| CONDITIONS TO BE IN SIMPLIFIED FORM | EXAMPLES OF VIOLATIONS |
|---|---|
| 1. Radicand contains no factor with a power greater or equal to the index | 1. $\sqrt[3]{x^5}$, $\sqrt{5 \cdot 2^3}$ |
| 2. Radicand contains no fractions | 2. $\sqrt{\dfrac{x}{y}}$, $\sqrt[2]{\dfrac{2}{7}}$ |
| 3. No denominator contains a radical | 3. $\dfrac{x}{\sqrt[3]{a}}$, $\dfrac{\sqrt{5}}{\sqrt{11}}$ |

To satisfy Condition 1, we must consider

$$\sqrt[n]{ab} = (ab)^{1/n} = a^{1/n} \cdot b^{1/n} = \sqrt[n]{a}\sqrt[n]{b}$$

Thus,

> ***THEOREM 6.9*** For any positive real numbers $a$ and $b$ and positive integer $n$,
>
> $$\sqrt[n]{ab} = \sqrt[n]{a}\sqrt[n]{b}$$

> To satisfy Condition 1 for simplified radicals:
>
> 1. Find the largest $n$th-power factor within the radicand.
> 2. Use Theorem 6.9 to separate terms.
> 3. Simplify.

**EXAMPLE 26** Simplify $\sqrt{40}$.

**SOLUTION** We factor out the largest perfect square, 4.

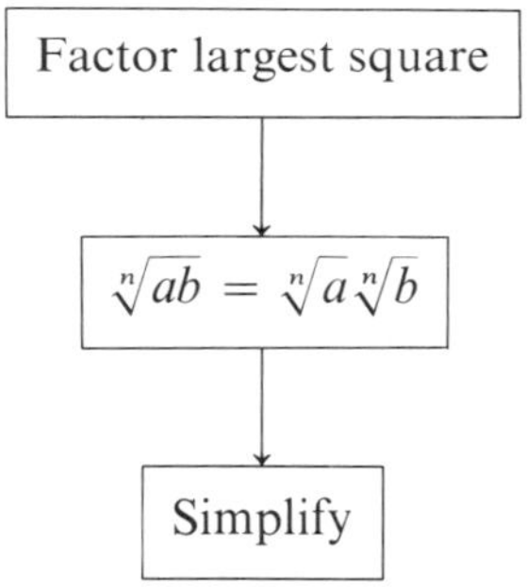


$$\sqrt{40} = \sqrt{4 \cdot 10}$$
$$= \sqrt{4}\sqrt{10}$$
$$= 2\sqrt{10}$$

**EXAMPLE 27** Simplify $\sqrt[3]{54}$.

**SOLUTION** We factor out the largest perfect third power, 27.

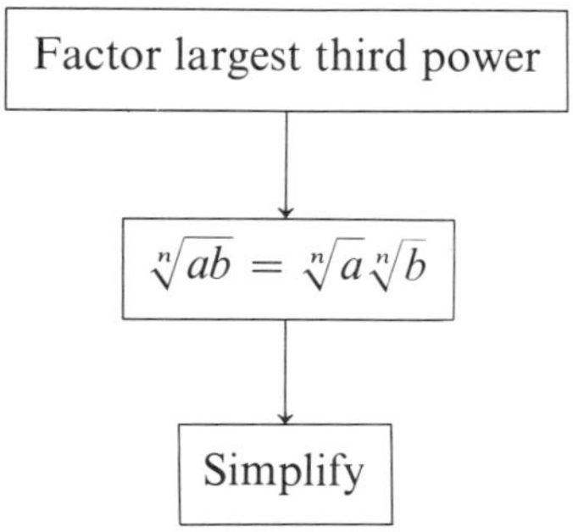


$$\sqrt[3]{54} = \sqrt[3]{27 \cdot 2}$$
$$= \sqrt[3]{27} \cdot \sqrt[3]{2}$$
$$= 3\sqrt[3]{2}$$

**EXAMPLE 28** Simplify $\sqrt{50a^3b^4c^5}$.

**SOLUTION** We factor out the largest perfect square, $25a^2b^4c^4$.

| | |
|---|---|
| Factor largest square | $\sqrt{50a^3b^4c^5} = \sqrt{25a^2b^4c^4 \cdot 2ac}$ |
| $\sqrt[n]{ab} = \sqrt[n]{a}\sqrt[n]{b}$ | $= \sqrt{25a^2b^4c^4}\sqrt{2ac}$ |
| Simplify | $= 5ab^2c^2\sqrt{2ac}$ |

**EXAMPLE 29** Simplify $\sqrt[4]{96x^7y^8z^9}$.

**SOLUTION** We factor out the largest perfect fourth power, $16x^4y^8z^8$.

| | |
|---|---|
| Factor largest fourth power | $\sqrt[4]{96x^7y^8z^9} = \sqrt[4]{16x^4y^8z^8 \cdot 6x^3z}$ |
| $\sqrt[n]{ab} = \sqrt[n]{a}\sqrt[n]{b}$ | $= \sqrt[4]{16x^4y^8z^8}\sqrt[4]{6x^3z}$ |
| Simplify | $= 2xy^2z^2\sqrt[4]{6x^3z}$ |

| YES | NO |
|---|---|
| $\sqrt{ab} = \sqrt{a}\sqrt{b}$ | ~~$\sqrt{a+b} = \sqrt{a} + \sqrt{b}$~~ |

Theorem 6.9 works for multiplication, but *not* for addition.

Now consider $\sqrt[n]{\dfrac{a}{b}}$.

$$\sqrt[n]{\frac{a}{b}} = \left(\frac{a}{b}\right)^{1/n} = \frac{a^{1/n}}{b^{1/n}} = \frac{\sqrt[n]{a}}{\sqrt[n]{b}}$$

> ***THEOREM 6.10*** For any positive real numbers $a$ and $b$ and positive integer $n$,
>
> $$\sqrt[n]{\frac{a}{b}} = \frac{\sqrt[n]{a}}{\sqrt[n]{b}} \qquad b \neq 0$$

> To remove a fraction from a radicand:
>
> 1. Use Theorem 6.10 to split the numerator and denominator.
> 2. Simplify.

**EXAMPLE 30** Simplify $\sqrt[3]{\frac{15}{8}}$.

**SOLUTION** We split the numerator and denominator.

| | |
|---|---|
| $\sqrt[n]{\frac{a}{b}} = \frac{\sqrt[n]{a}}{\sqrt[n]{b}}$ | $\sqrt[3]{\frac{15}{8}} = \frac{\sqrt[3]{15}}{\sqrt[3]{8}}$ |
| ↓ Simplify | $= \frac{\sqrt[3]{15}}{2}$ |

**EXAMPLE 31** Simplify $\sqrt[4]{\frac{y}{x^8}}$.

**SOLUTION** We again use Theorem 6.10.

| | |
|---|---|
| $\sqrt[n]{\frac{a}{b}} = \frac{\sqrt[n]{a}}{\sqrt[n]{b}}$ | $\sqrt[4]{\frac{y}{x^8}} = \frac{\sqrt[4]{y}}{\sqrt[4]{x^8}}$ |
| ↓ Simplify | $= \frac{\sqrt[4]{y}}{x^2}$ |

Sometimes, when we remove a fraction from the radicand, we end up with a radical in the denominator. This violates Condition 3 for simplified radicals.

> To remove an $n$th-root radical from a denominator:
>
> 1. Find a number $\sqrt[n]{F}$ that will multiply to make the denominator radicand a perfect $n$th power.
> 2. Multiply the fraction by $1 = \dfrac{\sqrt[n]{F}}{\sqrt[n]{F}}$.
> 3. Simplify.

**EXAMPLE 32** Simplify $\dfrac{3}{\sqrt{10}}$.

**SOLUTION** If we multiply the denominator by $\sqrt{10}$, it will become $\sqrt{100}$, a perfect square inside the radical.

| | |
|---|---|
| Multiply by $1 = \dfrac{\sqrt{10}}{\sqrt{10}}$ | $\dfrac{3}{\sqrt{10}} = \dfrac{3}{\sqrt{10}} \cdot \dfrac{\sqrt{10}}{\sqrt{10}}$ |
| ↓ $\sqrt[n]{ab} = \sqrt[n]{a}\sqrt[n]{b}$ | $= \dfrac{3\sqrt{10}}{\sqrt{100}}$ |
| ↓ Simplify | $= \dfrac{3\sqrt{10}}{10}$ |

**EXAMPLE 33** Simplify $\sqrt[3]{\dfrac{7}{2x^2}}$.

**SOLUTION** Using Theorem 6.10, we get $\dfrac{\sqrt[3]{7}}{\sqrt[3]{2x^2}}$.

To remove $\sqrt[3]{2x^2}$ from the denominator, we look for a multiplier that will make a perfect third power. Since $8x^3$ is a perfect third power, if we multiply $\sqrt[3]{2x^2}$ by $\sqrt[3]{4x}$ we get $\sqrt[3]{8x^3}$.

| | |
|---|---|
| $\sqrt[n]{\frac{a}{b}} = \frac{\sqrt[n]{a}}{\sqrt[n]{b}}$ | $\sqrt[3]{\frac{7}{2x^2}} = \frac{\sqrt[3]{7}}{\sqrt[3]{2x^2}}$ |
| Multiply by $1 = \frac{\sqrt[3]{4x}}{\sqrt[3]{4x}}$ | $= \frac{\sqrt[3]{7}}{\sqrt[3]{2x^2}} \cdot \frac{\sqrt[3]{4x}}{\sqrt[3]{4x}}$ |
| $\sqrt[n]{ab} = \sqrt[n]{a}\sqrt[n]{b}$ | $= \frac{\sqrt[3]{28x}}{\sqrt[3]{8x^3}}$ |
| Simplify | $= \frac{\sqrt[3]{28x}}{2x}$ |

Again, the $\sqrt[3]{4x}$ factor is used since it is the one that will produce a perfect third power, $\sqrt[3]{8x^3}$, in the denominator.

This process of eliminating radicals from the denominator is called **rationalizing the denominator.**

## PROBLEM SET 6.5

Write each of the following in simplified form. (Assume that all variables are positive.)

*See Example 26*

1. $\sqrt{50}$ 2. $\sqrt{18}$ 3. $\sqrt{12}$

4. $\sqrt{48}$ 5. $-\sqrt{1000}$ 6. $\sqrt{80}$

7. $\sqrt{98}$ 8. $-\sqrt{200{,}000}$ 9. $\sqrt{24{,}000}$

*See Example 27*

10. $\sqrt[3]{24}$ 11. $\sqrt[3]{250}$ 12. $\sqrt[3]{192}$

13. $\sqrt[3]{135}$ 14. $\sqrt[4]{48}$ 15. $\sqrt[4]{162}$

16. $\sqrt[5]{160}$ 17. $\sqrt[6]{640}$ 18. $\sqrt[8]{1024}$

*See Example 28*

19. $\sqrt{8x^3}$ 20. $\sqrt{20a^5}$ 21. $\sqrt{45r^7}$

22. $\sqrt{40u^2v^3}$ 23. $\sqrt{56p^3q^5r^6}$ 24. $\sqrt{125m^4n^5k^6}$

25. $\sqrt{27a^4b^7c^9}$ 26. $\sqrt{200x^3y^5z^7}$ 27. $\sqrt{8000h^3k^7m^9n^{11}}$

*See Example 29*

28. $\sqrt[3]{40y^4}$
29. $\sqrt[3]{500r^7}$
30. $\sqrt[3]{128d^{10}}$
31. $\sqrt[3]{10{,}000u^5v^7}$
32. $\sqrt[4]{80h^7k^8m^9}$
33. $\sqrt[4]{324x^5y^7z^9}$
34. $\sqrt[5]{320a^7b^{11}c^{16}}$
35. $\sqrt[6]{320m^{11}n^{13}k^{19}}$
36. $\sqrt[10]{2048p^7q^{19}r^{23}}$

*See Example 30*

37. $\sqrt{\frac{19}{4}}$
38. $\sqrt{\frac{27}{16}}$
39. $\sqrt{\frac{63}{100}}$
40. $\sqrt{\frac{48}{49}}$
41. $\sqrt{\frac{24}{25}}$
42. $\sqrt{\frac{45}{121}}$
43. $\sqrt[3]{\frac{11}{8}}$
44. $\sqrt[3]{\frac{16}{27}}$
45. $\sqrt[3]{\frac{81}{125}}$
46. $\sqrt[4]{\frac{23}{16}}$
47. $\sqrt[5]{\frac{64}{100{,}000}}$
48. $\sqrt[6]{\frac{5}{64}}$

*See Example 31*

49. $\sqrt{\frac{a^5}{b^4}}$
50. $\sqrt{\frac{x^7}{y^8}}$
51. $\sqrt{\frac{u}{9v^4}}$
52. $\sqrt{\frac{m^3}{81n^6}}$
53. $\sqrt{\frac{r^3s^2}{100p^2q^8}}$
54. $\sqrt{\frac{18a^3b^5}{25c^4d^{10}}}$
55. $\sqrt[3]{\frac{x^7}{y^9}}$
56. $\sqrt[3]{\frac{m^8}{n^{15}}}$
57. $\sqrt[3]{\frac{16a^4b^5}{27c^6d^9}}$
58. $\sqrt[4]{\frac{r^5s^7}{16p^8q^{12}}}$
59. $\sqrt[5]{\frac{x^2y^8}{243z^5t^{10}}}$
60. $\sqrt[6]{\frac{a^5b^8c^{13}}{64d^6e^{18}}}$

*See Example 32*

61. $\frac{4}{\sqrt{11}}$
62. $\frac{5}{\sqrt{15}}$
63. $\frac{3}{\sqrt{3}}$
64. $\frac{14}{\sqrt{7}}$
65. $\frac{50}{\sqrt{10a}}$
66. $\frac{18}{\sqrt{2z}}$
67. $\sqrt{\frac{3}{13}}$
68. $\sqrt{\frac{1}{2}}$
69. $\sqrt{\frac{3}{5}}$
70. $\sqrt{\frac{7}{10u}}$
71. $\sqrt{\frac{11}{2a^3}}$
72. $\sqrt{\frac{1}{5r^5}}$

*See Example 33*

73. $\frac{7}{\sqrt[4]{4}}$
74. $\frac{10}{\sqrt[4]{3}}$
75. $\frac{5}{\sqrt[3]{25x}}$
76. $\frac{6}{\sqrt[4]{8a^2}}$
77. $\frac{2}{\sqrt[3]{9u^4v}}$
78. $\frac{8t}{\sqrt[5]{4t^4w^3}}$
79. $\sqrt[3]{\frac{4}{9}}$
80. $\sqrt[4]{\frac{1}{2}}$
81. $\sqrt[3]{\frac{3a}{4b}}$

82. $\sqrt[4]{\dfrac{5x}{27y^2}}$ 83. $\sqrt[5]{\dfrac{10u}{81x^2y^3z}}$ 84. $\sqrt[10]{\dfrac{3a^2}{128b^6c^9d^{11}}}$

*Business applications*

85. The minimum cost for an inventory–production situation is given when

$$q = \sqrt{\frac{2krc}{a(k - r)}}$$

(see page 193). Simplify this radical.

86. Platter Records find that the average cost per album to produce $q$ albums is

$$C = \frac{5}{\sqrt[3]{q}}$$

Simplify this radical fraction.

*Electricity application*

87. In a circuit, the resonant frequency $f_0$ is given by

$$f_0 = \frac{1}{2\pi\sqrt{LC}}$$

where $L$ is the inductance and $C$ the capacitance. Simplify this radical fraction.

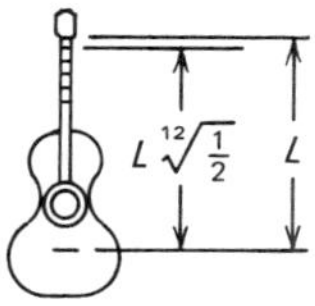


*Music application*

88. In a guitar, let the length from bridge to the neck be $L$. Then the length from the bridge to the first fret is $L\sqrt[12]{\frac{1}{2}}$. Simplify this radical.

*Chemical application*

89. Graham's law states that the rate of diffusion of a gas $R$ is related to its density $D$ by

$$R = \frac{k}{\sqrt{D}}$$

where $k$ is a constant. Simplify this.

## 6.6 RADICAL EXPRESSIONS

A **radical expression** is an algebraic expression that contains radicals. For example,

$$2 + \sqrt{7}, \qquad \frac{8}{3 - \sqrt{2}}, \qquad \text{and} \qquad \frac{\sqrt{6}}{\sqrt{10} - \sqrt{5}}$$

are radical expressions. In this section, we study how to add, subtract, multiply, divide, and simplify these expressions. In dealing with radical expressions, it is often helpful to think of an expression such as $2 + 5\sqrt{3}$ as if it were $2 + 5x$.

**EXAMPLE 34** We use the distributive law to combine like terms.

**(a)** $2\sqrt{7} + 8\sqrt{7} = (2 + 8)\sqrt{7} = 10\sqrt{7}$

**(b)** $5\sqrt[3]{10} - 2\sqrt[3]{10} + x\sqrt[3]{10} = (3 + x)\sqrt[3]{10}$

**EXAMPLE 35** Simplify $\sqrt{6x} + \sqrt{24x} + \sqrt{54x}$.

**SOLUTION** We first simplify each term as much as possible.

| | |
|---|---|
| Simplify each term | $\sqrt{6x} + \sqrt{24x} + \sqrt{54x} = \sqrt{6x} + \sqrt{4}\sqrt{6x} + \sqrt{9}\sqrt{6x}$ |
| | $= \sqrt{6x} + 2\sqrt{6x} + 3\sqrt{6x}$ |
| ↓ Combine like terms | $= 6\sqrt{6x}$ |

**EXAMPLE 36** Simplify $T = 2\sqrt[3]{2a^4} + 3\sqrt[3]{16a^7} - 4\sqrt[3]{54a}$.

**SOLUTION** We again simplify each term first; then we use the distributive law to combine like terms.

| | |
|---|---|
| Given | $T = 2\sqrt[3]{2a^4} + 3\sqrt[3]{16a^7} - 4\sqrt[3]{54a}$ |
| ↓ Simplify each term | $= 2\sqrt[3]{a^3}\sqrt[3]{2a} + 3\sqrt[3]{8a^6}\sqrt[3]{2a} - 4\sqrt[3]{27}\sqrt[3]{2a}$ |
| ↓ Simplify | $= 2 \cdot a\sqrt[3]{2a} + 3 \cdot 2a^2\sqrt[3]{2a} - 4 \cdot 3\sqrt[3]{2a}$ |
| ↓ Distributive law | $= (2a + 6a^2 - 12)\sqrt[3]{2a}$ |

> To multiply radical expressions:
>
> 1. Multiply them using the rules to multiply polynomials (especially FOIL and special-product rules).
> 2. Simplify the radicals and combine like terms.

**EXAMPLE 37** Multiply $(2 + \sqrt{5})(4 - 3\sqrt{2})$.

**SOLUTION** We use the FOIL method from Chapter 3.

| | |
|---|---|
| FOIL | $(2 + \sqrt{5})(4 - 3\sqrt{2}) = 8 - 6\sqrt{2} + 4\sqrt{5} - 3\sqrt{5}\sqrt{2}$ |
| ↓ Simplify | $= 8 - 6\sqrt{2} + 4\sqrt{5} - 3\sqrt{10}$ |

This cannot be simplified further. Notice that we treated this multiplication just as we would have treated the product $(2 + x)(4 - 3y)$.

**EXAMPLE 38** Multiply $(4 + \sqrt{3})^2$.

**SOLUTION** We use Theorem 3.4 for squaring binomials: $(a + b)^2 = a^2 + 2ab + b^2$.

| | |
|---|---|
| Theorem 3.4 | $(4 + \sqrt{3})^2 = 16 + 2 \cdot 4\sqrt{3} + \sqrt{3}\sqrt{3}$ |
| ↓ Simplify | $= 16 + 8\sqrt{3} + \sqrt{9}$ |
| ↓ Simplify | $= 16 + 8\sqrt{3} + 3 = 19 + 8\sqrt{3}$ |

Recall from Chapter 3 that $(a + b)(a - b) = a^2 - b^2$. This holds true with radical expressions. If we have the sum and difference of radicals, the product is the difference of their squares.

**EXAMPLE 39** The following all use $(a + b)(a - b) = a^2 - b^2$.

**(a)** $(\sqrt{6} + 5)(\sqrt{6} - 5) = \sqrt{36} - 25 = 6 - 25 = -19$
**(b)** $(\sqrt{10} + \sqrt{3})(\sqrt{10} - \sqrt{3}) = \sqrt{100} - \sqrt{9} = 10 - 3 = 7$
**(c)** $(8 + \sqrt{11})(8 - \sqrt{11}) = 64 - \sqrt{121} = 64 - 11 = 53$

Notice that the product is always an integer without a radical. We use this fact to simplify expressions with a radical in the denominator. The pairs, such as $8 + \sqrt{11}$ and $8 - \sqrt{11}$, are called **conjugates.**

> To simplify an expression with a sum or difference of radicals in the denominator:
>
> 1. Determine the conjugate of the denominator.
> 2. Multiply the fraction by $1 = \dfrac{\text{conjugate}}{\text{conjugate}}$.
> 3. Simplify.

**EXAMPLE 40** Simplify $\dfrac{2}{\sqrt{7} - 2}$.

**SOLUTION** We rationalize the denominator by multiplying top and bottom by its conjugate, $\sqrt{7} + 2$.

Multiply by $1 = \dfrac{\text{conjugate}}{\text{conjugate}}$

$$\frac{2}{\sqrt{7} - 2} = \frac{2}{\sqrt{7} - 2} \cdot \frac{(\sqrt{7} + 2)}{(\sqrt{7} + 2)}$$

↓

Distributive law

$$= \frac{2\sqrt{7} + 4}{\sqrt{49} - 4}$$

↓

Simplify

$$= \frac{2\sqrt{7} + 4}{7 - 4} = \frac{2\sqrt{7} + 4}{3}$$

**EXAMPLE 41** Simplify $\dfrac{\sqrt{6} - \sqrt{2}}{\sqrt{5} + \sqrt{2}}$.

**SOLUTION** The conjugate of the denominator is $\sqrt{5} - \sqrt{2}$.

| | |
|---|---|
| Multiply by 1 $= \dfrac{\text{conjugate}}{\text{conjugate}}$ | $\dfrac{\sqrt{6}-\sqrt{2}}{\sqrt{5}+\sqrt{2}} = \dfrac{\sqrt{6}-\sqrt{2}}{\sqrt{5}+\sqrt{2}} \cdot \dfrac{\sqrt{5}-\sqrt{2}}{\sqrt{5}-\sqrt{2}}$ |
| ↓ FOIL | $= \dfrac{\sqrt{30}-\sqrt{12}-\sqrt{10}+\sqrt{4}}{\sqrt{25}-\sqrt{4}}$ |
| ↓ Simplify | $= \dfrac{\sqrt{30}-\sqrt{4}\sqrt{3}-\sqrt{10}+\sqrt{4}}{5-2}$ |
| ↓ Simplify | $= \dfrac{\sqrt{30}-2\sqrt{3}-\sqrt{10}+2}{3}$ |

## PROBLEM SET 6.6

Simplify each of the following radical expressions.

*See Example 34*

1. $6\sqrt{2} + 7\sqrt{2} + \sqrt{2}$
2. $5\sqrt{3} - 8\sqrt{3} - 2\sqrt{3}$
3. $8\sqrt{3x} - 10\sqrt{3x} + 17\sqrt{3x}$
4. $4\sqrt{ab} - 10\sqrt{ab} - c\sqrt{ab}$
5. $5\sqrt[3]{x} - 7\sqrt[3]{x} + 3x\sqrt[3]{x}$
6. $x\sqrt[3]{2} + 2x\sqrt[3]{2} + 3x^2\sqrt[3]{2}$

*See Example 35*

7. $\sqrt{2} + \sqrt{8} + \sqrt{18}$
8. $\sqrt{3} + \sqrt{12} + \sqrt{27}$
9. $\sqrt{5} + 8\sqrt{20}$
10. $6\sqrt{28} - 9\sqrt{7}$
11. $4\sqrt{8} - 3\sqrt{18}$
12. $5\sqrt{12} - 7\sqrt{75}$
13. $2\sqrt{6} - 3\sqrt{24} + 5\sqrt{96}$
14. $3\sqrt{10} + 8\sqrt{90} - 5\sqrt{360}$
15. $\sqrt{x} + \sqrt{4x} + \sqrt{9x}$
16. $\sqrt{3a} + \sqrt{12a} + \sqrt{48a}$
17. $2\sqrt{2t} - 3\sqrt{8t} + 4\sqrt{50t}$
18. $4\sqrt{7w} - 2\sqrt{63w} + 5\sqrt{700w}$

*See Example 36*

19. $\sqrt[3]{2} + \sqrt[3]{16} + \sqrt[3]{128}$
20. $\sqrt[3]{10} + \sqrt[3]{80} + \sqrt[3]{270}$
21. $3\sqrt[3]{16} + 5\sqrt[3]{54}$
22. $10\sqrt[3]{320} - 5\sqrt[3]{135}$
23. $4\sqrt[3]{81x} + 6\sqrt[3]{24x}$
24. $10\sqrt[3]{54a} - 3\sqrt[3]{250a}$
25. $\sqrt[3]{x} + \sqrt[3]{x^4} + \sqrt[3]{x^7}$
26. $\sqrt[3]{r^2} + \sqrt[3]{r^5} + \sqrt[3]{r^8}$
27. $\sqrt[3]{2p} + 2\sqrt[3]{16p^4} + 3\sqrt[3]{54p^{10}}$
28. $\sqrt[3]{3k^2} - 2\sqrt[3]{24k^5} + 10\sqrt[3]{81k^{11}}$

Multiply the following radical expressions and simplify.

*See Example 37*

29. $3(\sqrt{7} + \sqrt{5})$
30. $5(\sqrt{3} - 2\sqrt{2})$
31. $\sqrt{6}(\sqrt{5} - \sqrt{2})$
32. $\sqrt{7}(\sqrt{5} + \sqrt{3})$

33. $\sqrt{3}(\sqrt{12} - \sqrt{27})$

34. $\sqrt{2}(3\sqrt{8} + \sqrt{18})$

35. $(1 + \sqrt{2})(2 + \sqrt{3})$

36. $(3 - \sqrt{5})(2 + \sqrt{3})$

37. $(8 - \sqrt{7})(4 + \sqrt{7})$

38. $(2 + \sqrt{6})(10 - \sqrt{6})$

39. $(\sqrt{10} + \sqrt{2})(\sqrt{10} - 4\sqrt{2})$

40. $2\sqrt{3} - \sqrt{7})(5\sqrt{3} + \sqrt{7})$

41. $(\sqrt{5} + \sqrt{6})(\sqrt{3} - \sqrt{2})$

42. $(\sqrt{7} - \sqrt{10})(\sqrt{5} - \sqrt{11})$

*See Example 38*

43. $(1 + \sqrt{5})^2$

44. $(2 - \sqrt{6})^2$

45. $(\sqrt{2} + \sqrt{3})^2$

46. $(\sqrt{10} - \sqrt{7})^2$

47. $(\sqrt{5} - \sqrt{6})^2$

48. $(\sqrt{11} + \sqrt{13})^2$

*See Example 39*

49. $(1 + \sqrt{2})(1 - \sqrt{2})$

50. $(3 - \sqrt{7})(3 + \sqrt{7})$

51. $(\sqrt{13} - 3)(\sqrt{13} + 3)$

52. $(\sqrt{30} - 5)(\sqrt{30} + 5)$

53. $(\sqrt{10} - \sqrt{2})(\sqrt{10} + \sqrt{2})$

54. $(\sqrt{7} + \sqrt{5})(\sqrt{7} - \sqrt{5})$

55. $(\sqrt{70} - \sqrt{30})(\sqrt{70} + \sqrt{30})$

56. $(\sqrt{83} - \sqrt{101})(\sqrt{83} + \sqrt{101})$

Simplify the following by rationalizing the denominators.

*See Example 40*

57. $\dfrac{5}{\sqrt{6} - 3}$

58. $\dfrac{7}{\sqrt{5} + 2}$

59. $\dfrac{8}{5 - \sqrt{10}}$

60. $\dfrac{\sqrt{3}}{\sqrt{7} - 5}$

61. $\dfrac{\sqrt{10}}{\sqrt{15} - 6}$

62. $\dfrac{\sqrt{11}}{4 - \sqrt{8}}$

63. $\dfrac{1 - \sqrt{2}}{1 + \sqrt{2}}$

64. $\dfrac{2 + \sqrt{5}}{3 - \sqrt{6}}$

65. $\dfrac{5 - \sqrt{7}}{2 + \sqrt{10}}$

*See Example 41*

66. $\dfrac{9}{\sqrt{10} - \sqrt{2}}$

67. $\dfrac{2}{\sqrt{11} + \sqrt{3}}$

68. $\dfrac{5}{\sqrt{8} - \sqrt{7}}$

69. $\dfrac{\sqrt{5}}{\sqrt{6} - \sqrt{5}}$

70. $\dfrac{\sqrt{7}}{\sqrt{13} - \sqrt{30}}$

71. $\dfrac{\sqrt{15} + 1}{\sqrt{17} - \sqrt{19}}$

72. $\dfrac{\sqrt{11} - 2}{\sqrt{13} - \sqrt{10}}$

73. $\dfrac{\sqrt{2} + \sqrt{3}}{\sqrt{5} + \sqrt{7}}$

74. $\dfrac{\sqrt{5} - 2\sqrt{3}}{\sqrt{5} + \sqrt{3}}$

*Social application*

75. The time $t$ (in days) for a rumor to spread to a fraction $p$ of a community might be given by

$$t = \frac{1}{1 - \sqrt{p}}$$

Rationalize this.

*Metals application* 76. The Brinell hardness number $N$ for metals is given by

$$N = \frac{191}{10 - \sqrt{100 - d^2}}$$

Rationalize this.

*Chemical application* 77. The increase in temperature, $\Delta T$, due to a catalyst of concentration $u$ is given by

$$\Delta T = \frac{10u}{\sqrt{u} + 1}$$

Rationalize this.

*Business application* 78. The time $t$ (in weeks) that it takes to expose fraction $p$ of the market to a new item might be given by

$$t = \frac{10\sqrt{p}}{1 - \sqrt{p}}$$

Rationalize this.

## 6.7 EQUATIONS WITH RADICALS

A **radical equation** is an equation that contains at least one radical. For example, $\sqrt{x + 1} + 8 = 11$ is a radical equation.

To solve radical equations that involve an $n$th root:

1. Get one radical alone on one side of the equation.
2. Raise both sides to the $n$th power. [*Recall:* $(\sqrt[n]{a})^n = a$.]
3. Solve for the unknown.
4. Check to see if this solution really solves the original equation.

Step 4 is very important. The foregoing procedure will give all the solutions, but it may also give results that are not solutions. We call these **extraneous solutions.** So we must check.

**EXAMPLE 42** Solve $\sqrt{z + 1} - 1 = 2$.

**SOLUTION** We get the radical term alone.

| | |
|---|---|
| Given | $\sqrt{z+1} - 1 = 2$ |
| ↓ Get radical alone | $\sqrt{z+1} = 3$ |
| ↓ Square both sides | $(\sqrt{z+1})^2 = 3^2$ |
| ↓ Simplify | $z + 1 = 9$ |
| ↓ Subtract 1 from both sides | $z = 8$ |

We now check by substituting $z = 8$ into the given equation.

$$\sqrt{8+1} - 1 = \sqrt{9} - 1 = 3 - 1 = 2$$

It checks; therefore, the solution set is $\{8\}$.

**EXAMPLE 43** Solve $\sqrt{2x+2} + \sqrt{3x-5} = 0$.

**SOLUTION** We put each radical on one side of the equation.

| | |
|---|---|
| Given | $\sqrt{2x+2} + \sqrt{3x-5} = 0$ |
| ↓ Get radicals alone | $\sqrt{2x+2} = -\sqrt{3x-5}$ |
| ↓ Square both sides | $(\sqrt{2x+2})^2 = (-\sqrt{3x-5})^2$ |
| ↓ Simplify | $2x + 2 = 3x - 5$ |
| ↓ Subtract $2x$ from both sides | $2 = x - 5$ |
| ↓ Add 5 to both sides | $7 = x$ |

We now check this solution by substituting $x = 7$ into the given equation.

$$\sqrt{2(7) + 2} + \sqrt{3(7) - 5} = \sqrt{16} + \sqrt{16} = 4 + 4 \neq 0$$

The solution does *not* check. Therefore, $x = 7$ is extraneous, and the solution set is $\varnothing$; that is, there is no solution.

**EXAMPLE 44** Solve $\sqrt[3]{3a - 7} = 2$.

**SOLUTION** Here we cube both sides to solve.

| | |
|---|---|
| Given | $\sqrt[3]{3a - 7} = 2$ |
| Cube both sides | $3a - 7 = 8$ |
| Add 7 to both sides | $3a = 15$ |
| Divide both sides by 3 | $a = 5$ |

We check by substituting $a = 5$ into the original equation.

$$\sqrt[3]{3(5) - 7} = \sqrt[3]{8} = 2$$

This checks, so the solution set is $\{5\}$.

## PROBLEM SET 6.7

Solve the following radical equations for the unknown. (Be sure to check for extraneous solutions—see Example 43.)

*See Example 42*

1. $\sqrt{x} = 5$
2. $\sqrt{a} = 2$
3. $\sqrt{r} = -3$
4. $\sqrt{t} = 10$
5. $\sqrt{b} - 4 = 2$
6. $\sqrt{y} + 3 = 2$
7. $\sqrt{x + 1} = 7$
8. $\sqrt{m - 5} = 1$
9. $\sqrt{2x + 1} = 3$
10. $\sqrt{5r + 14} = 2$
11. $\sqrt{4t + 9} - 2 = 5$
12. $\sqrt{10u - 5} + 8 = 5$

*See Example 43*

13. $\sqrt{4v + 5} = \sqrt{10v - 1}$
14. $\sqrt{4y - 19} = \sqrt{11 - 2y}$
15. $\sqrt{2x - 2} = \sqrt{25 - x}$
16. $\sqrt{6k + 3} = \sqrt{9k - 9}$
17. $\sqrt{7m + 1} - \sqrt{5m + 13} = 0$
18. $\sqrt{8a - 5} - \sqrt{5a + 4} = 0$
19. $\sqrt{9x - 7} + \sqrt{6x + 8} = 0$
20. $\sqrt{10p - 13} + \sqrt{6p + 15} = 0$

*See Example 44*

21. $\sqrt[3]{a} = 5$
22. $\sqrt[3]{k} = -2$
23. $\sqrt[4]{b} = 2$
24. $\sqrt[4]{z} = 3$
25. $\sqrt[3]{x + 1} = 4$
26. $\sqrt[3]{2r - 1} = 3$
27. $(5x - 2)^{1/3} = (x + 6)^{1/3}$
28. $(6m - 3)^{1/3} = (4m + 7)^{1/3}$
29. $(5g - 1)^{1/3} = 8^{2/3}$
30. $(5t + 14)^{1/6} = 4^{1/3}$

*Television applications*

31. The horizontal distance an antenna will carry a signal is given by

$$d = 1.22\sqrt{h}$$

(see Problem 89, page 194). Find the height $h$ of the antenna that will carry a signal $d = 30$ miles.

32. The noise voltage $E_R$ in a line is given by

$$E_R = 1.28 \times 10^{-10}\sqrt{R \cdot \Delta F}$$

where $R$ is the resistance and $\Delta F$ the bandwidth of the frequency. Find $R$ if $E_R = 5.4 \times 10^{-6}$ volt and $\Delta F = 4{,}000{,}000$ hertz.

*Music application*

33. The frequency $f$ of the tone of a stringed instrument (guitar, violin, bass, and so on) is given by

$$f = \frac{k}{L}\sqrt{\frac{T}{m}}$$

where $L$ is the length of string, $T$ the tension on the string, $m$ the mass density, and $k$ is a constant. Solve the equation for $T$.

# CHAPTER 6 SUMMARY

## Important Words and Phrases

conjugates (6.6)
extraneous solutions (6.7)
fractional exponent (6.4)
index (6.4)
negative exponent (6.2)
*n*th root (6.4)

radical (6.4)
radical equation (6.7)
radical expression (6.6)
radicand (6.4)
rationalizing the denominator (6.5)
scientific notation (6.3)
simplified form of a radical (6.5)
zero exponent (6.2)

## Important Definitions and Theorems

For any real numbers $a$ and $b$ and integers $m$ and $n$,

$$a^n \cdot a^m = a^{n+m}$$

$$(a^m)^n = a^{mn}$$

$$(ab)^m = a^m b^m$$

$$\left(\frac{a}{b}\right)^m = \frac{a^m}{b^m} \quad (b \neq 0)$$

$$a^0 = 1 \quad (a \neq 0)$$

$$a^{-n} = \frac{1}{a^n} \quad (a \neq 0)$$

$$\frac{1}{a^{-n}} = a^n \quad (a \neq 0)$$

$$\frac{a^m}{a^n} = a^{m-n} \quad (a \neq 0)$$

For any positive real numbers $a$, $b$, and $x$ and positive integers $m$ and $n$,

$$x = \sqrt[n]{a} \quad \text{means} \quad a = x^n$$

$$a^{1/n} = \sqrt[n]{a}$$

$$a^{m/n} = (a^m)^{1/n} = \sqrt[n]{a^m}$$

$$a^{m/n} = (a^{1/n})^m = (\sqrt[n]{a})^m$$

$$\sqrt[n]{ab} = \sqrt[n]{a}\sqrt[n]{b}$$

$$\sqrt[n]{\frac{a}{b}} = \frac{\sqrt[n]{a}}{\sqrt[n]{b}}$$

## Important Procedures

*To write a fraction with only positive exponents:* invert all factors with negative exponents and write with positive exponents.

*To express any number in scientific notation:*
1. If the number is less than 1: move the decimal point $q$ places to the right until the number is between 1 and 10; the power of 10 is $10^{-q}$.
2. If the number is between 1 and 10, leave it alone.
3. If the number is 10 or more: move the decimal point $p$ places to the left until the number is between 1 and 10; the power of 10 is $10^p$.

*To satisfy Condition 1 for simplified radicals:*
1. Find the largest $n$th-power factor within the radicand.
2. Use Theorem 6.9 to separate terms.
3. Simplify.

*To remove a fraction from a radical*, use Theorem 6.10 and simplify.

*To remove an nth-root radical from a denominator:*

1. Determine a number $\sqrt[n]{F}$ that will multiply to make the denominator radicand a perfect $n$th power.
2. Multiply by $1 = \dfrac{\sqrt[n]{F}}{\sqrt[n]{F}}$.
3. Simplify.

*To simplify an expressions with a sum or difference of radicals in the denominator:*

1. Determine the conjugate of the denominator.
2. Multiply by $1 = \dfrac{\text{conjugate}}{\text{conjugate}}$.
3. Simplify.

*To solve radical equations that involve an nth-root:*

1. Isolate one radical on one side of the equation.
2. Raise both sides to the $n$th power.
3. Solve for the unknown.
4. Check solution by substituting into the original equation. This procedure will give all the solutions, if any, but may also give extraneous solutions.

## CHAPTER 6 REVIEW EXERCISES

Simplify each of the following, so that each variable appears only once and all exponents are positive.

1. $(-6a^2b)(4a^5b^2)$
2. $\dfrac{-16x^4y}{8x^2y^3}$
3. $(-3u^2v^3)^4$
4. $\left(\dfrac{5m^2}{n^3k}\right)^3$
5. $(-6xy^2)^2(2x^2y)^3$
6. $\left[\left(\dfrac{p^2q}{r^3}\right)^2\left(\dfrac{p^3r^2}{q^4}\right)^3\right]^5$
7. $12^0$
8. $\dfrac{1}{3^{-2}}$
9. $a^{-4}b^5c^{-1}$
10. $a^{-6}a^7a^{-8}$
11. $\dfrac{k^3k^{-4}}{k^{-5}}$
12. $\left(\dfrac{x^{-2}y^{-3}}{x^4y^{-5}}\right)^{-2}$
13. $(x^{-1} - y^{-1})^{-1}$

Write the following in scientific notation.

14. 17,500,000,000
15. 0.0000621

Write the following in standard base-10 form.

16. $9.8 \times 10^4$

17. $3.7 \times 10^{-6}$

Compute the following.

18. $(1.7 \times 10^8)(6.2 \times 10^{-3})$

19. $\dfrac{2.4 \times 10^{-3}}{6 \times 10^{-8}}$

20. $(2 \times 10^{-3})^5$

Write the following as radicals.

21. $a^{1/7}$

22. $b^{4/5}$

23. $c^{-2/9}$

Write the following in fractional-exponent notation.

24. $\sqrt[3]{m}$

25. $\sqrt[7]{(a + b)^2}$

26. $\dfrac{1}{\sqrt[6]{x}}$

Simplify as much as possible.

27. $\sqrt[3]{125x^6}$

28. $-49^{1/2}$

29. $(-49)^{1/2}$

30. $32^{6/5}$

31. $\left(\dfrac{9}{16}\right)^{-3/2}$

32. $x^{1/3}(x^{1/2} - x^{1/4})$

33. $\left(\dfrac{a^{1/2}b^{-1/3}c^{1/4}}{a^{1/4}b^{-1/6}c^{-1/8}}\right)^{1/2}$

Write each of the following in a simplified form.

34. $\sqrt{108}$

35. $\sqrt[3]{800}$

36. $\sqrt{80x^3y^7}$

37. $\sqrt[3]{54p^4q^5r^6}$

38. $\sqrt{\dfrac{2}{5}}$

39. $\sqrt[3]{\dfrac{3a}{4x}}$

40. $6\sqrt{2} - 9\sqrt{2} - 12\sqrt{2}$

41. $2\sqrt{5} - 6\sqrt{20} - 5\sqrt{45}$

42. $2\sqrt[3]{135p} - \sqrt[3]{320p^4}$

Multiply the following.

43. $(2 + \sqrt{5})(3 - \sqrt{7})$

44. $(3 - \sqrt{2})^2$

45. $(\sqrt{6} - \sqrt{3})(\sqrt{6} + \sqrt{3})$

Rationalize the following denominators.

46. $\dfrac{\sqrt{10}}{1 - \sqrt{6}}$

47. $\dfrac{6 - \sqrt{2}}{\sqrt{3} + \sqrt{5}}$

Solve the following radical equations.

48. $\sqrt{2x + 1} = 7$

49. $\sqrt{4p + 7} - \sqrt{9p - 3} = 0$

50. $\sqrt[3]{r + 7} = 3$

# 7
# QUADRATIC EQUATIONS AND INEQUALITIES

In Chapter 2, we studied linear equations that have only terms of degree 1 or less. In this chapter, we consider equations with terms of degree 2 or less. Such an equation is called a **quadratic equation** and has the following **standard form:**

$$ax^2 + bx + c = 0$$

where $a$, $b$, and $c$ are constants ($a \neq 0$).

## 7.1 SOLVING QUADRATIC EQUATIONS BY FACTORING

The easiest way to solve quadratic equations is to factor (if possible) the polynomial on the left side of the equation above. We then use the **Zero-Product Theorem.** (See Theorem 1.4 of Chapter 1.)

> ***THEOREM 7.1*** ***(Zero-Product Theorem)*** For any real numbers $a$ and $b$, if $a \cdot b = 0$, then $a = 0$ or $b = 0$.

**EXAMPLE 1** Solve $(x + 8)(x - 5) = 0$.

**SOLUTION** By the Zero-Product Theorem, either $x + 8$ or $x - 5 = 0$. Now we solve each of these separately.

| | |
|---|---|
| Given | $(x + 8)(x - 5) = 0$ |
| ↓ | ↙ ↘ |
| Set each factor = 0 | $x + 8 = 0$ or $x - 5 = 0$ |
| ↓ | |
| Solve each equation | $x = -8$ or $x = 5$ |

Therefore, the solution set is $\{-8, 5\}$.

> To solve quadratic equations by factoring:
>
> 1. Put the equation into standard form.
> 2. Factor the polynomial.
> 3. Set each factor equal to zero.
> 4. Solve each of the first-degree equations.

**EXAMPLE 2** Solve $2x^2 - 9x - 35 = 0$.

**SOLUTION** This is already in standard form, so we start by factoring.

| | |
|---|---|
| Given | $2x^2 - 9x - 35 = 0$ |
| ↓ Factor | $(2x + 5)(x - 7) = 0$ |
| ↓ Set each factor $= 0$ | $2x + 5 = 0$ or $x - 7 = 0$ |
| ↓ Solve each equation | $x = -\frac{5}{2}$ or $x = 7$ |

The solution set is $\{-\frac{5}{2}, 7\}$.

**EXAMPLE 3** Solve $a^2 = 6a - 9$.

**SOLUTION** We first put this into standard form.

| | |
|---|---|
| Given | $a^2 = 6a - 9$ |
| ↓ Put into standard form | $a^2 - 6a + 9 = 0$ |
| ↓ Factor | $(a - 3)(a - 3) = 0$ |
| ↓ Set each factor $= 0$ | $a - 3 = 0$ or $a - 3 = 0$ |
| ↓ Solve each equation | $a = 3$ or $a = 3$ |

Notice that both equations have the same solution, $a = 3$. Therefore, the solution set is simply $\{3\}$.

| YES | NO |
|---|---|
| $(t - 7)(t + 6) = 0$ | ~~$(t - 7)(t + 6) = -22$~~ |
| $t - 7 = 0$ or $t + 6 = 0$ | ~~$t - 7 = -22$ or $t + 6 = -22$~~ |

The Zero-Product Theorem is true only when the product is equal to *zero*.

**EXAMPLE 4** Solve $(t - 7)(t + 6) = -22$.

**SOLUTION** We multiply this out and put it into standard form.

| | |
|---|---|
| Given | $(t - 7)(t + 6) = -22$ |
| FOIL | $t^2 - t - 42 = -22$ |
| Put into standard form | $t^2 - t - 20 = 0$ |
| Factor | $(t - 5)(t + 4) = 0$ |
| Set each factor $= 0$ | $t - 5 = 0$ or $t + 4 = 0$ |
| Solve each equation | $t = 5$ or $t = -4$ |

The solution set is $\{5, -4\}$.

**EXAMPLE 5** Solve $\frac{10}{k^2} - \frac{7}{k} = 3$.

**SOLUTION** We first multiply by the LCD, $k^2$, to put the equation into standard form.

|  |  |
|---|---|
| Given | $\frac{10}{k^2} - \frac{7}{k} = 3$ |
| Multiply both sides by LCD, $k^2$ | $k^2\left[\frac{10}{k^2} - \frac{7}{k}\right] = 3k^2$ |
| Distributive law | $10 - 7k = 3k^2$ |
| Subtract $3k^2$ from both sides | $-3k^2 - 7k + 10 = 0$ |
| Multiply both sides by $-1$ | $3k^2 + 7k - 10 = 0$ |
| Factor | $(3k + 10)(k - 1) = 0$ |
| Set each factor $= 0$ | $3k + 10 = 0$ or $k - 1 = 0$ |
| Solve each equation | $k = \frac{-10}{3}$ or $k = 1$ |

Therefore, the solution set is $\{-\frac{10}{3}, 1\}$.

**EXAMPLE 6** Solve $x = \sqrt{24 - 5x}$.

**SOLUTION** As in Chapter 6, we solve radical equations by squaring both sides.

| | |
|---|---|
| Given | $x = \sqrt{24 - 5x}$ |
| ↓ Square both sides | $x^2 = 24 - 5x$ |
| ↓ Put in standard form | $x^2 + 5x - 24 = 0$ |
| ↓ Factor | $(x - 3)(x + 8) = 0$ |
| ↓ Set each factor $= 0$ | $x - 3 = 0 \quad \text{or} \quad x + 8 = 0$ |
| ↓ Solve each equation | $x = 3 \quad \text{or} \quad x = -8$ |

Recall that with radical equations, we must check for extraneous solutions.

*Check* $x = 3$: Left side $= 3$

Right side $= \sqrt{24 - 15} = \sqrt{9} = 3$ Checks

*Check* $x = -8$: Left side $= -8$

Right side $= \sqrt{24 - (-40)} = \sqrt{64} = 8$ Does *not* check

Notice that $x = -8$ does not check; it is extraneous. Therefore, the solution set is $\{3\}$.

## PROBLEM SET 7.1

Solve the following equations.

*See Example 1*

1. $(x - 3)(x - 4) = 0$
2. $(a - 6)(a - 10) = 0$
3. $(k + 7)(k - 2) = 0$
4. $(r + 8)(r + 9) = 0$
5. $(2y - 7)(y - 8) = 0$
6. $(2z + 8)(5z - 1) = 0$

*See Example 2*

7. $x^2 - 5x + 6 = 0$
8. $a^2 - 8a + 7 = 0$
9. $t^2 + 3t - 40 = 0$
10. $r^2 + 18r + 17 = 0$

11. $2y^2 - 13y + 6 = 0$

12. $3x^2 + 13x - 10 = 0$

13. $6a^2 - 5a - 4 = 0$

14. $15m^2 - 29m + 8 = 0$

*See Example 3*

15. $x^2 + x = 12$

16. $y^2 + 2y = 35$

17. $z^2 = 16z - 63$

18. $t^2 = -18 - 11t$

19. $r = 15 - 6r^2$

20. $52a = 9 - 12a^2$

21. $b^2 - 10b + 6 = 3b - b^2$

22. $4m^2 + 10m = m^2 - 16m + 40$

*See Example 4*

23. $(x + 8)(x - 1) = -18$

24. $(k - 10)(k + 6) = -15$

25. $(m - 3)^2 = 49$

26. $(a - 2)^2 = 25$

27. $(x - 5)(x - 3) = 27 + 3x$

28. $(t - 4)(t + 6) = 6 - 5t$

29. $(2t + 1)(t - 5) = t(t - 13)$

30. $(3x - 4)(2x - 3) = 2x(2x - 21)$

*See Example 5*

31. $\dfrac{x^2}{4} + 4 = \dfrac{5x}{2}$

32. $\dfrac{a^2}{9} + \dfrac{2a}{3} = 3$

33. $t - 14 = \dfrac{-33}{t}$

34. $1 = \dfrac{13}{z} - \dfrac{36}{z^2}$

35. $\dfrac{3k}{4} = 5 + \dfrac{8}{k}$

36. $m = \dfrac{11}{2} + \dfrac{3}{m}$

37. $\dfrac{3}{x - 1} + \dfrac{14}{2x - 1} = 3$

38. $\dfrac{4}{a - 2} - \dfrac{10}{a - 3} = 1$

*See Example 6*

39. $r = \sqrt{r + 30}$

40. $y = \sqrt{13y - 40}$

41. $x = \sqrt{8x + 9}$

42. $m = \sqrt{14m - 49}$

43. $k = \sqrt{\dfrac{17k - 30}{2}}$

44. $a = \sqrt{\dfrac{11a + 7}{6}}$

*Physical application*

45. An object is thrown in the air, and its height $h$ at any time $t$ is given by $h = -16t^2 + 160t$. At what times is $h = 256$? (Substitute $h = 256$ and solve for $t$.)

*Health application*

46. It is known that a certain disease is spreading through a town at a rate $R$ given by

$$R = \frac{-y^2}{100{,}000} + \frac{y}{10}$$

where $y$ is the number of people with the disease. Compute $y$ if $R = 210$ people per day.

*Life science applications*

47. The yield $y$ in bushels per acre in a rice field is given by $y = 10x - \frac{x^2}{2}$, where $x$ is the density of the rice plant. Find the density $x$ that will give a yield of $y = 50$.

48. The probability $p$ that a certain plant will be dead after $x$ hours is given by

$$P = 1 - \frac{2}{x^2 + x + 2}$$

At what time $x$ is there a three-fourths chance that the plant will be dead? (If $p = \frac{3}{4}$, find $x$.)

*Business applications*

49. The supply $S$ and demand $D$ for a certain box of candy are given by

$$S = 1000p - 500 \qquad \text{and} \qquad D = \frac{14{,}000}{p}$$

where $p$ is the price. When $S = D$, the system is in equilibrium. Set the supply formula equal to the demand formula, and solve for $p$.

50. The minimum cost to keep an inventory is given when

$$Q - \frac{1600}{Q} = 0$$

where $Q$ is the order size. Find $Q$.

## 7.2 COMPLETING THE SQUARE

Solving quadratic equations by factoring is fine, when we can factor. But very often we cannot factor the polynomial. Therefore, we need to find other methods to solve quadratic equations. We start with the following theorem.

> ***THEOREM 7.2*** For any real number $a \geq 0$, if $x^2 = a$, then $x = \sqrt{a}$ or $x = -\sqrt{a}$

In words, this theorem tells us that *we can take the square roots of both sides of an equation* (if both sides are positive or zero).

**EXAMPLE 7** Solve $x^2 = 25$.

**SOLUTION** Using Theorem 7.2, we get $x = 5$ or $x = -5$. We could have also gotten this answer by writing the equation as $x^2 - 25 = 0$ and factoring $(x + 5)(x - 5) = 0$. Thus, the solution set is $\{-5, 5\}$.

**EXAMPLE 8** Solve $(a + 6)^2 = 49$.

**SOLUTION** This problem is only a little more difficult than Example 7. We split it into two single equations, and then solve each of them.

| | | | |
|---|---|---|---|
| Given | $(a + 6)^2 = 49$ | | |
| Take square roots of both sides | $a + 6 = 7$ | or | $a + 6 = -7$ |
| Solve each equation | $a = 1$ | or | $a = -13$ |

The solution set is thus $\{1, -13\}$.

**EXAMPLE 9** Solve $(3k - 1)^2 = 15$.

**SOLUTION** We solve this exactly as Example 8, except that the numbers will be a bit more awkward.

| | | | |
|---|---|---|---|
| Given | $(3k - 1)^2 = 15$ | | |
| Take square roots of both sides | $3k - 1 = \sqrt{15}$ | or | $3k - 1 = -\sqrt{15}$ |
| Add 1 to both sides | $3k = 1 + \sqrt{15}$ | or | $3k = 1 - \sqrt{15}$ |
| Divide both sides by 3 | $k = \dfrac{1 + \sqrt{15}}{3}$ | or | $k = \dfrac{1 - \sqrt{15}}{3}$ |

This is the solution set: $\left\{\dfrac{1 + \sqrt{15}}{3}, \dfrac{1 - \sqrt{15}}{3}\right\}$.

**EXAMPLE 10** Solve $x^2 + 6x = 13$.

**SOLUTION** If we rewrite this as $x^2 + 6x - 13 = 0$, we go nowhere since it cannot be factored. Instead, the technique that we use is to make the left side look like a square (as in Examples 8 and 9). To do this, recall from Chapter 3 that

$$(x + a)^2 = x^2 + 2ax + a^2$$

The left side of our equation, $x^2 + 6x$, is close to $x^2 + 2ax + a^2$ above. The $6x$ term corresponds to the $2ax$ term; so $a = 3$. Therefore, we need $a^2 = 9$ to make it a square.

| | |
|---|---|
| Given | $x^2 + 6x = 13$ |
| Add 9 to both sides to get square on left | $x^2 + 6x + 9 = 22$ |
| Rewrite left side | $(x + 3)^2 = 22$ |
| Take square roots of both sides | $x + 3 = \sqrt{22}$ or $x + 3 = -\sqrt{22}$ |
| Subtract 3 from both sides | $x = \sqrt{22} - 3$ or $x = -\sqrt{22} - 3$ |

The solution set is $\{\sqrt{22} - 3, -\sqrt{22} - 3\}$.

The method used is called **completing the square,** since that is exactly what we did: We added 9 to both sides to make the left side a perfect square. The trick is finding what to add to both sides to make the square. Here is the procedure.

> To solve a quadratic equation by completing the square:
>
> 1. Rewrite the equation in the form $x^2 + mx = n$. (Divide by the coefficient of $x^2$, if necessary.)

> 2. Find $\frac{m}{2}$ and add $\left(\frac{m}{2}\right)^2$ to both sides.
> 3. Rewrite the equations as $\left(x + \frac{m}{2}\right)^2 = n + \left(\frac{m}{2}\right)^2$.
> 4. Solve, taking the square roots of both sides.

**EXAMPLE 11** Solve $3x^2 - 4x - 10 = 0$.

**SOLUTION** We cannot factor this, so we will solve it by completing the square. We first rewrite it.

| | |
|---|---|
| Given | $3x^2 - 4x - 10 = 0$ |
| Rewrite as $x^2 + mx = n$ | $3x^2 - 4x = 10$ |
| $m = \frac{-4}{3}$, so $\frac{m}{2} = \frac{-2}{3}$ | $x^2 - \frac{4}{3}x = \frac{10}{3}$ |
| Add $\left(\frac{m}{2}\right)^2 = \frac{4}{9}$ to both sides | $x^2 - \frac{4}{3}x + \frac{4}{9} = \frac{10}{3} + \frac{4}{9} = \frac{34}{9}$ |
| Rewrite as a square | $\left(x - \frac{2}{3}\right)^2 = \frac{34}{9}$ |
| Take square roots of both sides | $x - \frac{2}{3} = \frac{\sqrt{34}}{3}$ or $x - \frac{2}{3} = \frac{-\sqrt{34}}{3}$ |
| Add $\frac{2}{3}$ to both sides | $x = \frac{2 + \sqrt{34}}{3}$ or $x = \frac{2 - \sqrt{34}}{3}$ |

Notice that we find $m$, then $\frac{m}{2}$, and finally we add $\left(\frac{m}{2}\right)^2$ to both sides. This will always give a perfect square $\left(x + \frac{m}{2}\right)^2$. In Example 11 we see that the solution set is

$$\left\{\frac{2 + \sqrt{34}}{3}, \frac{2 - \sqrt{34}}{3}\right\}.$$

## PROBLEM SET 7.2

Solve the following quadratic equations.

*See Example 7*

1. $x^2 = 81$
2. $a^2 = 4$
3. $k^2 = 15$
4. $g^2 = 32$
5. $r^2 = \frac{16}{25}$
6. $z^2 = \frac{4}{49}$

*See Example 8*

7. $(x + 2)^2 = 25$
8. $(a - 3)^2 = 16$
9. $\left(r - \frac{2}{5}\right)^2 = 9$
10. $\left(t + \frac{7}{3}\right)^2 = 81$
11. $(y - 5)^2 = 7$
12. $(k + 7)^2 = 5$

*See Example 9*

13. $(2m + 1)^2 = 16$
14. $(5y - 3)^2 = 9$
15. $(3a - 2)^2 = 3$
16. $(8u - 5)^2 = 10$
17. $\left(10v - \frac{1}{3}\right)^2 = \frac{2}{9}$
18. $\left(5f + \frac{1}{7}\right)^2 = \frac{15}{49}$

Solve the following equations by completing the square.

*See Example 10*

19. $x^2 + 4x = 77$
20. $a^2 - 6a = 40$
21. $z^2 - 8z = 21$
22. $t^2 + 10t = 1$
23. $y^2 - y = 7$
24. $k^2 + 3k = 5$
25. $p^2 - \frac{3}{4}p = \frac{5}{4}$
26. $u^2 + \frac{7}{2}u = \frac{9}{2}$
27. $x^2 - \frac{5}{3}x = \frac{8}{3}$

*See Example 11*

28. $m^2 - \frac{2}{5}m = \frac{11}{5}$
29. $2x^2 - 4x = 5$
30. $3a^2 - 12a = 10$
31. $5k^2 - 2k = 4$
32. $10r^2 + 4r = 7$
33. $4m^2 - 3m - 5 = 0$

*Health application*

34. The amount of sugar in the blood after a certain ingestion of food is given by $S = 3.8 + 0.2t - 0.1t^2$, where $t$ is the time in hours. Find the time $t$ at which the sugar level is zero by setting $S = 0$ and completing the square.

(*Hint:* Multiply the formula by $-10$ and take the larger of the two solutions.)

*Business application* 35. For a certain company, its inventory costs are minimized if the quantity $Q$ that it orders satisfies

$$\frac{C_1}{2} - \frac{C_3 r}{Q^2} = 0$$

Find $Q$ if $C_1 = 1$, $C_3 = 1000$, and $r = 2000$.

## 7.3 COMPLEX NUMBERS

Suppose that we have an equation such as $x^2 = -16$. Is the solution $x = 4$? No, because $4^2 = +16$. Is the solution $x = -4$? No, because $(-4)^2 = +16$. There is no real-number solution to $x^2 = -16$. We use a new set of numbers called the **complex numbers** to solve equations such as this.

We start by defining the following.

**DEFINITION 7.1** $i = \sqrt{-1}$ and $i^2 = -1$

The symbol $i$ is not a real number; rather, it is called an **imaginary number.** It is used to simplify

$$\sqrt{-a} = \sqrt{-1}\sqrt{a} = i\sqrt{a}$$

**DEFINITION 7.2** For any real number $a \geq 0$,

$$\sqrt{-a} = i\sqrt{a}$$

**EXAMPLE 12** **(a)** $\sqrt{-16} = i\sqrt{16} = 4i$

**(b)** $\sqrt{-7} = i\sqrt{7}$

**(c)** $-\sqrt{-25} = -i\sqrt{25} = -5i$

We now define the **complex numbers, $C$,** to be the set of all numbers of the form $a + bi$, where $a$ and $b$ are real numbers. For example, $2 + 7i$, $4.3 - 9.7i$, and $\frac{1}{2} - i\sqrt{3}$ are complex numbers. Given a complex number $a + bi$ (or $a + ib$), $a$ is the **real part** and $b$ is the **imaginary part.**

**EXAMPLE 13**

| | Complex Number | Real Part | Imaginary Part |
|---|---|---|---|
| (a) | $3 + 4i$ | 3 | 4 |
| (b) | $5i$ | 0 | 5 |
| (c) | $\dfrac{7 - i\sqrt{2}}{3}$ | $\dfrac{7}{3}$ | $\dfrac{-\sqrt{2}}{3}$ |

Complex numbers can be added, subtracted, multiplied, and divided. (One suggestion is to think of complex numbers like first-degree polynomials. For instance, treat $3 - 2i$ like $3 - 2x$. They are *not* the same, but their algebra is very similar.)

**DEFINITION 7.3** For any real numbers $a$ and $b$,

$$(a + bi) + (c + di) = (a + c) + (b + d)i$$

To add complex numbers:

1. Add the real parts.
2. Add the imaginary parts.

**EXAMPLE 14** **(a)** $(5 + 7i) + (4 - 3i) = (5 + 4) + (7 - 3)i = 9 + 4i$

**(b)** $(8 - 12i) + (6 + 5i) + (10 - i) = (8 + 6 + 10) + (-12 + 5 - 1)i = 24 - 8i$

To subtract complex numbers:

1. Replace the number to be subtracted by its opposite (change the signs).
2. Add the results.

**EXAMPLE 15** Subtract $(4 + 3i) - (-10 + 5i)$.

**SOLUTION** As with polynomials, we change subtraction to addition of the opposite.

| | |
|---|---|
| Change signs | $(4+3i)-(-10+5i)=(4+3i)+(10-5i)$ |
| Add | $=14-2i$ |

> To multiply complex numbers:
>
> 1. Use FOIL or the distributive law to multiply out the terms.
> 2. Use $i^2 = -1$ to simplify the result to the form $a + bi$.

**EXAMPLE 16** Multiply $(2 - 3i)(5 + 4i)$.

**SOLUTION** We multiply this out using the FOIL method.

| | |
|---|---|
| FOIL | $(2-3i)(5+4i)=10+8i-15i-12i^2$ |
| $i^2 = -1$ | $=10+8i-15i+12$ |
| Simplify | $=22-7i$ |

Notice that we multiply these almost as though they were the polynomials $(2 - 3x)(5 + 4x)$, except that we use the fact that $i^2 = -1$.

If $a + bi$ is a complex number, we say that $a - bi$ is its **conjugate,** and vice versa. For example, $7 + 5i$ and $7 - 5i$ are conjugates of each other. (Compare this with the conjugates $3 + \sqrt{5}$ and $3 - \sqrt{5}$.) When we multiply conjugates, we always obtain a positive real number. We use the formula $(x + y)(x - y) = x^2 - y^2$.

**EXAMPLE 17** **(a)** $(7 + 5i)(7 - 5i) = 49 - 25i^2 = 49 + 25 = 74$

**(b)** $(4 - 10i)(4 + 10i) = 16 - 100i^2 = 16 + 100 = 116$

Again we used $i^2 = -1$ to simplify these.

Dividing complex numbers is similar to dividing or simplifying radical expressions: we multiply by $1 = \dfrac{\text{conjugate}}{\text{conjugate}}$.

> To divide complex numbers, $\dfrac{x + yi}{u + vi}$:
>
> 1. Find the conjugate of the denominator; that is $u - vi$.
> 2. Multiply the fraction by $1 = \dfrac{\text{conjugate}}{\text{conjugate}} = \dfrac{u - vi}{u - vi}$.
> 3. Simplify to the form $a + bi$.

**EXAMPLE 18** Divide $\dfrac{8 - 3i}{5 - 2i}$.

**SOLUTION** The conjugate of the denominator is $5 + 2i$.

| | |
|---|---|
| Multiply by $1 = \dfrac{\text{conjugate}}{\text{conjugate}}$ | $\dfrac{8 - 3i}{5 - 2i} = \dfrac{8 - 3i}{5 - 2i} \cdot \dfrac{5 + 2i}{5 + 2i}$ |
| ↓ FOIL | $= \dfrac{40 + 16i - 15i - 6i^2}{25 - 4i^2}$ |
| ↓ $i^2 = -1$ | $= \dfrac{40 + 16i - 15i + 6}{25 + 4}$ |
| ↓ Simplify | $= \dfrac{46 + i}{29}$ |
| ↓ Write as $a + bi$ | $= \dfrac{46}{29} + \dfrac{1}{29}i$ |

**EXAMPLE 19** Divide $\dfrac{5}{1 + i}$.

**SOLUTION** The conjugate of the denominator is $1 - i$.

| | |
|---|---|
| Multiply by $1 = \dfrac{\text{conjugate}}{\text{conjugate}}$ | $\dfrac{5}{1+i} = \dfrac{5}{1+i} \cdot \dfrac{1-i}{1-i}$ |
| ↓ Distributive law | $= \dfrac{5-5i}{1-i^2}$ |
| ↓ $i^2 = -1$ | $= \dfrac{5-5i}{1+1}$ |
| ↓ Simplify | $= \dfrac{5-5i}{2}$ |
| ↓ Write as $a + bi$ | $= \dfrac{5}{2} - \dfrac{5}{2}i$ |

Finally, let us look at the powers of $i$.

$$\begin{aligned}
i &= i \\
i^2 &= -1 \\
i^3 &= i^2 \cdot i = (-1)i = -i \\
i^4 &= i^2 \cdot i^2 = (-1)(-1) = 1 \\
i^5 &= i^4 \cdot i = 1 \cdot i = i \\
i^6 &= i^4 \cdot i^2 = 1 \cdot (-1) = -1 \\
&\vdots
\end{aligned}$$

If we continue this, we notice that the sequence repeats in multiples of 4 in the exponent.

***THEOREM 7.3***

$$\begin{aligned}
i &= i^1 = i^5 = i^9 = \cdots \\
-1 &= i^2 = i^6 = i^{10} = \cdots \\
-i &= i^3 = i^7 = i^{11} = \cdots \\
1 &= i^4 = i^8 = i^{12} = \cdots
\end{aligned}$$

**EXAMPLE 20** To simplify powers of $i$, we use Theorem 7.3.

**(a)** $i^{23} = i^{20} \cdot i^3 = (1)(-i) = -i$

**(b)** $i^{18} = i^{16}i^2 = (1)(-1) = -1$

**(c)** $i^{-5} = \frac{1}{i^5} = \frac{1}{i} = \frac{i}{i \cdot i} = \frac{i}{-1} = -i$

Since $i^{4n} = 1$ (where $4n$ is a multiple of 4), we always separate out an $i^{4n}$ term.

## PROBLEM SET 7.3

Write each of the following in terms of $i$ and simplify as much as possible.

*See Example 12*

1. $\sqrt{-49}$
2. $\sqrt{-100}$
3. $\sqrt{-25}$
4. $\sqrt{-121}$
5. $\sqrt{\frac{-4}{9}}$
6. $\sqrt{-\frac{16}{25}}$
7. $\sqrt{-24}$
8. $\sqrt{-18}$
9. $-\sqrt{-16}$
10. $-\sqrt{-36}$
11. $-\sqrt{-12}$
12. $-\sqrt{-50}$

Identify the real and imaginary parts of the following complex numbers.

*See Example 13*

13. $4 + 6i$
14. $8 - 2i$
15. $\frac{14 - i\sqrt{2}}{2}$
16. $\frac{6 + i\sqrt{3}}{8}$
17. $12$
18. $-6i$

Combine the following complex numbers and simplify to the form $a + bi$ or $a + ib$.

*See Example 14*

19. $(7 + i) + (8 - 3i)$
20. $(16 + 2i) + (4 + 7i)$
21. $(9 + 7i) + (-6 + 2i)$
22. $(4 - 5i) + (-10 - 2i)$
23. $i + 2i + 3i + 6$
24. $6 + 8i - 9i - 7i$
25. $(4 + i) + (8 - 2i) + (-6 - 3i)$
26. $(9 - i) + (-8 - i) + (2 + 6i)$

*See Example 15*

27. $(10 + 3i) - (6 + i)$
28. $(12 + 10i) - (8 - 3i)$
29. $(14 - 7i) - (-8 + 12i)$
30. $(3 - 10i) - (-14 - 6i)$
31. $i - 3i - 5i$
32. $8i - 5i - 10i$
33. $(2 - 5i) - (3 + 7i) + (8 - 2i)$
34. $(5 - 8i) + (4 + i) - (-10 + 3i)$

35. $(3 - i) - (8 - 4i) - (-6 - 2i)$ 36. $(16 - i) - (9 + 7i) - (-3 + 2i)$

*See Example 16*

37. $2i(2 + 3i)$ 38. $7i(3 - 4i)$

39. $(2 + i)(3 - i)$ 40. $(5 - i)(4 + i)$

41. $(2 - 3i)(4 + 5i)$ 42. $(5 + 7i)(2 + 3i)$

43. $(-3 + 4i)(2 - i)$ 44. $(7 - i)(-4 + i)$

45. $(-1 - i)(-6 + 7i)$ 46. $(-2 + i)(-9 - 2i)$

47. $(2 + 3i)^2$ 48. $(1 - 7i)^2$

49. $(5 - 2i)^2$ 50. $(6 + 10i)^2$

*See Example 17*

51. $(4 - 3i)(4 + 3i)$ 52. $(7 + i)(7 - i)$

53. $(2 + 5i)(2 - 5i)$ 54. $(9 - 6i)(9 + 6i)$

55. $(1 - 10i)(1 + 10i)$ 56. $(3 + 5i)(3 - 5i)$

*See Examples 18 and 19*

57. $\frac{4 + i}{3 - i}$ 58. $\frac{6 - 2i}{2 + i}$ 59. $\frac{7 - 5i}{6 + 3i}$ 60. $\frac{-8 + 3i}{5 - 2i}$

61. $\frac{10 + 3i}{2 + 4i}$ 62. $\frac{4 - 5i}{6 - i}$ 63. $\frac{-3 - i}{7 - i}$ 64. $\frac{4 - 2i}{-3 - 5i}$

65. $\frac{6 - 5i}{4i}$ 66. $\frac{2 + 5i}{-3i}$ 67. $\frac{10 - i}{i}$ 68. $\frac{-3 + i}{2i}$

69. $\frac{7}{2 + i}$ 70. $\frac{6}{4 - 2i}$ 71. $\frac{-4i}{1 - 6i}$ 72. $\frac{i}{1 - i}$

Simplify the following.

*See Example 20*

73. $i^3$ 74. $i^{10}$ 75. $i^{25}$ 76. $i^{100}$

77. $i^{17}$ 78. $i^{31}$ 79. $i^{-6}$ 80. $i^{-9}$

*Electricity application*

In electricity, the vector impedence in an alternating-current circuit is given by the complex number $R + jX$. (Here, $j$ is used instead of $i$, since $i$ usually denotes the current.) $X$ is given by $L\omega - 1/\omega C$. In a certain circuit:

$R$ = resistance = 1000 ohms $C$ = capacitance = 0.0000002 farad

$L$ = inductance = 0.5 henry $\omega$ = angular frequency = 2000 radians/sec

81. Find $X$. 82. Find $R + jX$.

83. Find the impedance $Z = \sqrt{(R + jX)(R - jX)}$.

84. If $L$ and $C$ are as given, find another $\omega$ that will make the imaginary part $X = 0$. (This is called *resonance*.)

## 7.4 THE QUADRATIC FORMULA

In Section 7.1, we saw how to solve a quadratic equation by factoring. This method is always our first choice; however, sometimes we cannot solve by factoring. In this section, we derive a formula for solving quadratic equations that will always work. We start with the equation $ax^2 + bx + c = 0$ and complete the square. (Review Section 7.2, if necessary.)

| | |
|---|---|
| Given | $ax^2 + bx + c = 0$ |
| Subtract $c$ from both sides | $ax^2 + bx = -c$ |
| Divide both sides by $a$ | $x^2 + \frac{b}{a}x = -\frac{c}{a}$ |
| $m = \frac{b}{a}$; so $\frac{m}{2} = \frac{b}{2a}$<br>Add $\left(\frac{m}{2}\right)^2 = \frac{b^2}{4a^2}$ to both sides | $x^2 + \frac{b}{a}x + \frac{b^2}{4a^2} = \frac{b^2}{4a^2} - \frac{c}{a}$ |
| Add fractions on the right | $x^2 + \frac{b}{a}x + \frac{b^2}{4a^2} = \frac{b^2 - 4ac}{4a^2}$ |
| Rewrite as a square | $\left(x + \frac{b}{2a}\right)^2 = \frac{b^2 - 4ac}{4a^2}$ |
| Take square roots of both sides | $x + \frac{b}{2a} = \frac{\pm\sqrt{b^2 - 4ac}}{2a}$ |
| Add $\frac{-b}{2a}$ to both sides | $x = \frac{-b}{2a} \pm \frac{\sqrt{b^2 - 4ac}}{2a}$ |
| Combine fractions | $x = \frac{-b \pm \sqrt{b^2 - 4ac}}{2a}$ |

Summarizing this, we have the **quadratic formula.**

> ***THEOREM 7.4*** The solutions to $ax^2 + bx + c = 0$, where $a \neq 0$ are
>
> $$x = \frac{-b \pm \sqrt{b^2 - 4ac}}{2a}$$
>
> This formula is a compact way of saying that the solution set is $\left\{\frac{-b + \sqrt{b^2 - 4ac}}{2a}, \frac{-b - \sqrt{b^2 - 4ac}}{2a}\right\}$.

**EXAMPLE 21** Solve $12x^2 + 13x + 3 = 0$.

**SOLUTION** We use Theorem 7.4, the quadratic formula.

| | |
|---|---|
| Identify terms | $a = 12; \quad b = 13; \quad c = 3$ |
| Quadratic formula | $x = \frac{-b \pm \sqrt{b^2 - 4ac}}{2a}$ |
| Substitute | $x = \frac{-13 \pm \sqrt{13^2 - 4(12)(3)}}{2(12)}$ |
| Simplify | $x = \frac{-13 \pm \sqrt{169 - 144}}{24}$ |
| Simplify | $x = \frac{-13 \pm \sqrt{25}}{24} = \frac{-13 \pm 5}{24}$ |
| Find + and − solutions | $x = \frac{-13 + 5}{24}$ or $x = \frac{-13 - 5}{24}$ |
| Simplify | $x = \frac{-1}{3}$ or $x = \frac{-3}{4}$ |

The solution set is $\{-\frac{1}{3}, -\frac{3}{4}\}$.

**EXAMPLE 22** Solve $\dfrac{1}{15} - \dfrac{1}{3x} + \dfrac{1}{5x^2} = 0$

**SOLUTION** We first multiply by the LCD, $15x^2$.

| | |
|---|---|
| Multiply by LCD, $15x^2$ | $15x^2\left[\dfrac{1}{15} - \dfrac{1}{3x} + \dfrac{1}{5x^2}\right] = 15x^2\,[0]$ |
| ↓ Distributive law | $x^2 - 5x + 3 = 0$ |
| ↓ Identify terms | $a = 1;\quad b = -5;\quad c = 3$ |
| ↓ Quadratic formula | $x = \dfrac{-b \pm \sqrt{b^2 - 4ac}}{2a}$ |
| ↓ Substitute | $x = \dfrac{-(-5) \pm \sqrt{(-5)^2 - 4(1)(3)}}{2(1)}$ |
| ↓ Simplify | $x = \dfrac{5 \pm \sqrt{25 - 12}}{2}$ |
| ↓ Simplify | $x = \dfrac{5 \pm \sqrt{13}}{2}$ |

Therefore, the solution set is $\left\{\dfrac{5 + \sqrt{13}}{2}, \dfrac{5 - \sqrt{13}}{2}\right\}$.

**EXAMPLE 23** Write the solutions to Example 22 as decimals.

**SOLUTION** We compute them using a hand calculator.

| | PUNCH | DISPLAY | MEANING |
|---|---|---|---|
| ⊕: | C | 0. | Clear |
| | 5 | 5. | 5 |
| | + 1 3 $\sqrt{\ }$ | 3.6055513 | Plus $\sqrt{13}$ |
| | = | 8.6055513 | $5 + \sqrt{13}$ |
| | ÷ 2 | 2. | Divided by 2 |
| | = | 4.3027756 | "+" solution |
| ⊖: | C | 0. | Clear |
| | 5 | 5. | 5 |
| | − 1 3 $\sqrt{\ }$ | 3.6055513 | Minus $\sqrt{13}$ |
| | = | 1.3944487 | $5 - \sqrt{13}$ |
| | ÷ 2 | 2. | Divided by 2 |
| | = | 0.6972244 | "−" solution |

Therefore, a set of approximate solutions is $\{4.303, 0.697\}$.

**EXAMPLE 24** Solve $2(m^2 + 4) = 3m$.

**SOLUTION** We must first put this into standard form, $2m^2 - 3m + 8 = 0$. Now we can substitute into the quadratic formula.

| | |
|---|---|
| Identify terms | $a = 2;\quad b = -3;\quad c = 8$ |
| Quadratic formula | $m = \dfrac{-b \pm \sqrt{b^2 - 4ac}}{2a}$ |
| Substitute | $m = \dfrac{-(-3) \pm \sqrt{(-3)^2 - 4(2)(8)}}{2(2)}$ |
| Simplify | $m = \dfrac{3 \pm \sqrt{9 - 64}}{4}$ |
| Simplify | $m = \dfrac{3 \pm \sqrt{-55}}{4}$ |
| Find the + and − solutions | $m = \dfrac{-3 + i\sqrt{55}}{4}$ or $m = \dfrac{-3 - i\sqrt{55}}{4}$ |

Notice that we get complex-number solutions. This is why it is so important to study complex numbers. The solution set is $\left\{\frac{-3}{4} + i\frac{\sqrt{55}}{4}, \frac{-3}{4} - i\frac{\sqrt{55}}{4}\right\}$.

**EXAMPLE 25** Solve $p = \sqrt{5p + 9}$

**SOLUTION** We square both sides first, and then solve.

| | |
|---|---|
| Given | $p = \sqrt{5p + 9}$ |
| Square both sides | $p^2 = 5p + 9$ |
| Put in standard form | $p^2 - 5p - 9 = 0$ |
| Identify terms | $a = 1; \quad b = -5; \quad c = -9$ |
| Quadratic formula | $p = \dfrac{-b \pm \sqrt{b^2 - 4ac}}{2a}$ |
| Substitute | $p = \dfrac{-(-5) \pm \sqrt{(-5)^2 - 4(1)(-9)}}{2(1)}$ |
| Simplify | $p = \dfrac{5 \pm \sqrt{25 + 36}}{2}$ |
| Simplify | $p = \dfrac{5 \pm \sqrt{61}}{2}$ |

The solutions appear to be $\frac{5 + \sqrt{61}}{2}$ and $\frac{5 - \sqrt{61}}{2}$, but we know that with radical equations we must check for extraneous solutions. But substituting these solutions into the original radical could be very difficult. Instead, we approximate the solutions using a hand calculator as we did in Example 23. When we do, we find that

$$\frac{5+\sqrt{61}}{2} \approx 6.4 \qquad \text{and} \qquad \frac{5-\sqrt{61}}{2} \approx -1.4$$

We now substitute these approximates into the original equation, $p = \sqrt{5p + 9}$.

1. $p \approx 6.4$ and $\sqrt{5(6.4) + 9} = \sqrt{32 + 9} = \sqrt{41} \approx 6.4$ Checks
2. $p \approx -1.4$ and $\sqrt{5(-1.4) + 9} = \sqrt{-7 + 9} = \sqrt{2} \approx 1.4$ Does *not* check

Therefore, the solution set is simply $\left\{\frac{5+\sqrt{61}}{2}\right\}$ or approximately $\{6.4\}$.

The number $b^2 - 4ac$ within the radical of the quadratic formula is called the **discriminant, *D*.**

We can use the discriminant to help summarize the results of this section:

1. If $D$ is positive, there are two real solutions.
2. If $D$ is zero, there is only one real solution.
3. If $D$ is negative, there are two complex solutions.

## PROBLEM SET 7.4

Find the solutions to the following equations using the quadratic formula.

*See Examples 21, 22, and 24*

1. $2x^2 - x - 15 = 0$
2. $3a^2 + 7a - 6 = 0$
3. $4k^2 + 5k - 3 = 0$
4. $m^2 - 10m + 2 = 0$
5. $r^2 - 3r - 5 = 0$
6. $3z^2 - 5z - 3 = 0$
7. $5x^2 - 4 = 2x$
8. $y^2 - y = 10$
9. $3p^2 - p = 5$
10. $3u + 8 = u^2$
11. $6w = 1 - 4w^2$
12. $5t^2 + t = -7$
13. $v(4v + 3) = 2$
14. $2k(1 - k) = 3$
15. $(3n + 1)(3n - 1) = 5$
16. $(2m - 3)(2m + 3) = 4$
17. $(x + 7)(x - 9) = x - 8$
18. $(u - 6)(u - 5) = u - 4$
19. $\frac{x^2}{2} + \frac{x}{3} + \frac{5}{6} = 0$
20. $\frac{a^2}{3} - \frac{a}{5} - \frac{7}{15} = 0$

21. $\dfrac{3}{2z^2} - \dfrac{5}{4z} + \dfrac{1}{8} = 0$

22. $\dfrac{4}{3u^2} + \dfrac{2}{9u} - \dfrac{1}{18} = 0$

23. $\dfrac{x}{x+1} = \dfrac{x+1}{2x}$

24. $\dfrac{t}{2t+1} = \dfrac{t+2}{t}$

Solve this following equations and write the solutions as decimals. (Use a hand calculator.)

*See Examples 22 and 23*

25. $x^2 - 5x - 8 = 0$

26. $a^2 + a - 5 = 0$

27. $2u^2 - 9u - 8 = 0$

28. $w^2 + 10w + 15 = 0$

29. $3r^2 - 2r = 4$

30. $5k^2 + 2 = 7k$

31. $m^2 = 8m - 6$

32. $2z^2 = 1 - 7z$

Solve the following radical equations. Check for extraneous solutions by first approximating the solutions on a hand calculator.

*See Examples 23 and 25*

33. $x = \sqrt{3 - 5x}$

34. $p = \sqrt{6p - 2}$

35. $r = \sqrt{10r - 3}$

36. $u = \sqrt{4 - 7u}$

37. $2m = \sqrt{7m + 6}$

38. $3t = \sqrt{5t + 7}$

39. $z = \sqrt{\dfrac{3 - 5z}{2}}$

40. $w = \sqrt{\dfrac{5 - 4w}{7}}$

*Art application*

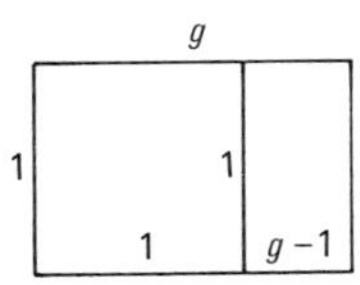


41. The "golden rectangle" shown at the left has dimensions 1 by $g$. If a 1 by 1 square is cut out, the remaining rectangle has dimensions $(g - 1)$ by 1, which is related to the original rectangle by

$$\frac{\text{Length}}{\text{Width}} = \frac{g}{1} = \frac{1}{g - 1}$$

Find $g$. (Consider only the positive solution.)

*Physical application*

42. An object is thrown in the air, and its height $h$ at any time $t$ is given by $h = -16t^2 + 80t$. At what times is the height equal to 70?

*Business applications*

43. The supply $S$ and demand $D$ for a certain men's shirt are given by

$$S = 20{,}000p - 150{,}000 \quad \text{and} \quad D = \frac{100{,}000}{p}$$

where $p$ is the price. Set $S = D$ and find the equilibrium price.

44. The price $p$ of a item depends on the quantity demanded $x$ by the equation $p = 10{,}000 - 200x - x^2$. What demand $x$ will correspond to a price $p = 2000$?

45. A company finds that its marginal profit (in thousands) is given by

$$MP = -x^2 + 50x - 100$$

where $x$ is number (in thousands) of lamps produced. The profit is maximum where MP = 0. Set MP = 0 and solve for $x$.

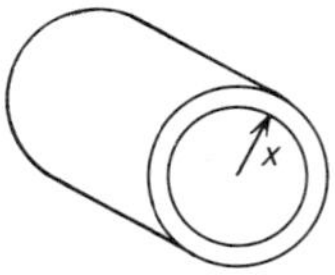


*Health applications*

46. Blood flows through blood vessels with a velocity $v = 1.28 - 20{,}000x^2$, where $x$ is the distance (in centimeters) from the center of blood vessel. Find the distance $x$ when $v = 0.28$ centimeters per second.

47. The population of a bacteria culture is given by

$$p = 10{,}000 + 1000t - 125t^2$$

where $t$ is the time in hours. Find the times at which the population will be $p = 11{,}000$.

*Photographic application*

48. The formula that gives the magnification of a lens is

$$L = \left(M + \frac{1}{M} + 2\right)f + n$$

Find the magnification $M$ after substituting the following data:

$$L = \text{length from lens to image} = 49$$

$$f = \text{focal length of the lens} = 4$$

$$n = \text{separation of nodal points} = 1$$

## 7.5 WORD PROBLEMS

In this section, we study word problems that lead to quadratic equations. The approach will be very similar to those used in Chapters 2 and 5.

> To solve word problems leading to quadratic equations:
>
> 1. Identify the unknown.
> 2. Translate the problem into an equation.
> 3. Put the equation into standard form.
> 4. Solve by factoring or using the quadratic formula.

**EXAMPLE 26** The sum of two numbers is 9. The sum of their squares is 45. Find the two numbers.

**SOLUTION** We must first identify the unknowns. Since the numbers add up to 9, we can call them $x$ and $9 - x$.

| | |
|---|---|
| Identify unknowns | Let $x$ = one unknown<br>$9 - x$ = other unknown |
| Given | Sum of squares is 45 |
| Translate | $x^2 + (9 - x)^2 = 45$ |
| Square binomial | $x^2 + 81 - 18x + x^2 = 45$ |
| Put into standard form | $2x^2 - 18x + 36 = 0$ |
| Divide both sides by 2 | $x^2 - 9x + 18 = 0$ |
| Factor | $(x - 6)(x - 3) = 0$ |
| Set each factor = 0 | $x - 6 = 0$ or $x - 3 = 0$ |
| Solve each equation | $x = 6$ or $x = 3$ |

Therefore, the integers are 3 and 6.

**EXAMPLE 27** The grassy mall at Raccoon State College is 40 meters by 30 meters. There is a walkway of uniform width around it, and the total area of the walkway is 296 square meters. What is the width of the walkway?

**SOLUTION** It is helpful to start with a sketch of this situation (shown at the top of page 243). The unknown width is $x$. Notice that the dimensions of the large rectangle are $40 + 2x$ by $30 + 2x$, since there is a walkway of width $x$ on *both* ends of the length and width.

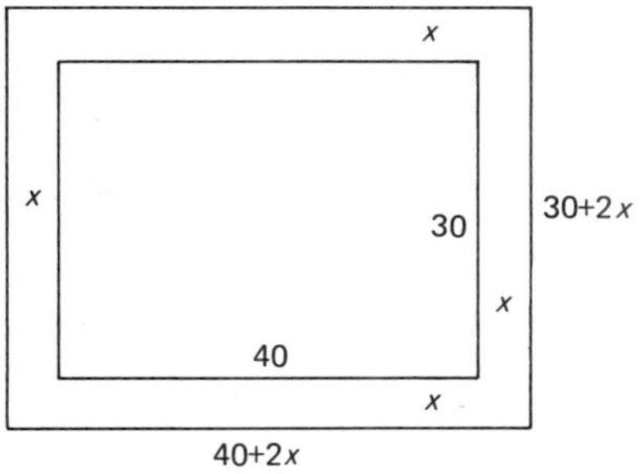


| | |
|---|---|
| Set up area equation | Area (total) $=$ $A$(mall) $+$ $A$(walkway) |
| ↓ Substitute | $(40 + 2x)(30 + 2x) = 40 \cdot 30 + 296$ |
| ↓ Simplify | $1200 + 140x + 4x^2 = 1200 + 296$ |
| ↓ Put into standard form | $4x^2 + 140x - 296 = 0$ |
| ↓ Divide by 4 | $x^2 + 35 - 74 = 0$ |
| ↓ Factor | $(x - 2)(x + 37) = 0$ |
| ↓ Set each factor = 0 | $x - 2 = 0$ or $x + 37 = 0$ |
| ↓ Solve each equation | $x = 2$ or $x = -37$ |

Since a path $-37$ meters wide makes no sense, the width is simply 2 meters.

**EXAMPLE 28** There is a 10-mile-per-hour wind. Amy rides her bicycle 40 miles into the wind and 40 miles back with the wind, in 3 total hours. At what velocity (on a still day) was she pedaling?

**SOLUTION** Recall from Chapter 5 that we have the formula: *distance is velocity times time* ($d = vt$). Let $x$ be the unknown velocity; then $x - 10$ is the velocity against the wind and $x + 10$ is the velocity with the wind. Let us complete a $d$–$v$–$t$ table.

| | $d$ | $v$ | $t = \frac{d}{v}$ |
|---|---|---|---|
| Going | 40 | $x - 10$ | $\frac{40}{x - 10}$ |
| Coming | 40 | $x + 10$ | $\frac{40}{x + 10}$ |

| Step | |
|---|---|
| Given | Total time is 3 hours |
| Translate | $\frac{40}{x - 10} + \frac{40}{x + 10} = 3$ |
| Multiply by the LCD, $(x - 10)(x + 10)$ | $(x - 10)(x + 10)\left[\frac{40}{x - 10} + \frac{40}{x + 10}\right] = 3(x - 10)(x + 10)$ |
| Distributive law | $40x + 400 + 40x - 400 = 3x^2 - 300$ |
| Put into standard form | $-3x^2 + 80x + 300 = 0$ |
| Multiply by $-1$ | $3x^2 - 80x - 300 = 0$ |
| Factor | $(3x + 10)(x - 30) = 0$ |
| Set each factor $= 0$ | $3x + 10 = 0$ or $x - 30 = 0$ |
| Solve each equation | $x = \frac{-10}{3}$ or $x = 30$ |

Since $-10/3$ makes no sense here, the answer is 30 miles per hour.

## PROBLEM SET 7.5

See Example 26

1. The sum of two numbers is 14. Their product is 33. Find the two numbers.
2. The sum of two numbers is 19. Their product is 60. Find the two numbers.
3. The sum of two numbers is 10. The sum of their squares is 68. Find the two numbers.
4. The sum of two numbers is 11. The sum of the squares is 61. Find the two numbers.
5. The sum of a number and its reciprocal is $\frac{10}{3}$. Find the number.
6. The sum of a number and its reciprocal is $\frac{13}{6}$. Find the number.
7. Find two consecutive odd integers whose product is 143.
8. Find two consecutive even integers whose product is 120.
9. A number plus twice its reciprocal is $\frac{9}{2}$. Find the number.
10. A number plus three times its reciprocal is $\frac{28}{3}$. Find the number.

See Example 27

11. The length of a rectangle is 5 more than its width. If the area is 66, find the dimensions.
12. The length of a rectangle is 7 more than its width. If the area is 60, find the dimensions.
13. The perimeter of a rectangle is 28, and the area is 48. Find the dimensions. (*Hint:* Write the width in terms of the length and perimeter.)
14. The perimeter of a rectangle is 36, and the area is 72. Find the dimensions. (See the hint in Problem 13.)
15. A rectangle 15 by 20 is increased all the way around by a uniform border of width $x$. If the new area is 336, find the width of the border.
16. A rectangle 12 by 18 is increased all the way around by a uniform border of width $x$. If the area of the border is 64, find the width of the border.
17. A swimming pool measures 20 feet by 25 feet. A concrete walkway of uniform width is built around the pool. The area of the walkway is 250 square feet. Find the width of the pathway.
18. A rectangular mall is 15 meters by 25 meters. There is a pathway of uniform width around it. If the total area (mall and pathway) is 416 square meters, find the width of the pathway.

See Example 28

19. Lisa made a hike of 24 miles (12 miles each way). Her velocity going was 1 mile per hour faster than her velocity returning. The whole hike took 7 hours. Find her velocities, going and coming.
20. Mike made a boat trip, 12 miles upstream and 12 miles downstream. The current was 1 mile per hour and the trip took 5 hours. Find his velocity (as if he were in still water).

21. Willie made a 140-mile trip, 40 miles at one speed and 100 miles at 10 miles per hour faster. The entire trip took 3 hours. Find the two speeds.

22. Jackie drove 5 miles in rush-hour traffic and 20 miles at 20 miles per hour faster. The entire ride took $\frac{3}{4}$ of an hour. Find the two speeds.

## 7.6 QUADRATIC INEQUALITIES

A **quadratic inequality** is an inequality in one of the following standard forms:

$$ax^2 + bx + c > 0$$

$$ax^2 + bx + c \geq 0$$

$$ax^2 + bx + c < 0$$

$$ax^2 + bx + c \leq 0$$

The technique we use to solve these is graphing on the number line.

**EXAMPLE 29** Solve $(x - 2) \cdot (x - 5) > 0$.

**SOLUTION** This is already factored for us. How do we make a product of two terms positive? We find where the factors are both positive or both negative. When we find the values where a factor is positive, we can easily find the values where it is negative.

Find where each factor is positive

$(x - 2) > 0$ if $x > 2$

$(x - 5) > 0$ if $x > 5$

Graph the signs of each factor

$(x - 2)$: - - - - - -O+ + + + + + + + + + + + + + + + + +

$(x - 5)$: - - - - - - - - - - - - - - - - - - -O+ + + + + + +

Determine the signs of the product

*Product* + + + + + O- - - - - - - - - - - -O + + + + + + +

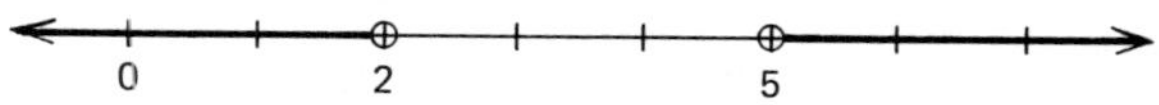


Solution is where product is +

Solution set $= \{x \mid x < -2 \quad \text{or} \quad x > 5\}$

> To solve a quadratic inequality:
>
> 1. Put it into standard form.
> 2. Factor the polynomial.
> 3. Determine where each factor is positive. (The zero and negative regions are then obvious.)
> 4. Draw in the signs of each factor on the number line.
> 5. Determine the signs of the product.
> 6. From the graph, take the region where the product has the desired sign.

**EXAMPLE 30** Solve $x^2 + 5x - 6 \le 0$.

**SOLUTION** We use the procedure stated above. We want the product on the left to be negative or zero.

| | |
|---|---|
| Factor | $(x + 6)(x - 1) \le 0$ |
| Find where each factor is positive | $(x + 6) > 0$ if $x > -6$<br>$(x - 1) > 0$ if $x > 1$ |
| Graph the sign of each factor | $(x + 6)$ - - - ○ + + + + + + + + + + + + + + + + + + + + +<br><br>$(x - 1)$ - - - - - - - - - - - - - - - - - - ○ + + + + + + |
| Determine the signs of the product | *Product* + + [○ - - - - - - - - - - - - - ○] + + + + + |
| | 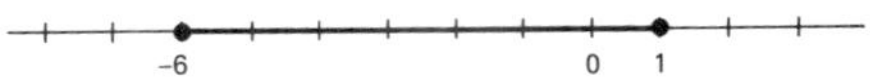 |
| Solution is where product is – or 0 | Solution set $= \{x \mid -6 \le x \le 1\}$ |

**EXAMPLE 31** Solve $3a^2 + 10a \geq 8$.

**SOLUTION** In standard form, this is $3a^2 + 10a - 8 \geq 0$. When we factor, we will want the product to be positive or zero.

| | |
|---|---|
| Factor | $(3a - 2)(a + 4) \geq 0$ |
| Find where each factor is positive | $3a - 2 > 0$ if $a > \frac{2}{3}$<br>$a + 4 > 0$ if $a > -4$ |
| Graph the sign of each factor | $(3a - 2)$ – – – – – – – – – – – – – – – – – – – ○ + + + + +<br>$(a + 4)$ – – – – – ○ + + + + + + + + + + + + + + + + + |
| Determine the sign of the product | *Product* + + + + + ○ – – – – – – – – – – – – – ○ + + + + + |
| | 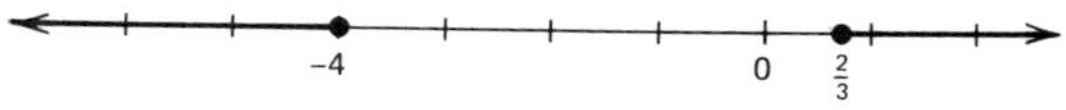 |
| Solution is where product is 0 or + | Solution set $= \{x \mid x \leq -4 \text{ or } x \geq \frac{2}{3}\}$ |

## PROBLEM SET 7.6

Solve the following quadratic inequalities and graph the solution set.

*See Example 29*

1. $(x - 1)(x - 2) > 0$
2. $(a + 5)(a - 2) < 0$
3. $(2r + 1)(r - 3) \leq 0$
4. $(3t - 1)(t + 2) \geq 0$
5. $(3z - 2)(4z + 5) < 0$
6. $(4x + 3)(5x - 2) > 0$

*See Example 30*

7. $u^2 - 7u + 12 \geq 0$
8. $m^2 - m - 2 \leq 0$
9. $k^2 + 7k - 18 > 0$
10. $a^2 - 2a - 15 < 0$
11. $2x^2 + 11x + 5 \leq 0$
12. $3r^2 + 11r - 4 \geq 0$
13. $6t^2 + t - 2 < 0$
14. $10m^2 + 11m - 6 > 0$

*See Example 31*

15. $x^2 \geq 9$
16. $a^2 \leq 25$
17. $z^2 - 3z > 10$
18. $p^2 + 3p < 28$
19. $k^2 + 10 \leq 11k$
20. $r^2 - 44 \geq 7r$
21. $2x^2 + x < 15$
22. $3t^2 + 16t > 12$

*Physical application*

23. An object is shot into the air, and the height $h$ is given by $h = -16t^2 + 80t$, where $t$ is time in seconds. At what times $t$ will the height be greater than 96 feet?

*Life science application*

24. The yield $y$ in bushels per acre in a rice field is given by $y = 10x - x^2/2$, where $x$ is the density of the rice plant. At what densities $x$ will the yield be greater than 48 bushels per acre?

*Health application*

25. The population of a bacteria culture is given by $p = 10{,}000 + 1000t - 100t^2$, where $t$ is the time in hours. At what times $t$ will the population be less than 12,100?

# CHAPTER 7 SUMMARY

## Important Words and Phrases

completing the square (7.2)
complex numbers (7.3)
conjugate (7.3)
discriminant (7.4)
extraneous solutions (7.1)
imaginary number (7.3)
imaginary part (7.3)
quadratic equation (introduction)
quadratic formula (7.4)
quadratic inequality (7.6)
real part (7.3)
standard form (introduction)
Zero-Product Theorem (7.1)

## Important Theorems and Definitions

If $a \cdot b = 0$, then $a = 0$ or $b = 0$, or both.

If $x^2 = a$, then $x = \sqrt{a}$ or $x = -\sqrt{a}$. $(a \geq 0)$

The **quadratic formula** gives the solutions to $ax^2 + bc + c = 0$ $(a \neq 0)$ as

$$x = \frac{-b \pm \sqrt{b^2 - 4ac}}{2a}$$

$i = \sqrt{-1}$, $i^2 = -1$, $\sqrt{-a} = i\sqrt{a}$. $(a \geq 0)$

## Important Procedures

*To solve a quadratic equation by factoring:*
1. Put the equation into standard form; factor the polynomial.
2. Set each factor equal to zero.
3. Solve each of the first-degree equations.

*To solve a quadratic equation by completing the square:*
1. Rewrite the equation in the form $x^2 + mx = n$.
2. Add $\left(\frac{m}{2}\right)^2$ to both sides.
3. Rewrite the equation as $\left(x + \frac{m}{2}\right)^2 = n + \left(\frac{m}{2}\right)^2$.
4. Solve by taking the square root of both sides.

*To solve word problems leading to quadratic equations:*
1. Identity the unknown(s).
2. Translate the problem into an equation.
3. Put the equation into standard form.
4. Solve by factoring or using the quadratic formula.

*To add, subtract, and multiply complex numbers:* we follow the usual rules for binomials such as $a + bx$, except that $i^2 = -1$.

*To divide complex numbers:*
1. Multiply the numerator and denominator by the conjugate of the denominator.
2. Simplify.

*To solve a quadratic inequality:*
1. Put it into standard form.
2. Factor the polynomial.
3. Determine where the factors are positive and negative.
4. Mark the signs on a number line.
5. Determine the signs of the product.
6. Take the region with the desired sign.

## CHAPTER 7 REVIEW EXERCISES

Solve the following quadratic equations by factoring.

1. $(x - 5)(2x + 7) = 0$
2. $t^2 + 2t - 15 = 0$

3. $2a^2 + a = 12a + 6$

4. $(m + 4)(m - 5) = -14$

5. $z + 6 = \frac{7}{z}$

6. $y = \sqrt{18 - 7y}$

Solve the following quadratic equations completing the square.

7. $x^2 = \frac{4}{81}$

8. $(3k - 1)^2 = \frac{3}{16}$

9. $r^2 + 8r = 1$

10. $3u^2 - 5u - 2 = 0$

Write the following in terms of $i$.

11. $\sqrt{-64}$

12. $-\sqrt{-25}$

13. Identify the real and imaginary parts of $-6 + 10i$.

Combine the following complex numbers and simplify to the form $a + bi$ or $a + ib$.

14. $(7 - 2i) + (3 + 5i)$

15. $(9 - 3i) - (-5 - 4i) - i$

16. $(3 - 4i)(2 + 7i)$

17. $(8 - i)(8 + i)$

18. $\frac{5}{2 - i}$

19. $\frac{2 - 3i}{4 + i}$

20. $i^3$

21. $i^{20}$

Solve the following equations using the quadratic formula. (Use a hand calculator to help with problems 24 and 25.)

22. $p^2 - 6p - 10 = 0$

23. $(5x - 1)(x + 3) = 5$

24. $3r^2 - r = 5$

25. $a = \sqrt{1 - 3a}$

26. The sum of two numbers is 8. The sum of their squares is 50. Find the numbers.

27. A swimming pool is 10 feet by 20 feet. There is a border of uniform width. The area of the border is 64 square feet. Find the width.

28. The McCoys drove 120 miles at one speed and 30 miles at 10 miles per hour slower. The entire 150-mile trip took 4 hours. Find the two speeds.

Solve the following quadratic inequalities.

29. $(2x + 5)(x - 7) > 0$

30. $2a^2 - 9a - 5 \leq 0$

# 8

# LINEAR EQUATIONS AND INEQUALITIES IN TWO VARIABLES

In Chapter 2, we asked when $x + 2 = 6$ is true; of course, $x = 4$ makes this true. In this chapter, we might ask when $x + y = 10$. Now the answer will depend on both $x$ and $y$ and will be written as a pair $(x, y)$. Here, there will be infinitely many solutions, such as (1, 9), (2, 8), (10, 0), (−5, 15), (18, −8), and $(\frac{1}{2}, 9\frac{1}{2})$. (Each of these pairs adds up to 10.)

Since there are too many solutions to write out, it is actually easier to draw (or sketch) the solution. We call this drawing the **graph** of the equation.

## 8.1 GRAPHING THE STRAIGHT LINE

Equations such as $x + y = 10$ will have solutions that are pairs $(x, y)$. We picture these pairs on the two-dimensional **Cartesian** (or **rectangular**) **coordinate system.** This coordinate system is formed by two number lines, called **coordinate axes** at right angles to each other, as shown in Fig. 8.1. From the illustration we can see the following six facts:

1. The positive directions are up and to the right.
2. The negative directions are down and to the left.

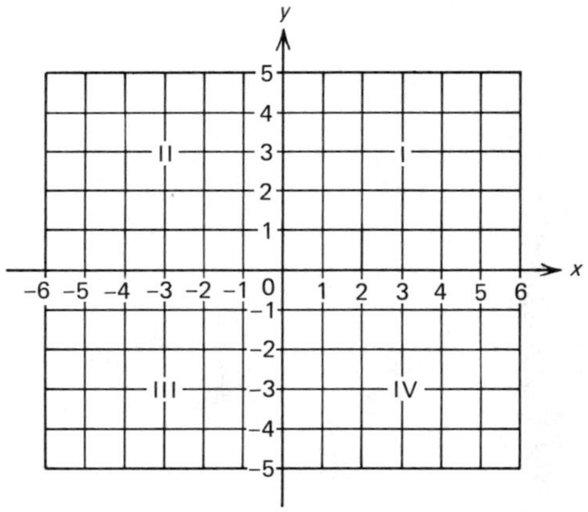


**Figure 8.1**

3. The horizontal coordinate axis is called the ***x*-axis.**
4. The vertical coordinate axis is called the ***y*-axis.**
5. The plane is divided into four **quadrants,** labeled I, II, III, and IV.
6. The point where the axes meet is called the **origin.** This point has coordinate (0, 0).

> For the pair $(x, y)$:
>
> 1. $x$ means the horizontal movement from the origin (right for positive, left for negative). This is called the ***x*-coordinate.**
> 2. $y$ means the vertical movement from the origin (up for positive, down for negative). This is called the ***y*-coordinate.**

**EXAMPLE 1** Locate the points $(3, 4)$, $(1, -5)$, $(-2, 0)$, $(-3, -1)$, and $(4, 3)$ on a graph.

**SOLUTION** We determine what each pair means and put the points on the graph in Fig. 8.2.

**(a)** $(3, 4)$ means 3 to the right, 4 up.

**(b)** $(1, -5)$ means 1 to the right, 5 down.

**(c)** $(-2, 0)$ means 2 to the left, and no units up or down.

**(d)** $(-3, -1)$ means 3 to the left, 1 down.

**(e)** $(4, 3)$ means 4 to the right, 3 down.

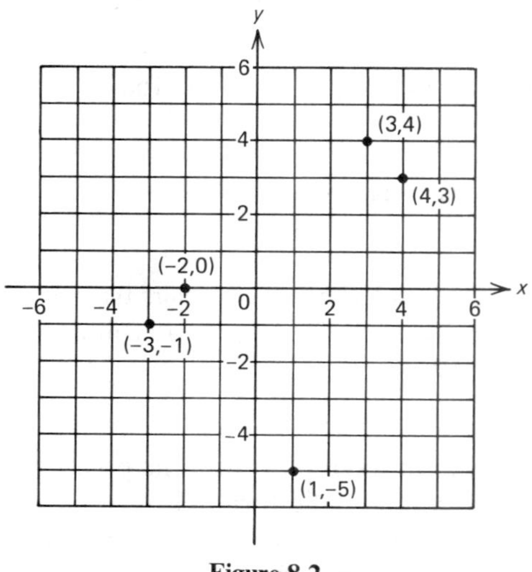


**Figure 8.2**

Notice that (3, 4) is *not* the same as (4, 3). That is why $(x, y)$ is called an **ordered pair.** The order *does* matter.

**EXAMPLE 2** Give the coordinates of the points shown in Fig. 8.3.

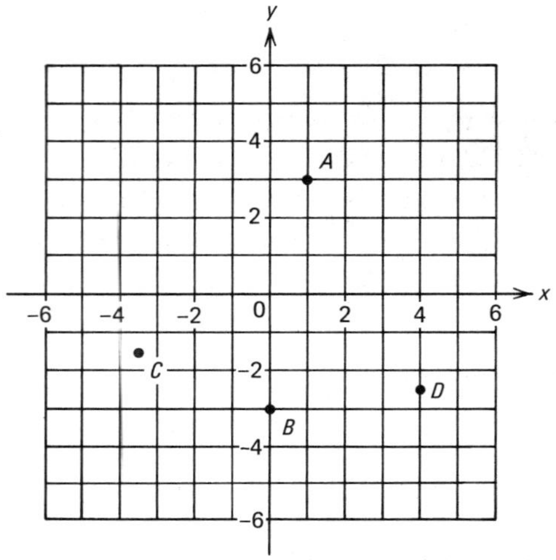


**Figure 8.3**

**SOLUTION** We start at the origin and "walk" to the points.

**(a)** $A$ is 1 to the right, 3 up. Thus, $A = (1, 3)$.

**(b)** $B$ is no units right or left, 3 down. Thus, $B = (0, -3)$.

**(c)** $C$ is 3.5 to the left, 1.5 down. Thus, $C = (-3.5, -1.5)$.

**(d)** $D$ is 4 to the right, 2.5 down. Thus, $D = (4, -2.5)$.

**EXAMPLE 3** Graph the equation $x + y = 10$.

**SOLUTION** We plot some of the ordered pairs that we know satisfy this equations. For example, we know (0, 10), (1, 9), (2, 8), (3, 7), (4, 6), and $(9\frac{1}{2}, \frac{1}{2})$ are solutions. We put this on Fig. 8.4 (on page 257).

Notice in Example 3 that the points form a line which we draw in. In fact, any equation in the form $ax + by = c$, with $a$ and $b$ not both 0, is called a **linear equation** and will form a line when graphed. The form $ax + by = c$ is called the **standard form** of a linear equation.

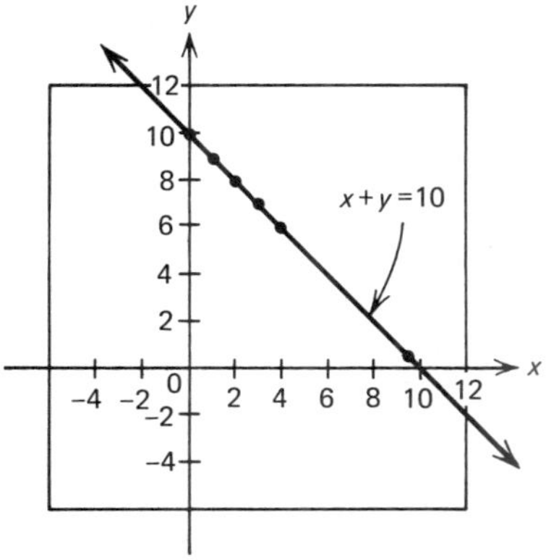


**Figure 8.4**

The ***x*-intercept** of the graph of an equation is the $x$-coordinate of the point where the graph crosses the $x$-axis. The ***y*-intercept** is the $y$-coordinate of the point where the graph crosses the $y$-axis.

> To find the $x$-intercept, set $y = 0$ and solve the equation for $x$.
>
> To find the $y$-intercept, set $x = 0$ and solve the equation for $y$.

**EXAMPLE 4** Find the $x$- and $y$-intercepts of $40x - 5y = 200$ and graph the equation.

**SOLUTION** We first find the $x$-intercept by setting $y = 0$.

| | |
|---|---|
| Given | $40x - 5y = 200$ |
| ↓ Set $y = 0$ | $40x - 5(0) = 200$ |
| ↓ Simplify | $40x = 200$ |
| ↓ Divide both sides by 40 | $x = 5$ |

The $x$-intercept is 5, and the line crosses the $x$-axis at (5, 0). Now we find the $y$-intercept by setting $x = 0$.

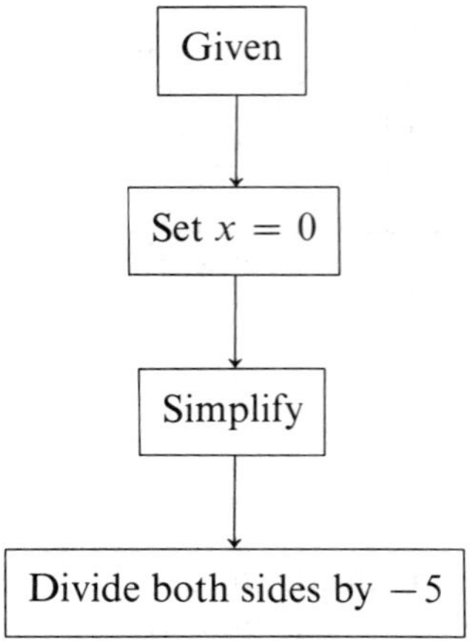


$$40x - 5y = 200$$

$$40(0) - 5y = 200$$

$$-5y = 200$$

$$y = -40$$

The $y$-intercept is $-40$, and the line crosses the $y$-axis at $(0, -40)$. Since two points determine a line, the intercepts (5, 0) and $(0, -40)$ will give us the line. Notice from this graph that *the axes need not be scaled the same. We scale each axis to fit the points of the problem.* Also, it is usually a good practice to find a third point on the line. This serves as a check and also helps us to graph. For instance, a third point here is $(2, -24)$.

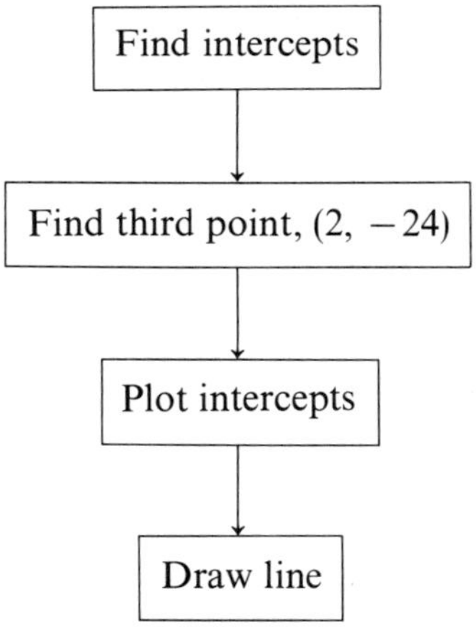


| $x$ | $y$ |
|---|---|
| 0 | $-40$ |
| 5 | 0 |
| 2 | $-24$ |

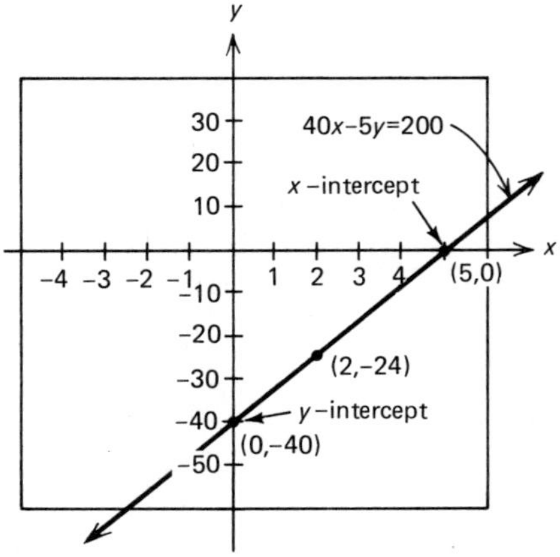

Summarizing, we have the following procedure.

To graph the equation $ax + by = c$, where $a$, $b$, and $c$ are not zero:

1. Find the $x$- and $y$-intercepts.
2. Find a third solution to the equation.
3. Plot these intercepts on the proper axes.
4. Draw the straight line through these points.

**EXAMPLE 5** Graph the equations $x = 3$ and $y = -1$.

**SOLUTION** The equation $x = 3$ seems strange since there is no $y$-term. This means that no matter what $y$ is, $x$ will be 3. Thus, (3, 1), (3, 7), (3, −2) are all points on the line. In Fig. 8.5, we see that this is a vertical line.

Similarly, $y = -1$ means that the $y$-component will be −1 no matter what $x$ is. We see in Fig. 8.5 that this is a horizontal line.

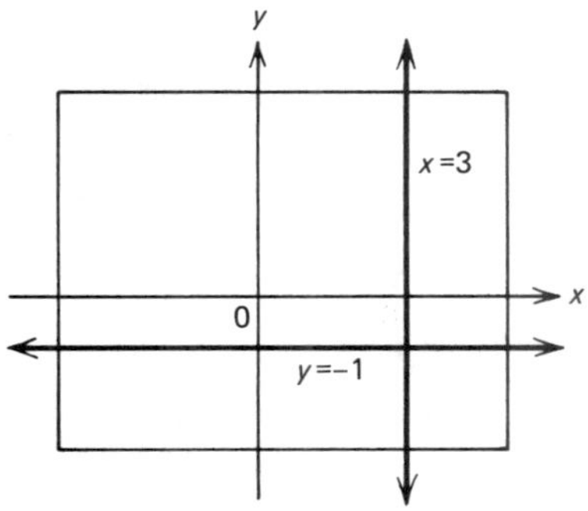


**Figure 8.5**

## PROBLEM SET 8.1

Locate the following points on a rectangular coordinate system.

*See Example 1*

1. $A = (2, 5)$
2. $B = (4, 1)$
3. $C = (3, -2)$
4. $D = (6, -1)$
5. $E = (-7, 4\frac{1}{2})$
6. $F = (-1, 5)$
7. $G = (-2, -3)$
8. $H = (-4, -5\frac{1}{4})$
9. $I = (0, 6)$
10. $J = (-2, 0)$
11. $K = (-4.5, 3)$
12. $L = (2, -3.5)$

Identify the coordinates of the points shown in Fig. 8.6.

*See Example 2*

13. $A$ 14. $B$ 15. $C$ 16. $D$
17. $E$ 18. $F$ 19. $G$ 20. $H$
21. $I$ 22. $J$ 23. $K$ 24. $L$

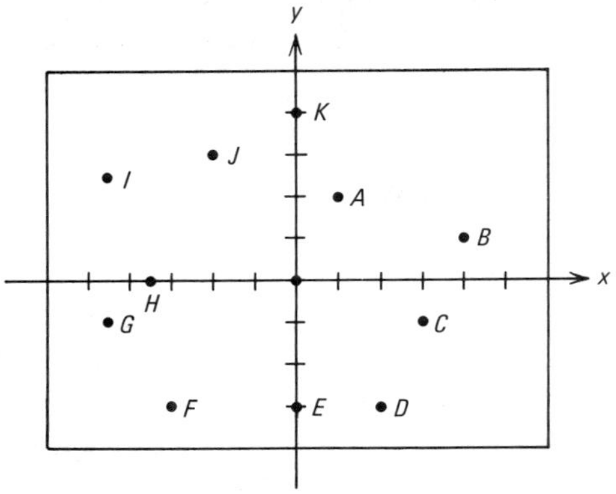


**Figure 8.6**

For the following equations, identify the $x$- and $y$-intercepts; then graph.

*See Examples 3 and 4*

25. $x + y = 6$
26. $x + y = -2$
27. $x - y = 7$
28. $x - y = -4$
29. $2x + y = 10$
30. $x - 3y = 6$
31. $2x - 3y = 6$
32. $4x + 3y = 12$
33. $5x + 2y = 13$
34. $2x - 7y = 6$
35. $3x - 2x = 8$
36. $6x + 5y = 8$
37. $x - y = 0$
38. $x + y = 0$
39. $2x + y = 0$
40. $x - 2y = 0$

Graph the following equations of lines.

*See Example 5*

41. $y = 7$
42. $y = -6$
43. $x = 2$
44. $x = 4$
45. $x = 0$
46. $y = 0$

*Physical application*

47. The relation between hydrostatic pressure $p$ (in pounds per square inch) and water depth $d$ is given by $33p + 15d = 495$. Graph the equation (with $p$ as the $x$-axis and $d$ as the $y$-axis).

*Health application*

48. A patient in a hospital must get 1 milligram of thiamine. One ounce of prune juice provides 0.05 mg thiamine; 1 ounce of orange juice provides

0.08 mg. Let $x$ be the amount of prune juice and $y$ be the amount of orange juice. Then to get 1 mg of thiamine from these juices, we have the relation $0.05x + 0.08y = 1$. Graph this equation. (*Hint:* First multiply both sides by 100 to remove the decimals.)

*Business applications*

49. A company buys a $36,000 truck. As the truck ages and depreciates, its value $v$ is related to its age $a$ by $4000a + v = 36{,}000$. Graph this equation with $a$ on the $x$-axis and $v$ on the $y$-axis. (*Note:* the axes will be scaled very differently, so find the intercepts first.)

50. The relation between the price $p$ of an item and the quantity $q$ sold is given by $q + 1000p = 6000$. Graph this equation. (See the note in Problem 49.)

## 8.2 DISTANCE AND SLOPE

If we have two points in the plane, we often ask the following questions:

1. What is the distance between the points?
2. What is the steepness of the line through the points?

To find the **distance** between points, we use the Pythagorean theorem, which states the following: for right triangles, if $a$ and $b$ are the shorter legs and $c$ is the longer hypotenuse, then $c^2 = a^2 + b^2$, or $c = \sqrt{a^2 + b^2}$ (Fig. 8.7).

Now suppose that we have two points, $(x_1, y_1)$ and $(x_2, y_2)$. See Fig. 8.8. (It does not matter which point is labeled with the "1" subscript and which with the "2".) We let $a = x_2 - x_1$ and $b = y_2 - y_1$. Here, $a$ is the change

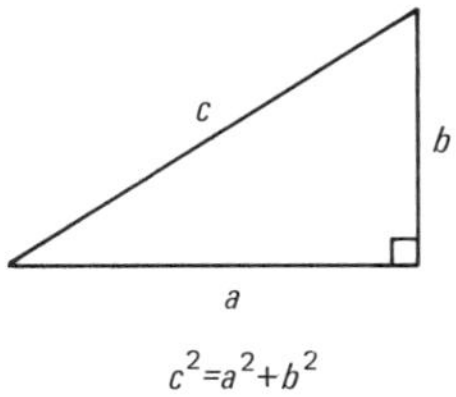


$c^2=a^2+b^2$

Figure 8.7

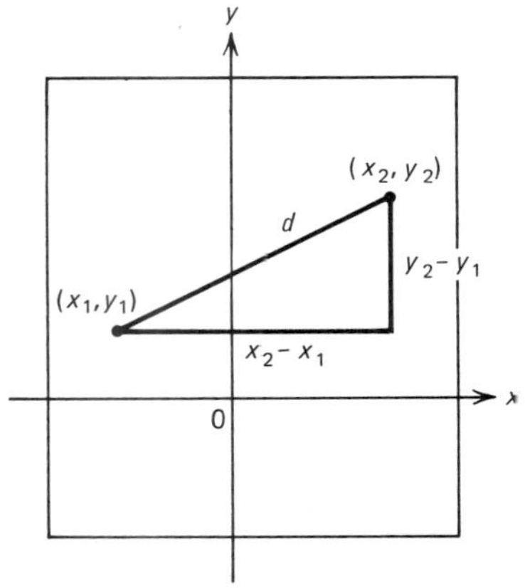


Figure 8.8

in the $x$-coordinate, $b$ is the change in the $y$-coordinate, and $d$ is the distance between the points. Now we use the Pythagorean Theorem.

| | |
|---|---|
| Pythagorean Theorem | $c^2 = a^2 + b^2$ |
| ↓ Substitute | $d^2 = (x_2 - x_1)^2 + (y_2 - y_1)^2$ |
| ↓ Take square root | $d = \sqrt{(x_2 - x_1)^2 + (y_2 - y_1)^2}$ |

This gives us our distance theorem.

> ***THEOREM 8.1*** For any points $(x_1, y_1)$ and $(x_2, y_2)$ in the plane, the distance $d$ between them is given by
>
> $$d = \sqrt{(x_2 - x_1)^2 + (y_2 - y_1)^2}$$

**EXAMPLE 6** Find the distance between the points $(-1, 3)$ and $(3, 6)$.

**SOLUTION** See Fig. 8.9. We let $(-1, 3) = (x_1, y_1)$ and $(3, 6) = (x_2, y_2)$. Now we use Theorem 8.1. (Be careful with the subtraction.)

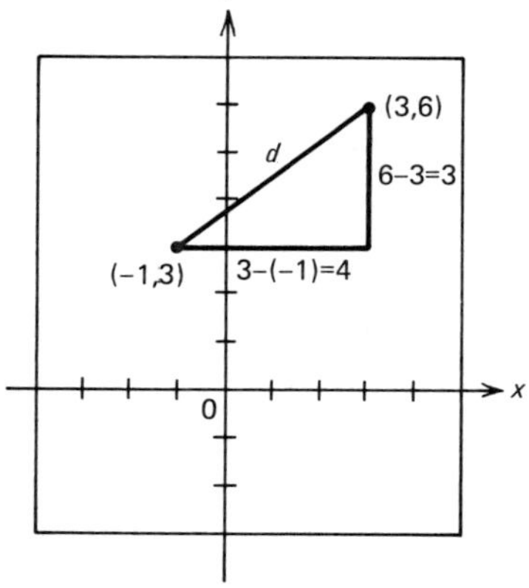


Figure 8.9

| | |
|---|---|
| Theorem 8.1 | $d = \sqrt{(x_2 - x_1)^2 + (y_2 - y_1)^2}$ |
| ↓ Substitute | $= \sqrt{(3 - (-1))^2 + (6 - 3)^2}$ |
| ↓ Simplify | $= \sqrt{4^2 + 3^2} = \sqrt{25} = 5$ |

Thus, the distance between the points is 5.

| YES | NO |
|---|---|
| $\sqrt{4^2 + 3^2} = \sqrt{25} = 5$ | ~~$\sqrt{4^2 + 3^2} = 4 + 3 = 7$~~ |

We must simplify the inside term first.

**EXAMPLE 7** Find the distance between $(1, -3)$ and $(-2, -9)$.

**SOLUTION** We substitute these values into Theorem 8.1 and subtract carefully.

| | |
|---|---|
| Theorem 8.1 | $d = \sqrt{(x_2 - x_1)^2 + (y_2 - y_1)^2}$ |
| ↓ Substitute | $= \sqrt{(-2 - 1)^2 + (-9 - (-3))^2}$ |
| ↓ Simplify | $= \sqrt{(-3)^2 + (-6)^2}$ |
| ↓ Simplify | $= \sqrt{9 + 36} = \sqrt{45} = 3\sqrt{5}$ |

Thus, the distance between the points is $3\sqrt{5}$.

The steepness of a line is measured by a quantity called the **slope** (written ***m***), which is defined as follows.

> **DEFINITION 8.1** For points $(x_1, y_1)$ and $(x_2, y_2)$ in the plane:
>
> $$m = \text{Slope} = \frac{\text{Rise}}{\text{Run}} = \frac{y_2 - y_1}{x_2 - x_1}$$

The slope compares the change in $y$ (the vertical rise) to the change in $x$ (the horizontal run).

**EXAMPLE 8** Find the slopes of the lines shown in Fig. 8.10.

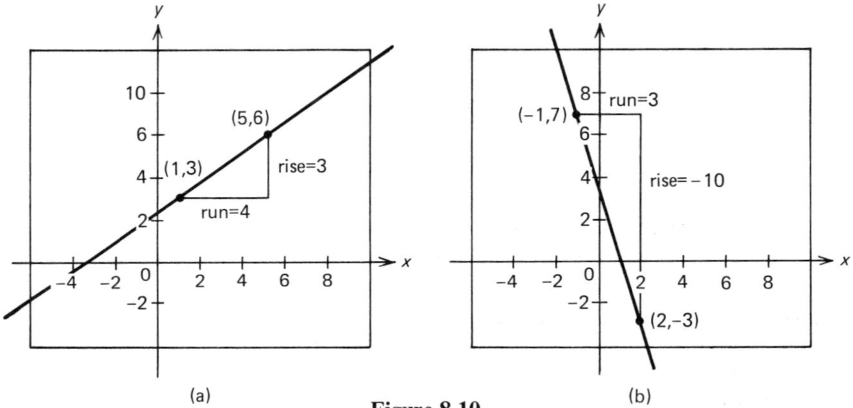


**Figure 8.10**

**SOLUTION** In Fig. 8.10(a), we see that $x_2 - x_1 = 5 - 1 = 4$ and $y_2 - y_1 = 6 - 3 = 3$. Therefore,

| Definition 8.1 |
|---|

$$m = \frac{\text{Rise}}{\text{Run}} = \frac{y_2 - y_1}{x_2 - x_1} = \frac{3}{4}$$

In Fig. 8.10(b), we see that $x_2 - x_1 = 2 - (-1) = 3$ and $y_2 - y_1 = -3 - 7 = -10$. Therefore,

| Definition 8.1 |
|---|

$$m = \frac{\text{Rise}}{\text{Run}} = \frac{y_2 - y_1}{x_2 - x_1} = \frac{-10}{3}$$

Notice that the first line went up to the right and had a positive slope, while the second line went down and had a negative slope.

**EXAMPLE 9** Find the slopes of the lines shown in Fig. 8.11 (on page 265).

**SOLUTION** In Fig. 8.11(a), we see $x_2 - x_1 = 3 - (-1) = 4$, but $y_2 - y_1 = 3 - 3 = 0$ ($y$ does not change). Therefore,

| Definition 8.1 |
|---|

$$m = \frac{\text{Rise}}{\text{Run}} = \frac{0}{4} = 0$$

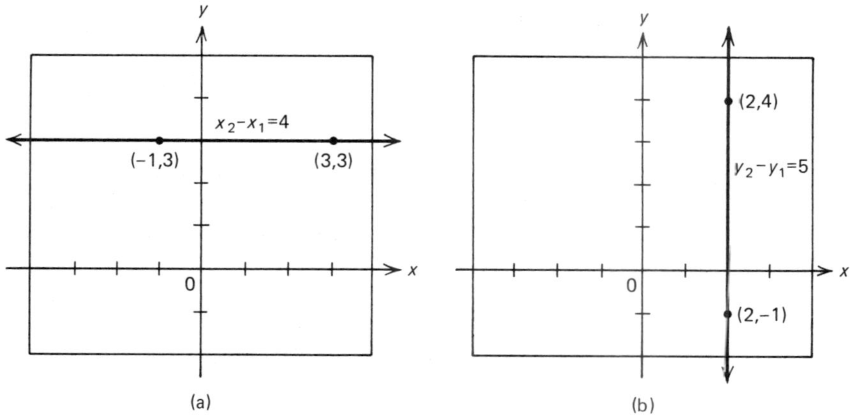


**Figure 8.11**

In Fig. 8.11(b), we see $y_2 - y_1 = 4 - (-1) = 5$, but $x_2 - x_1 = 2 - 2 = 0$ ($x$ does not change). Therefore,

Definition 8.1

$$m = \frac{\text{Rise}}{\text{Run}} = \frac{y_2 - y_1}{x_2 - x_1}$$

$$= \frac{5}{0} \quad \text{is undefined}$$

The horizontal line in Fig. 8.11(a) has a slope zero (as do all horizontal lines). The vertical line in Fig. 8.11(b) has no slope (as is the case with all vertical lines) since we cannot divide by zero.

We summarize these examples in Table 8.1.

**Table 8.1**

| LINE | SLOPE |
|---|---|
| Rising (left to right) | Positive |
| Falling (left to right) | Negative |
| Horizontal | Zero |
| Vertical | Undefined |
| Steep | Large |
| Gradual | Small |

Finally, we have the following relation that can be shown about slopes.

***THEOREM 8.2*** Two lines are parallel if and only if they have the same slope: $m_1 = m_2$.

> ***THEOREM 8.3*** Two lines are perpendicular if and only if their slopes satisfy $m_1 m_2 = -1$ or $m_1 = \dfrac{-1}{m_2}$.

Figure 8.12 shows how these theorems appear. In Fig. 8.12(a), the parallel lines all have the same slope, $\frac{3}{2}$. In Fig. 8.12(b), the perpendicular lines have slopes that are the negative reciprocals of each other, $\frac{7}{4}$ and $-\frac{4}{7}$.

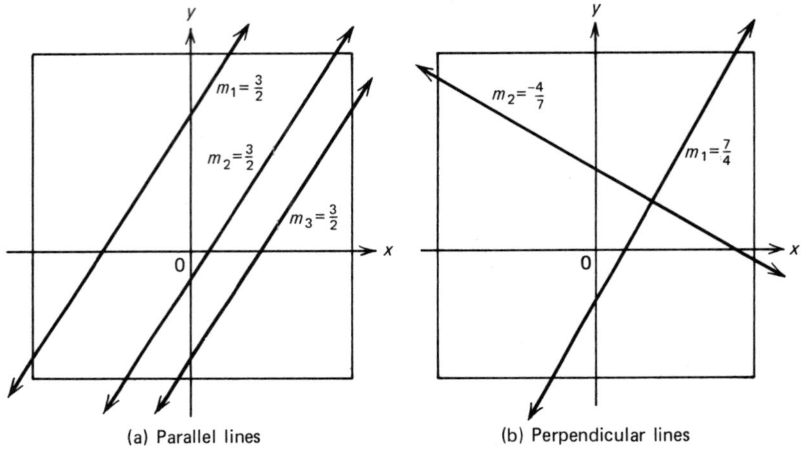


**Figure 8.12**

**EXAMPLE 10** Graph the points (1, 2), (4, 3), and (3, 6). Find the slopes of all the line segments. Are any two of them parallel or perpendicular?

**SOLUTION** The graph appears in Fig. 8.13 (on page 267).

We use Definition 8.1 to find the slopes.

| Slope of $l_1$ | $m_1 = \dfrac{3-2}{4-1} = \dfrac{1}{3}$ |
|---|---|
| Slope of $l_2$ | $m_2 = \dfrac{3-6}{4-3} = \dfrac{-3}{1} = -3$ |
| Slope of $l_3$ | $m_3 = \dfrac{6-2}{3-1} = \dfrac{4}{2} = 2$ |

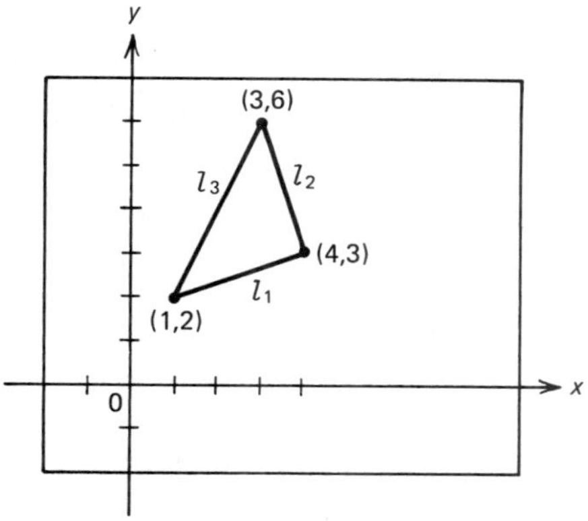


**Figure 8.13**

None of these three are equal, so none are parallel. Notice that $l_1$ and $l_2$ have slopes $\frac{1}{3}$ and $-3$, which are negative reciprocals. Thus, $l_1$ is perpendicular to $l_2$.

## PROBLEM SET 8.2

Find the distances between the following points.

*See Examples 6 and 7*

1. (1, 6) and (3, 10)
2. (2, 8) and (5, 20)
3. (3, 7) and (1, 15)
4. (4, −6) and (2, 2)
5. (−10, 3) and (1, −7)
6. (−2, 5) and (−6, −1)
7. (5, 8) and (10, 8)
8. (6, −2) and (5, −8.5)
9. (1, 9) and (1, −8)
10. (6, −4) and (−1, −4)
11. (−3, 8) and (−3, 2)
12. (9, −3) and (6, 0)

Find the slopes of the line segments through the following points.

*See Examples 8 and 9*

13. (1, 6) and (3, 10)
14. (2, 8) and (5, 20)
15. (3, 7) and (1, 15)
16. (4, −6) and (2, 2)
17. (−10, 3) and (1, −7)
18. (−2, 5) and (−6, −1)
19. (5, 8) and (10, 8)
20. (6, −2) and (5, −8.5)
21. (1, 9) and (1, −8)
22. (6, −4) and (−1, −4)
23. (−3, 8) and (−3, 2)
24. (9, −3) and (6, 0)

For each set of four points in the problems below, determine if any pair of lines through points $A$, $B$, $C$, and $D$ are parallel or perpendicular. (*Hint:* Sketch the points first and look for likely possibilities.)

*See Example 10*

25. $A = (1, 5)$, $B = (3, 7)$, $C = (6, 2)$, and $D = (9, 5)$
26. $A = (3, 9)$, $B = (6, 1)$, $C = (8, 0)$, and $D = (16, 3)$
27. $A = (1, -3)$, $B = (5, 1)$, $C = (5, -1)$, and $D = (11, 4)$
28. $A = (2, 8)$, $B = (-6, 3)$, $C = (4, -2)$, and $D = (-8, -3)$
29. $A = (1, -5)$, $B = (-2, 4)$, $C = (8, 1)$, and $D = (11, 2)$
30. $A = (6, -4)$, $B = (0, 3)$, $C = (1, -2)$, and $D = (-7, 5)$

*Business applications*

The *marginal profit* is basically the extra profit brought in by manufacturing extra items. It is given by the formula

$$\text{MP} = \frac{P_2 - P_1}{Q_2 - Q_1}$$

Calculate the marginal revenue MP for the following sets of $(Q, P)$ pairs, where $Q$ is the quantity produced and $P$ the profit brought in.

31. (100, \$7000) and (120, \$8300)
32. (2000, \$10,000) and (2300, \$9400)
33. (85, \$550) and (90, \$530)
34. (6400, \$120,000) and (6500, \$125,000)

*Science application*

The average velocity of a moving object is given by

$$v = \frac{s_2 - s_1}{t_2 - t_1}$$

where $s$ is the distance an object travels in time $t$. Calculate the average velocity for the following sets of $(t, s)$ pairs.

35. (10 sec, 250 m) and (14 sec, 290 m)
36. (6 min, 1400 ft) and (6.5 min, 1500 ft)
37. (5 hr, 240 mi) and (6.5 hr, 305 mi)
38. (1 day, 640 mi) and (1.5 day, 900 mi)

## 8.3 THE EQUATION OF A STRAIGHT LINE

The line is a concept from geometry. As we will see, when we put a line on a Cartesian coordinate plane, there is always a simple equation that relates the $x$- and $y$-coordinates. We need *both* the slope and a point on the line. Figure 8.14(a) and (b) shows what happens if we have only the slope or only a point. There are infinitely many such lines. Figure 8.14(c) shows that with *both* the slope and a point we get only one line.

What is the equation of the line through the given point $(x_1, y_1)$ and with slope $m$? In other words, if we have a typical point $(x, y)$ on the line, what is the relation between the coordinates $x$ and $y$? To answer this we look at Fig. 8.15. This shows a line with slope $m$ and through the given point $(x_1, y_1)$ and the typical point $(x, y)$.

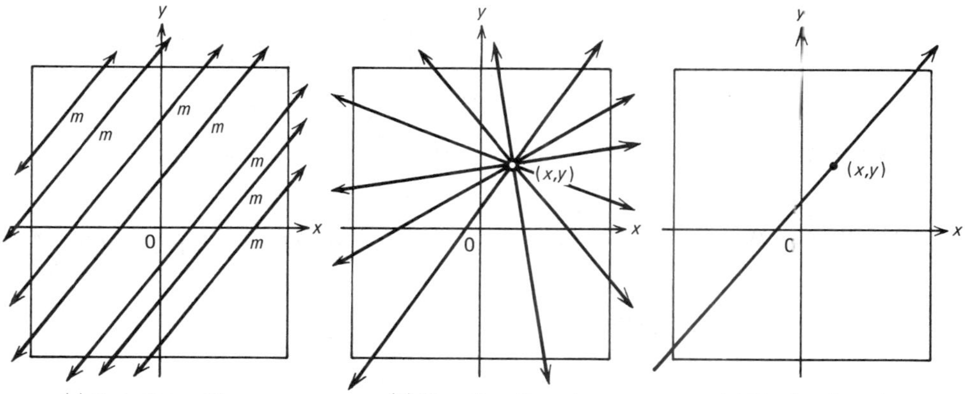


(a) Many lines with the same slope, $m$

(b) Many lines through the same point, $(x,y)$

(c) *The* line through $(x,y)$ with slope $m$

**Figure 8.14**

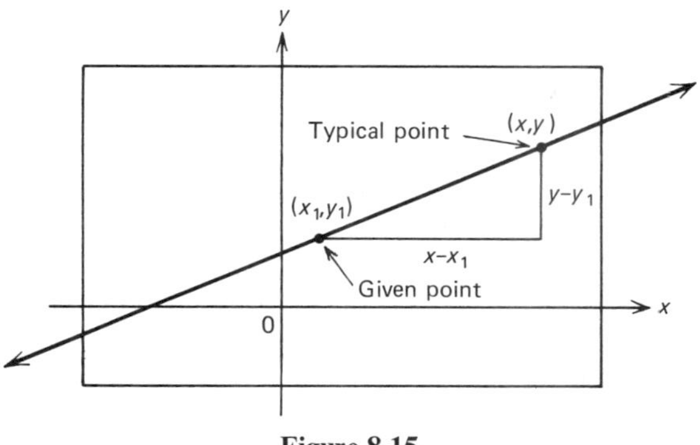


**Figure 8.15**

We compute the slope using the slope formula.

| | |
|---|---|
| Definition 8.1 | $m = \dfrac{y - y_1}{x - x_1}$ |
| ↓ Multiply both sides by $(x - x_1)$ | $m(x - x_1) = y - y_1$ |

This becomes a very valuable theorem.

> ***THEOREM 8.4*** The equation of the line with slope $m$ through point $(x_1, y_1)$ is given by
>
> $y - y_1 = m(x - x_1)$
>
> This form is called the **point-slope form** of a line because a point and the slope are given.

**EXAMPLE 11** Find the equation of the line with slope 3 through point $(1, -4)$.

**SOLUTION** We simply substitute these values into Theorem 8.4. (Be careful with the subtractions.)

| | |
|---|---|
| Identify terms | $m = 3;\quad x_1 = 1;\quad y_1 = -4$ |
| ↓ Theorem 8.4 | $y - y_1 = m(x - x_1)$ |
| ↓ Substitute | $y - (-4) = 3(x - 1)$ |
| ↓ Simplify | $y + 4 = 3x - 3$ |
| ↓ Subtract 4 from both sides | $y = 3x - 7$ |

Notice that this is a slightly different form than $ax + by = c$. In general, it is as follows.

> **DEFINITION 8.2** $y = mx + b$
>
> is the **slope-intercept form** of a line. It is called this since $m$ (the coefficient of $x$) is the slope, and $b$ is the $y$-intercept. See Fig. 8.16.

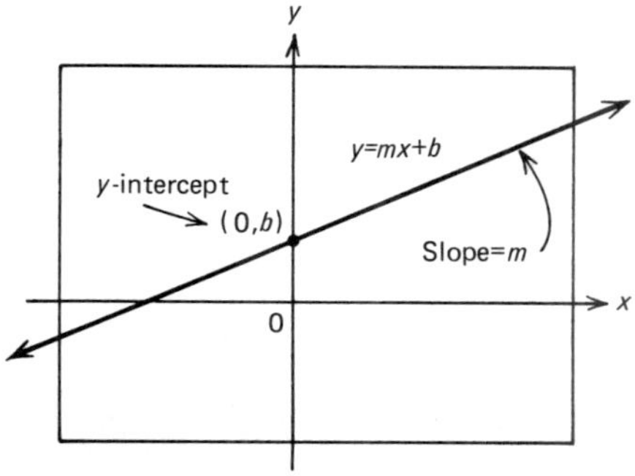


**Figure 8.16**

**EXAMPLE 12** Find the equation of the line through (1, 9) and (3, 4). Leave the answer in both forms.

**SOLUTION** To find the equation we need a point and the slope. We have a point, so we must find the slope. Then we substitute into Theorem 8.4, with the given point $(x_1, y_1) = (1, 9)$.

| | |
|---|---|
| Definition 8.1 | $m = \dfrac{y_2 - y_1}{x_2 - x_1} = \dfrac{4 - 9}{3 - 1} = \dfrac{-5}{2}$ |
| ↓ Theorem 8.4 | $y - y_1 = m(x - x_1)$ |
| ↓ Substitute | $y - 9 = -\dfrac{5}{2}(x - 1)$ |
| ↓ Distributive law | $y - 9 = -\dfrac{5}{2}x + \dfrac{5}{2}$ |
| ↓ Add $9 = \dfrac{18}{2}$ to both sides | $y = -\dfrac{5}{2}x + \dfrac{23}{2}$ |

Notice that we used (1, 9) as the point $(x_1, y_1)$ in the formula. Could we have used (3, 4) as the point? Yes, and we would get the same answer. To put this equation into standard form, we multiply both sides by 2 and move the $x$-term to the left.

| | |
|---|---|
| Multiply both sides by 2 | $2[y] = 2\left[\frac{-5}{2}x + \frac{23}{2}\right]$ |
| ↓ Distributive law | $2y = -5x + 23$ |
| ↓ Add $5x$ to both sides | $5x + 2y = 23$ |

We can use the slope-intercept form to help find the slope and $y$-intercept of a line.

> To find the slope and $y$-intercept of a line:
>
> 1. Write the equation in slope-intercept form, $y = mx + b$.
> 2. The slope is the coefficient $m$.
> 3. The $y$-intercept is $b$.

**EXAMPLE 13** **(a)** For $y = 6x - 3$, the slope is 6. The $y$-intercept is $-3$.

**(b)** For $y = \frac{-7}{3}x + 8$, the slope is $\frac{-7}{3}$. The $y$-intercept is 8.

**(c)** For $y = 4$, the slope is 0. The $y$-intercept is 4.

**EXAMPLE 14** Find the slope and $y$-intercept of $2x + 7y = 3$.

**SOLUTION** To find the slope, we put this in the form $y = mx + b$. To do this, we solve for $y$.

| | |
|---|---|
| Given | $2x + 7y = 3$ |
| ↓ Subtract $2x$ from both sides | $7y = -2x + 3$ |
| ↓ Divide both sides by 7 | $y = \frac{-2}{7}x + \frac{3}{7}$ |

Therefore, the slope is $\frac{-2}{7}$, the coefficient of $x$, and the $y$-intercept is $\frac{3}{7}$.

**EXAMPLE 15** Find the equation of the line through (1, 6) and perpendicular to $3x + 5y = 7$.

**SOLUTION** To find the equation of any line, we need a *point* and the *slope*. We have a point, (1, 6), so we need the slope. We know that the line we are seeking is perpendicular to $3x + 5y = 7$, so the slope we need will be the negative reciprocal of the slope $3x + 5y = 7$.

| | |
|---|---|
| Given line | $3x + 5y = 7$ |
| ↓ Subtract $3x$ from both sides | $5y = -3x + 7$ |
| ↓ Divide both sides by 5 | $y = \frac{-3}{5}x + \frac{7}{5}$ |

Therefore, the slope of the given line is $\frac{-3}{5}$. Thus, the slope of the desired perpendicular is $\frac{5}{3}$, the negative reciprocal. Now we use Theorem 8.4.

| | |
|---|---|
| Identify terms | $m = \frac{5}{3}$; $(x_1, y_1) = (1, 6)$ |
| ↓ Substitute into Theorem 8.4 | $y - 6 = \frac{5}{3}(x - 1)$ |
| ↓ Distributive law | $y - 6 = \frac{5}{3}x - \frac{5}{3}$ |
| ↓ Add $6 = \frac{18}{3}$ to both sides | $y = \frac{5}{3}x + \frac{13}{3}$ |

**EXAMPLE 16** Find the equation of the line parallel to $2x + 3y = 8$ and through (1, 4).

**SOLUTION** We have a point, so we must now find the slope.

| | |
|---|---|
| Given | $2x + 3y = 8$ |
| Subtract $2x$ from both sides | $3y = -2x + 8$ |
| Divide both sides by 3 | $y = \frac{-2}{3}x + \frac{8}{3}$ |

Since the slope of the given line is $\frac{-2}{3}$, the slope of the parallel line that we are seeking is also $\frac{-2}{3}$. Now we use Theorem 8.4.

| | |
|---|---|
| Identify terms | $m = \frac{-2}{3}$ and $(x_1, y_1) = (1, 4)$ |
| Substitute in Theorem 8.4 | $y - 4 = \frac{-2}{3}(x - 1)$ |
| Distributive law | $y - 4 = \frac{-2}{3}x + \frac{2}{3}$ |
| Add $4 = \frac{12}{3}$ to both sides | $y = \frac{-2}{3}x + \frac{14}{3}$ |

If we multiply both sides of the equation by 3 and rearrange terms, we get $2x + 3y = 14$.

## PROBLEM SET 8.3

Find the equations of the lines with the following slopes and points.

*See Example 11*

1. $m = 5$; $(3, 7)$
2. $m = 6$; $(4, 2)$
3. $m = 10$; $(1, -3)$
4. $m = 2$; $(-4, 0)$
5. $m = 6$; $(1, 8)$
6. $m = -3$; $(5, 1)$
7. $m = -1$; $(8, -3)$
8. $m = -5$; $(-2, -2)$
9. $m = \frac{2}{5}$; $(6, -1)$
10. $m = \frac{-7}{2}$; $(-5, 2)$

Find the equations of the lines through the following pairs of points.

*See Example 12*

11. (1, 6) and (3, 10)
12. (2, 8) and (5, 20)
13. (3, 7) and (1, 15)
14. (4, −6) and (2, 2)
15. (−10, 3) and (1, −7)
16. (−2, 5) and (−6, −1)
17. (5, 8) and (10, 8)
18. (6, 2) and (5, −8)
19. (6, −4) and (−1, −4)
20. (9, −3) and (6, 0)

For each of the following equations, find the slope and $y$-intercept of the corresponding line.

*See Example 13*

21. $y = 7x - 5$
22. $y = 4x + 6$
23. $y = -6x - 2$
24. $y = -4x + 5$
25. $y = \frac{2}{3}x - \frac{7}{3}$
26. $y = \frac{-3}{10}x + 12$
27. $y = 6$
28. $y = -14$

*See Example 14*

29. $2x + y = 5$
30. $6x + 2y = 13$
31. $6x + 5y = 2$
32. $3x - 4y = 8$
33. $9x - y = 7$
34. $12x - 5y = 16$
35. $2x + 5y - 7 = 0$
36. $8x - 3y - 12 = 0$

Find the equations of the lines with the following conditions.

*See Examples 15 and 16*

37. Parallel to $y = 7x - 5$, through (6, 2)
38. Parallel to $y = -3x + 4$, through (7, −2)
39. Perpendicular to $y = 2x - 6$, through (1, 7)
40. Perpendicular to $y = -\frac{1}{3}x - 8$, through (−2, −3)
41. Parallel to $2x - y = 12$, through (−3, 1)
42. Parallel to $5x + 2y = 6$, through (−2, 5)
43. Perpendicular to $4x + 5y = 12$, through (−3, 0)
44. Perpendicular to $3x - 7y = 1$, through (1, −2)

*Ecology applications*

45. In a certain European town, the weekly death rate $r$ is related to the sulfur dioxide air content $C$. Find the slope and the equation of the line through the $(C, r)$ data points: (129 $\mu g/m^3$, 98 deaths) and (323 $\mu g/m^3$, 104 deaths).

46. In a certain city, the air pollution is 10 ppm at 7 A.M., increases to 146 ppm at 3 P.M., and decreases back to 10 ppm at 7 A.M. the next day. Set 7 A.M. = 7, 3 P.M. = 15, and 7 A.M. the next day = 31. Find the equations of the line segments between (7, 10) and (15, 146) and between (15, 146) and (31, 10). (Let $x$ be the time and $y$ the pollution.) Graph both of these segments together.

*Business applications*

47. A company finds it can sell 10,000 shirts at \$7 each and 6000 of the same shirt at \$11 each. Find the equation of the line segment between (10,000 shirts, \$7) and (6000 shirts, \$11). This is called a *demand line.*

48. At the start of a loan, a business owes \$8000. After 12 payments, they owe \$5000. Find the equation of the line between (0 payments, \$8000) and (12 payments, \$5000).

*Consumer application*

49. To run her car, Ms. Morison notices that it cost \$1350 to drive 5000 miles a year, and \$1800 to drive 10,000 miles per year. Find the equation of the line segment between (5000 miles, \$1350) and (10,000 miles, \$1800).

## 8.4 LINEAR INEQUALITIES

Recall that $ax + by = c$ is a linear equation. In this section, we study **linear inequalities** which have the standard forms

$$ax + by < c \qquad ax + by > c$$

$$ax + by \leq c \qquad ax + by \geq c$$

Just as $x < 2$ is a half-line cut at the point $x = 2$, a linear inequality such as $2x + 5y \leq 10$ is a half-plane cut by the line $2x + 5y = 10$. We now see how to graph these linear inequalities.

To graph a linear inequality:

1. Graph the corresponding line, $ax + by = c$.
2. Determine if a test point $P$ (not on the line) satisfies the inequality. [Any point will do, but (0, 0) is usually a very convenient point if it is not on the line.]

3a. If $P$ satisfies the inequality, shade the region of the plane *containing* $P$.

> 3b. If $P$ does not satisfy the inequality, shade the region of the plane *not containing P*.
>
> 4a. If the inequality is $\leq$ or $\geq$, include the line in the graph.
>
> 4b. If the inequality is $<$ or $>$, exclude the line from the graph (use a dashed line).

**EXAMPLE 17** Graph $2x + 5y \leq 10$.

**SOLUTION** We start by graphing $2x + 5y = 10$.

Find intercepts

| $x$ | $y$ |
|---|---|
| 0 | 2 |
| 5 | 0 |

Graph the line $2x + 5y = 10$

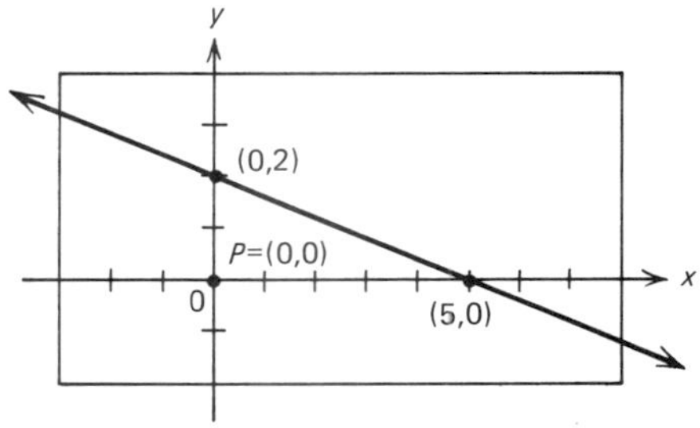


Choose $P = (0, 0)$ not on line

Check $P = (0, 0)$ in inequality

Is $2(0) + 5(0) \leq 10$? YES

"YES" means shade region containing $P$

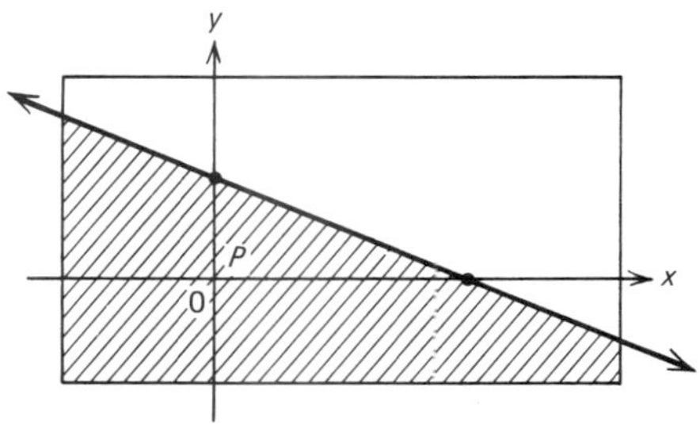


$\leq$ means include line

**EXAMPLE 18** Graph $3x + 4y > 24$.

**SOLUTION** Again, we first graph the line $3x + 4y = 24$.

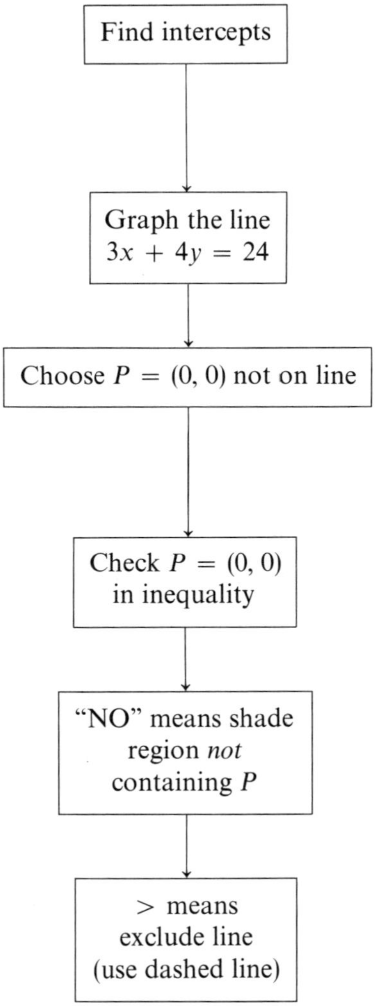


| $x$ | $y$ |
|---|---|
| 0 | 6 |
| 8 | 0 |

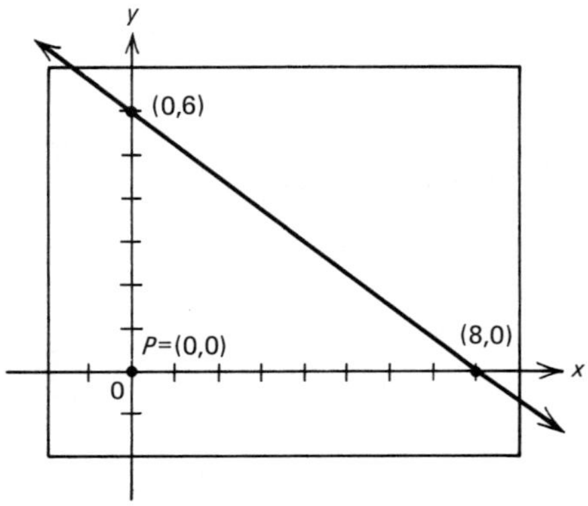


Is $3(0) + 4(0) > 24$? NO

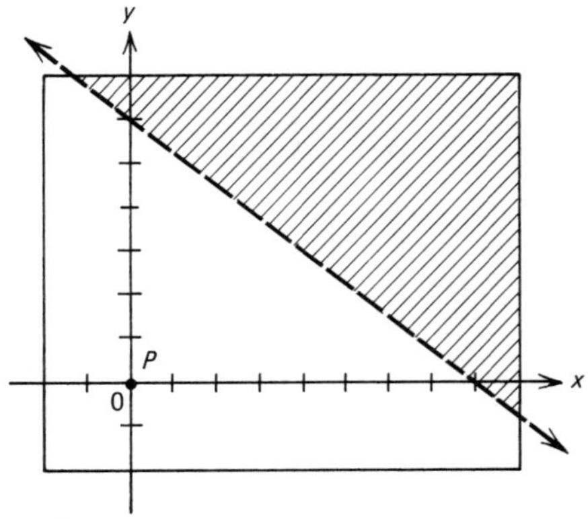


**EXAMPLE 19** Graph $y \leq 2x$.

**SOLUTION** We graph the line $y = 2x$ by picking $x$-values and solving for the $y$-values.

Choose $x$-values, solve for $y$-values

| $x$ | $y$ |
|---|---|
| 0 | 0 |
| 1 | 2 |
| 2 | 4 |

↓

Graph the line $y = 2x$

↓

Choose $P = (5, 2)$ not on line

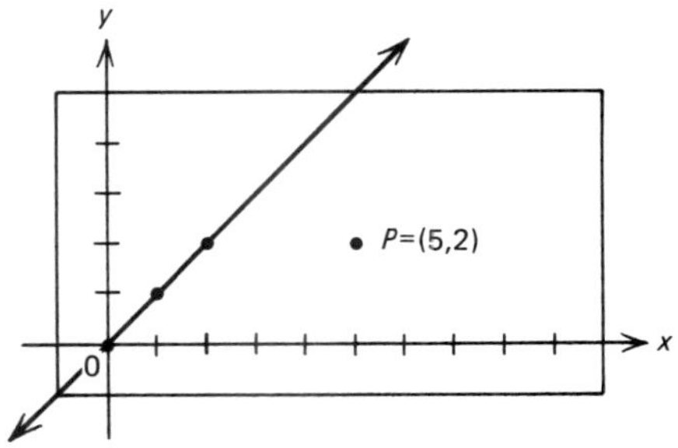


↓

Check $P = (5, 2)$ in inequality

Is $2 \geq 2(5)$? NO

↓

"NO" means shade region *not* containing $P$

↓

$\geq$ means include line

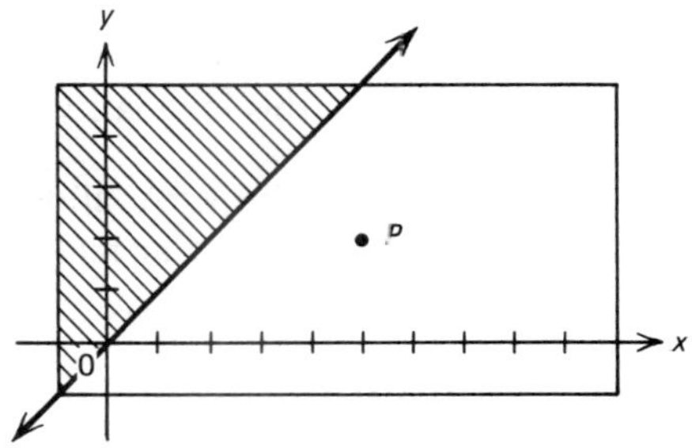


Notice that we had to let $P$ be another point, such as (5, 2), since (0, 0) was already on the line.

## PROBLEM SET 8.4

Graph the following linear inequalities.

*See Examples 17 and 18*

1. $x + y < 5$
2. $x - y > 4$
3. $2x + 3y \geq 6$
4. $4x + 2y \leq 16$
5. $x - 5y > 10$
6. $6x + y < 18$

| | | |
|---|---|---|
| 7. $5x + 3y \leq 15$ | 8. $7x - 2y \geq 14$ | 9. $4x - 3y < 8$ |
| 10. $2x + 5y > 11$ | 11. $6x + 7y \geq 1$ | 12. $x - 6y \leq 2$ |
| 13. $y > x$ | 14. $y < 6x$ | 15. $y \leq 2x + 1$ |
| 16. $y \geq 1 - x$ | 17. $y < 3$ | 18. $y > -4$ |
| 19. $y \geq 1$ | 20. $y \leq 10$ | 21. $x > -5$ |
| 22. $x < -6$ | 23. $x \leq 2$ | 24. $x \geq -1$ |

*See Example 19*

*Health application*

25. A patient is not allowed to have over 330 milligrams of cholesterol from his diet of eggs and liver. The relation is

$$165x + 90y \leq 330$$

where $x$ is the number of eggs and $y$ the amount of liver in ounces. Graph this inequality.

26. In order to get at least 1 milligram of thiamine (see page 260), a patient's intake of prune and orange juice must satisfy

$$0.05x + 0.08y \geq 1$$

where $x$ is ounces of prune juice and $y$ is ounces of orange juice. Graph this inequality.

*Ecology application*

27. The EPA allows a manufacturing plant no more than 8000 grams of particular matter. Product 1 produces 50 grams per gallon while product 2 produces 80 grams per gallon. The relation is

$$50x + 80y \leq 8000$$

where $x$ is gallons of product 1 and $y$ is gallons of product 2. Graph this inequality.

*Business application*

28. A company has 1200 employee-hours to use to manufacture a new line of chairs and desks. A chair takes 2 hours to make, and the desk takes 6 hours to make. They satisfy the relation

$$2x + 6y \leq 1200$$

where $x$ is the number of chairs and $y$ the number of desks. Graph this inequality.

## 8.5 VARIATION

In science, health, and business, we see two types of variation between quantities: direct and inverse variation. We say that $x$ **varies directly** with an expression $t$ if $x$ increases as $t$ increases (and vice versa) and is written as follows.

> **DEFINITION 8.3** *$x$ varies directly* with $t$ if
>
> $$x = kt$$
>
> where $k$ is a **constant of variation.**

We say that $x$ **varies inversely** with an expression $t$ if $x$ decreases as $t$ increases (and vice versa) and is written as follows.

> **DEFINITION 8.4** *$x$ varies inversely* with $t$ if
>
> $$x = \frac{k}{t}$$

**EXAMPLE 20** The following are examples of direct variation.

| STATEMENT | EQUATION |
|---|---|
| $x$ varies directly with $t$ | $x = kt$ |
| $y$ varies directly with the square root of $x$ | $y = k\sqrt{x}$ |
| $z$ varies directly with the cube of $w$ | $z = kw^3$ |

**EXAMPLE 21** The following are examples of inverse variation.

| STATEMENT | EQUATION |
|---|---|
| $x$ varies inversely with $y$ | $x = \frac{k}{y}$ |
| $F$ varies inversely with the square of $r$ | $F = \frac{k}{r^2}$ |
| $t$ varies inversely with the cube root of $s$ | $t = \frac{k}{\sqrt[3]{s}}$ |

> To find the constant of variation, $k$:
>
> 1. Translate the statement into an algebraic equation.
> 2. Substitute given information about the quantities into the equation.
> 3. Solve for $k$.

**EXAMPLE 22** According to Boyle's law, the pressure of a gas varies inversely to its volume. If the volume is 50 and the pressure is 25, find the constant of variation.

**SOLUTION** Let $p$ = pressure and V = volume. We translate the statement into an equation and then substitute V = 50 and $p$ = 25.

| | |
|---|---|
| Given | Pressure varies inversely to volume |
| ↓ Translate | $p = \dfrac{k}{V}$ |
| ↓ Substitute | $25 = \dfrac{k}{50}$ |
| ↓ Multiply both sides by 50 | $1250 = k$ |

Therefore, $k = 1250$, and the relation is $p = \dfrac{1250}{V}$.

**EXAMPLE 23** For a certain product, the cost varies directly as the square root of the number produced. If it costs \$1600 to produce 400 items, how much will it cost to produce 10,000 items?

**SOLUTION** Let $C$ = cost and $x$ = number produced. We first translate and solve for $k$ using $C = 1600$ and $x = 400$. Then we substitute $x = 10{,}000$ into this equation and find $C$.

| | |
|---|---|
| Given | Cost varies directly as square root of number |
| Translate | $C = k\sqrt{x}$ |
| Substitute | $1600 = k\sqrt{400}$ |
| Simplify | $1600 = k \cdot 20$ |
| Divide both sides by 20 | $80 = k$ |
| Rewrite original equation with $k = 80$ | $C = 80\sqrt{x}$ |
| Substitute $x = 10{,}000$ | $C = 80\sqrt{10{,}000}$ |
| Simplify | $C = 80 \quad 100 = 8000$ |

When a quantity varies with two or more other variables, we say that the first variable **varies jointly** with the others.

**EXAMPLE 24** The following are examples of joint variation.

| STATEMENT | EQUATION |
|---|---|
| $x$ varies directly with $y$ and the square of $z$ | $x = kyz^2$ |
| $T$ varies directly with the cube of $x$ and inversely to the square root of $r$ | $T = \dfrac{kx^3}{\sqrt{r}}$ |
| $z$ varies directly with $a$, $b$, and $c$ and inversely to the cube root of $d$ | $z = \dfrac{kabc}{\sqrt[3]{d}}$ |

**EXAMPLE 25** The force of repulsion between two negatively charged particles varies directly with the amount of charge on each and inversely with the square of the distance between them. Two particles of charges 10 and 5 at a distance of 20 repel each other with a force of 1000. Use this to find the force for charges 20 and 15 at a distance of 10.

**SOLUTION** We solve this as we did in Example 23. Let $F$ be force, $Q_1$ and $Q_2$ be the charges, and $r$ be the distance. We use $Q_1 = 10$, $Q_2 = 5$, $r = 20$, and $F = 1000$ to find $k$. Then we substitute $Q_1 = 20$, $Q_2 = 15$, and $r = 10$ to find $F$.

| | |
|---|---|
| Given | Force varies directly with amounts of charges and inversely with square of distance |
| Translate | $F = \frac{kQ_1Q_2}{r^2}$ |
| Substitute | $1000 = \frac{k(10)(5)}{20^2}$ |
| Simplify | $1000 = \frac{50k}{400}$ |
| Multiply by $\frac{400}{50}$ | $8000 = k$ |
| Rewrite original equation with $k = 8000$ | $F = \frac{8000Q_1Q_2}{r^2}$ |
| Substitute | $F = \frac{8000(20)(15)}{10^2}$ |
| Simplify | $F = \frac{2{,}400{,}000}{100} = 24{,}000$ |

## PROBLEM SET 8.5

Translate the following statements into algebraic equations.

*See Example 20*

1. $x$ varies directly as $y$.
2. $t$ varies directly as the square of $r$.
3. $a$ varies directly as the fourth power of $b$.
4. $u$ varies directly as the cube root $v$.

*See Example 21*

5. $z$ varies inversely to $x$.
6. $m$ varies inversely to the cube of $n$.
7. $L$ varies inversely to the square root $g$.
8. $s$ varies inversely to the fifth power of $t$.

*See Example 24*

9. $a$ varies directly with $x$, $y$, and the cube of $z$.
10. $r$ varies directly with the square of $s$ and cube root of $t$.
11. $x$ varies inversely to $a$ and the fourth power of $b$.
12. $u$ varies directly with $v$ and the cube of $w$ and inversely to $x$ and square root of $y$.

Find the constants of variation in the following situations.

*See Example 22*

13. Problem 1 with $x = 20$ and $y = 4$.
14. Problem 2 with $t = 75$ and $r = 5$.
15. Problem 3 with $a = 80$ and $b = 2$.
16. Problem 4 with $u = 20$ and $v = 8$.
17. Problem 5 with $z = 20$ and $x = 2$.
18. Problem 6 with $m = 3$ and $n = 4$.
19. Problem 7 with $L = 6$ and $g = 9$.
20. Problem 8 with $s = 10$ and $t = 2$.

*See Example 25*

21. Problem 9 with $a = 480$, $x = 4$, $y = 5$, and $z = 2$.
22. Problem 10 with $r = 72$, $s = 2$, and $t = 27$.
23. Problem 11 with $x = 3$, $a = 5$, and $b = 3$.
24. Problem 12 with $u = 12$, $v = 3$, $w = 2$, $x = 4$, and $y = 25$.

In the following cases, solve for the indicated variable.

*See Example 23*

25. In Problem 13, find $x$ if $y = 6$.
26. In Problem 14, find $t$ if $r = 2$.
27. In Problem 15, find $a$ if $b = 3$.
28. In Problem 16, find $u$ if $v = 27$.
29. In Problem 17, find $z$ if $x = 5$.
30. In Problem 18, find $m$ if $n = 3$.
31. In Problem 19, find $L$ if $g = 25$.
32. In Problem 20, find $s$ if $t = 1$.

*See Example 25*

33. In Problem 21, find $a$ if $x = 5$, $y = 1$, and $z = 3$.
34. In Problem 22, find $r$ if $s = 3$ and $t = 8$.
35. In Problem 23, find $x$ if $a = 4$ and $b = 2$.
36. In Problem 24, find $u$ if $v = 3$, $w = 1$, $x = 6$, and $y = 9$.

*Business applications*

37. The demand for a quantity varies inversely to its price. At \$0.50 each, the company sold 200,000 pens. How many pens would they sell at \$1.50 each?
38. A company's sales varies directly with the square root of its advertising budget. If a company had 1000 sales with a budget of \$100, how many sales will it have with a budget of \$900?

*Health applications*

39. A child's dosage of medicine varies directly with the child's weight and the adult dosage. Suppose that one medicine with a 2-tablespoon adult dosage called for 1 tablespoon for a 75-pound child. How much of a 3-tablespoon adult dosage should be given to a 50-pound child?
40. The rate at which an infection spreads through a community varies jointly with the number who have the infection and the number who do not. In a 100,000-person town, 1000 are infected and 99,000 are not, and the infection rate is 99 per week. What will the infection rate be when 2000 are infected and 98,000 are not?

*Science applications*

41. The volume of a gas varies directly with its absolute temperature and inversely to its pressure. Suppose that the gas occupies a volume of 20 when the temperature is 300°K and the pressure is 45. Find the volume when the temperature is 350°K and the pressure is 30.
42. The frequency of the pitch of a guitar string varies directly with the square root of the tension and inversely to the length. One string under 121 pounds of tension and 26 inches long has a frequency of 220. What is the frequency if the tension is 144 and the length is 24 inches?

43. The speed at which a falling object hits the ground varies directly with the square root of the height of the fall. If a fall of 9 feet produces a speed of 24 feet per second, what speed is attained after a 64-foot fall?

44. The intensity of light varies inversely with the square of the distance from the source of the light. If the intensity is 60 candle power at a distance of 3 feet, what is the intensity at 4 feet?

45. The critical load of a cylindrical column varies directly as the fourth power of its diameter and inversely to the square of the height. If a 200-centimeter-tall column with a 40-centimeter diameter can support 19,200 kilograms, how much load can a column 30 centimeters in diameter and 300 centimeters in height support?

# CHAPTER 8 SUMMARY

## Important Words and Phrases

Cartesian coordinate system (8.1)
constant of variation (8.5)
coordinate axis (8.1)
direct variation (8.5)
distance (8.2)
graph (introduction)
inverse variation (8.5)
joint variation (8.5)
linear equation (8.1)
linear inequality (8.4)
ordered pair (8.1)
origin (8.1)
parallel lines (8.2)
perpendicular lines (8.2)
point-slope form (8.3)
quadrant (8.1)
rectangular coordinate system (8.1)
slope (8.2)
slope-intercept form (8.3)
standard form (8.1)
$x$-intercept (8.1)
$y$-intercept (8.1)

## Important Definitions and Theorems

For points $(x_1, y_1)$ and $(x_2, y_2)$ in the plane,

$$d = \sqrt{(x_2 - x_1)^2 + (y_2 - y_1)^2}$$

$$m = \frac{y_2 - y_1}{x_2 - x_1} = \frac{\text{Rise}}{\text{Run}}$$

For parallel lines, $m_1 = m_2$

For perpendicular lines, $m_1 = \dfrac{-1}{m_2}$

Point-slope form: $y - y_1 = m(x - x_1)$

Slope-intercept form: $y = mx + b$

For direct variation, $x = kT$

For inverse variation, $x = \frac{k}{T}$

## Important Procedures

*To graph the equation* $ax + by = c$, where $a$, $b$, and $c$ are not zero:

1. Find the $x$- and $y$-intercepts.
2. Find a third point.
3. Plot these points.
4. Draw the line through these points.

*To find the equation of a line:*

1. Determine the slope $m$ and a point $(x_1, y_1)$.
2. Substitute these values into the equation $y - y_1 = m(x - x_1)$.

*To find the slope and y-intercept of a line:*

1. Write the equation in the form $y = mx + b$.
2. The slope is the coefficient $m$.
3. The $y$-intercept is $b$.

*To graph a linear inequality:*

1. Graph the corresponding line $ax + by = c$.
2. Determine if a test point $P$ (not on the line) satisfies the inequality.

3(a). If $P$ satisfies the inequality, shade the region of the plane containing $P$.
3(b). If $P$ does not satisfy the inequality, shade the region of the plane *not* containing $P$.
4(a). If the inequality is $\leq$ or $\geq$, include the line.
4(b). If the inequality is $<$ or $>$, exclude the line.

## CHAPTER 8 REVIEW EXERCISES

1. Locate the point $A = (3, -7)$ on a graph.
2. Locate the point $B = (-6, -4)$ on a graph.
3. Identify the coordinates of point $A$ in Fig. 8.17.
4. Identify the coordinates of point $B$ in Fig. 8.17.

For the equation $2x - 7y = 14$:

5. Find the $x$-intercept.
6. Find the $y$-intercept.

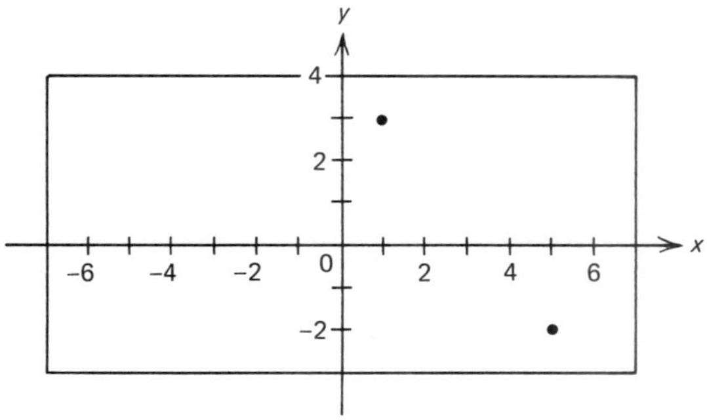


**Figure 8.17**

7. Graph the equation.
8. Graph the equation $x = -4$.

For the points $A = (1, -7)$ and $B = (-2, -3)$:

9. Find the distance between $A$ and $B$.
10. Find the slope of the line through $A$ and $B$.
11. If $A = (1, 2)$, $B = (3, 8)$, $C = (5, 2)$, and $D = (11, 0)$, are the lines through $A$ and $B$ and through $C$ and $D$ parallel, perpendicular, or neither?
12. Find the equation of the line with slope $-8$ and through point $(-1, 6)$.
13. Find the equation of the line through $(1, 4)$ and $(3, -6)$.
14. Find the slope of the line $2x - 3y = 7$.
15. Find the equation of the line through $(2, -3)$ and perpendicular to $4x - 2y = 7$.
16. Graph $2x - 3y \geq 6$.
17. Graph $y < 5x$.
18. Graph $x \leq 5$.

$x$ varies directly with $y$ and the square root $z$ and inversely to the square of $a$ and cube root of $b$.

19. Translate this into an algebraic equation.
20. Find the constant of variation if $x = 21$, $y = 2$, $z = 9$, $a = 1$, and $b = 8$.
21. Find $x$ if $y = 3$, $z = 4$, $a = 2$, and $b = 27$.

# 9

# FUNCTIONS AND SECOND-DEGREE RELATIONS

## 9.1 RELATIONS AND FUNCTIONS

A **relation** is any set of ordered pairs $(x, y)$. In a relation, the **domain** is the set of first coordinates $x$, and the **range** is the set of second coordinates $y$.

**EXAMPLE 1** The following set is a relation and is graphed in Fig. 9.1: $\{(1, 10), (2, 9), (3, 8), (4, 7), (3, 6), (2, 5), (1, 4)\}$.

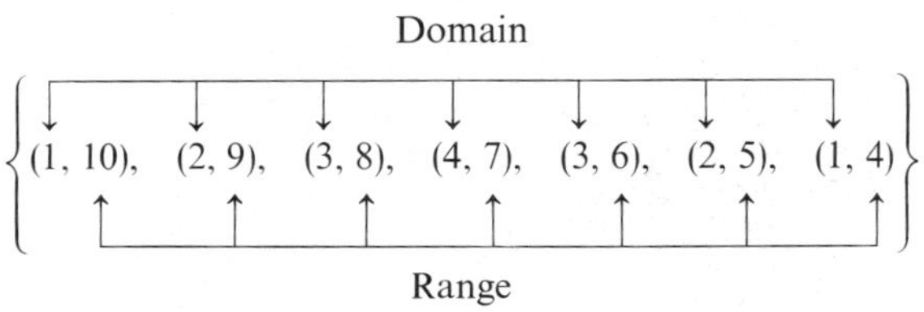


Domain = {first coordinates} = $\{1, 2, 3, 4\}$

Range = {second coordinates} = $\{4, 5, 6, 7, 8, 9, 10\}$

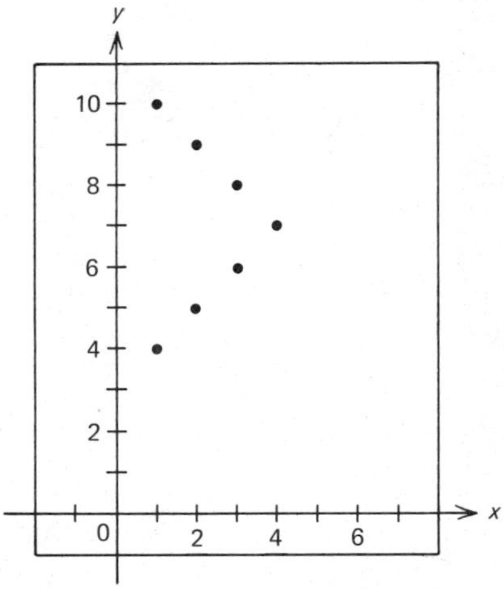


**Figure 9.1**

A special type of relation is a **function** in which every element of the domain is paired with *only one* element of the range. Example 1 is *not* a function since 1 is paired with both 10 and 4.

Although this is the proper definition of function, it is more useful to think of a **function** as a *rule* that assigns each element of the domain to a single element of the range (each $x$ goes with only one $y$).

**EXAMPLE 2** The following functions are rules that pair each element in the domain to one in the range.

| DOMAIN | | RANGE | DOMAIN | | RANGE |
|---|---|---|---|---|---|
| ⋮ | | ⋮ | ⋮ | | ⋮ |
| $-3$ | $\longrightarrow$ | 9 | $-3$ | $\longrightarrow$ | $-10$ |
| $-2$ | $\longrightarrow$ | 4 | $-2$ | $\longrightarrow$ | $-7$ |
| $-1$ | $\longrightarrow$ | 1 | $-1$ | $\longrightarrow$ | $-4$ |
| 0 | $\longrightarrow$ | 0 | 0 | $\longrightarrow$ | $-1$ |
| 1 | $\longrightarrow$ | 1 | 1 | $\longrightarrow$ | 2 |
| 2 | $\longrightarrow$ | 4 | 2 | $\longrightarrow$ | 5 |
| 3 | $\longrightarrow$ | 9 | 3 | $\longrightarrow$ | 8 |
| 4 | $\longrightarrow$ | 16 | 4 | $\longrightarrow$ | 11 |
| ⋮ | | ⋮ | ⋮ | | ⋮ |
| $x$ | $\longrightarrow$ | $x^2$ | $x$ | $\longrightarrow$ | $3x - 1$ |

In the first example, we square the first number to get the second. In the other example, we triple the first number and subtract one to get the second. In both cases, we had a rule or formula.

We usually write functions with letters such as $f$, $g$, $h$, $F$, or $G$. If $x$ is an element of the domain, we write

$$y = f(x)$$

(read "$f$ of $x$") as the number in the range that $x$ is assigned to. The notation is *not* multiplication, but rather a rule or formula with $x$. (We used this notation in Chapter 3 for polynomial functions—see page 77.)

**EXAMPLE 3** Let $f(x) = x^2 + 1$ and $g(x) = 5x + 2$. The function notation works as follows.

$$f(-1) = (-1)^2 + 1 = 2 \qquad g(-1) = 5(-1) + 2 = -3$$

$$f(0) = 0^2 + 1 = 1 \qquad g(0) = 5(0) + 2 = 2$$

$$f(1) = 1^2 + 1 = 2 \qquad g(1) = 5(1) + 2 = 7$$

$$f(2) = 2^2 + 1 = 5 \qquad g(2) = 5(2) + 2 = 12$$

$$f(3) = 3^2 + 1 = 10 \qquad g(3) = 5(3) + 2 = 17$$

$$f(10) = 10^2 + 1 = 101 \qquad g(10) = 5(10) + 2 = 52$$

$$f(t) = t^2 + 1 \qquad g(t) = 5t + 2$$

$$f(a + b) = (a + b)^2 + 1 = a^2 + 2ab + b^2 + 1 \qquad g(a + b) = 5(a + b) + 2 = 5a + 5b + 2$$

$$f[f(2)] = f[5] = 5^2 + 1 = 26 \qquad g[g(2)] = g[12] = 5(12) + 1 = 61$$

In this case, $f(x) = x^2 + 1$ means square and add 1 for whatever is within the parentheses, whether it is a number like 10, a symbol like $t$, or an expression like $a + b$. Similarly, $g(x) = 5x + 2$ means multiply by 5 and add 2 for whatever is within the parentheses.

**EXAMPLE 4** If $f(x) = x^2 + 2x + 5$, find

**(a)** $f(x + h)$ **(b)** $f(x + h) - f(x)$ **(c)** $\dfrac{f(x + h) - f(x)}{h}$

**SOLUTION** In addition to showing how functions operate, this example is important to other areas of study such as calculus.

**(a)**

Substitute

$$f(x + h)^2 = (x + h)^2 + 2(x + h) + 5$$

↓

Distributive law

$$= x^2 + 2xh + h^2 + 2x + 2h + 5$$

**(b)**

Substitute

$$f(x+h) - f(x) = (x^2 + 2xh + h^2 + 2x + 2h + 5) - (x^2 + 2x + 5)$$

↓

Subtract

$$= 2xh + h^2 + 2h$$

**(c)** Substitute → Divide by $h$

$$\frac{f(x+h)-f(x)}{h} = \frac{2xh + h^2 + 2h}{h}$$

$$= 2x + h + 2$$

We generally assume that the domains and ranges of functions and relations to be as large a set of real numbers as possible. The domains will be seen as a set of $x$-values on the $x$-axis, and the range as a set of $y$-values on the $y$-axis

We graph functions by plotting the pairs $(x, f(x))$. Since each $x$ is paired with *only one y*, each vertical line through some $x$-value will hit the graph of the function in at most one $y$-value.

The **vertical-line test** states that if any vertical line hits a curve more than once, it is not a function. On the other hand, if all vertical lines hit the curve at most once, it is a function.

| YES | NO |
|---|---|
| $f(-2) = 5$<br>$f(3) = 5$ | $f(4) = 10$<br>$f(4) = 6$ (crossed out) |

Two different $x$-values can go to the same $y$; but an $x$ cannot go to two different $y$-values.

**EXAMPLE 5** In Fig. 9.2, curves (a) and (b) are functions, but (c) and (d) are not functions since a vertical line hits them more than once. Also, we will indicate the domains on the $x$-axes and the ranges on the $y$-axes.

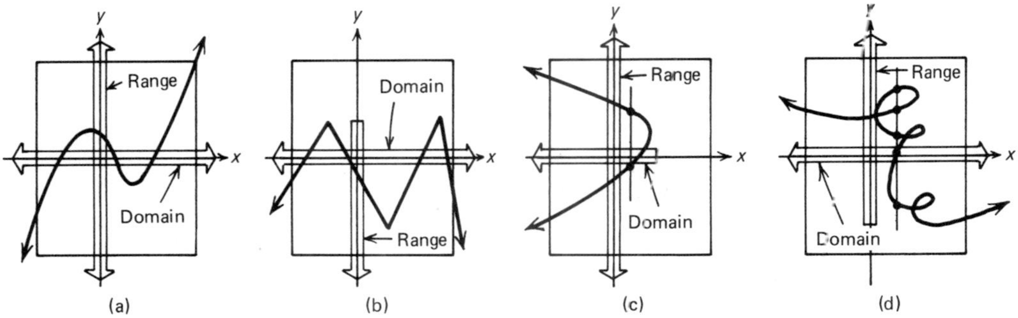


**Figure 9.2**

## PROBLEM SET 9.1

For the following relations, give the domains and ranges, graph, and state whether or not they are functions.

*See Example 1*

1. $\{(1, 5), (2, 7), (3, 10), (4, 14), (5, 19), (6, 25)\}$
2. $\{(10, 3), (20, 3), (30, 3), (40, 3)\}$
3. $\{(1, 5), (1, 6), (1, 7), (1, 8), (1, 9)\}$
4. $\{(2, 3), (4, 3), (3, 7), (1, 9), (5, 4), (2, 6)\}$
5. $\{(7, 4), (1, 6), (3, 4), (4, 4), (2, 6), (10, 4)\}$
6. $\{(1, 5), (2, 5), (3, 5), (4, 5), (5, 5), (3, 4), (3, 3)\}$

For each of the following rules, complete the assignments.

*See Example 2*

7.
$$\begin{array}{rcl} x & \longrightarrow & x^3 \\ 1 & \longrightarrow & 1 \\ 2 & \longrightarrow & 8 \\ 3 & \longrightarrow & \\ 4 & \longrightarrow & \\ -2 & \longrightarrow & \\ -5 & \longrightarrow & \end{array}$$

8.
$$\begin{array}{rcl} x & \longrightarrow & 4x + 3 \\ 0 & \longrightarrow & 3 \\ 1 & \longrightarrow & 7 \\ 2 & \longrightarrow & \\ 3 & \longrightarrow & \\ -4 & \longrightarrow & \\ -5 & \longrightarrow & \end{array}$$

9.
$$\begin{array}{rcl} x & \longrightarrow & \frac{1}{x} \\ 1 & \longrightarrow & 1 \\ 2 & \longrightarrow & \frac{1}{2} \\ 3 & \longrightarrow & \\ 4 & \longrightarrow & \\ -8 & \longrightarrow & \\ -10 & \longrightarrow & \end{array}$$

10.
$$\begin{array}{rcl} x & \longrightarrow & \sqrt{x} \\ 1 & \longrightarrow & 1 \\ 4 & \longrightarrow & 2 \\ 9 & \longrightarrow & \\ 25 & \longrightarrow & \\ 81 & \longrightarrow & \\ 400 & \longrightarrow & \end{array}$$

For the following functions, find:

(a) $f(0)$ (b) $f(-2)$ (c) $f(3)$
(d) $f(10)$ (e) $f[f(3)]$ (f) $f(a + b)$

*See Example 3*

11. $f(x) = x + 5$
12. $f(x) = 2x - 3$
13. $f(x) = 6 - x$
14. $f(x) = 4 - 2x$
15. $f(x) = 2x^2$
16. $f(x) = 3x^2 - 1$
17. $f(x) = x^2 - 3x + 4$
18. $f(x) = 2x^2 + 5x - 3$

For the following functions, find:

(a) $f(x + h)$ (b) $f(x + h) - f(x)$ (c) $\dfrac{f(x + h) - f(x)}{h}$

*See Example 4*

19. $f(x) = 2x + 5$
20. $f(x) = 4x - 1$
21. $f(x) = 2x^2$
22. $f(x) = 3x^2 - 1$
23. $f(x) = x^2 - 3x + 4$
24. $f(x) = 2x^2 + 5x - 3$

State whether the following sketches represent functions. Also, indicate the domains on the $x$-axes and the ranges on the $y$-axes.

*See Example 5*

25.

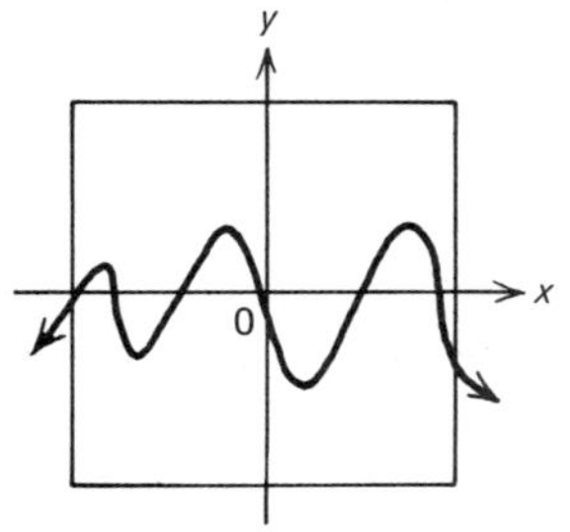


26.

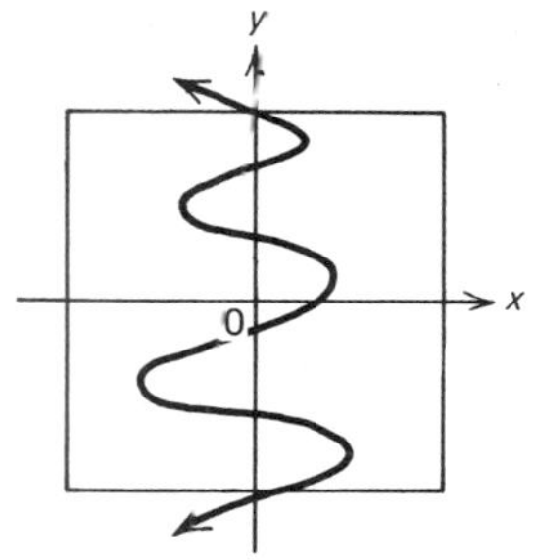


27.

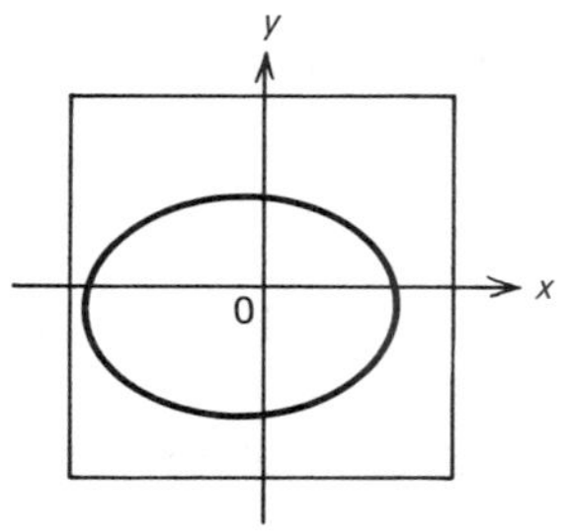


28.

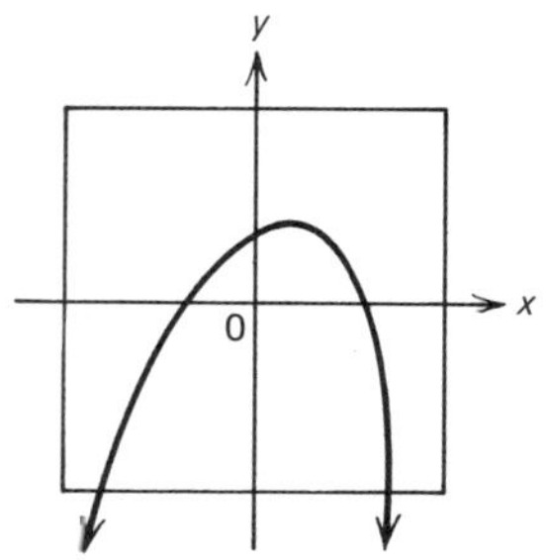

29.

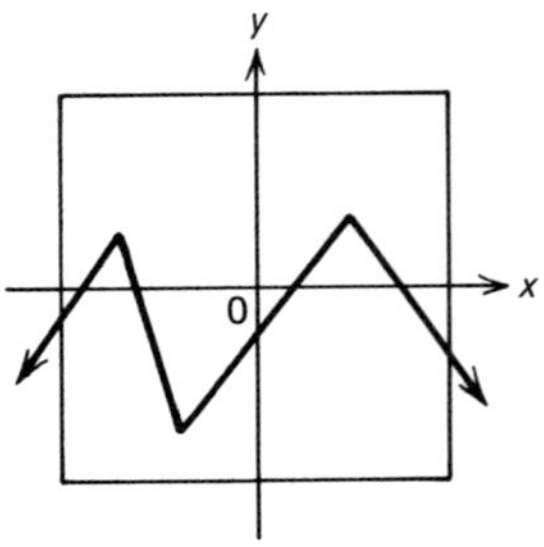


30.

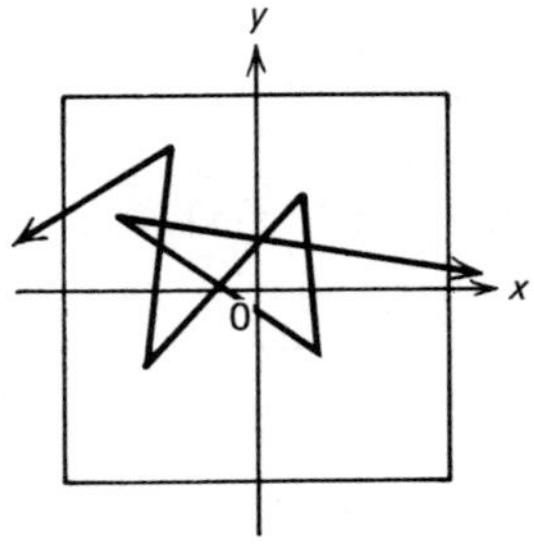


*Business application*

For a certain product, the *demand function* is given by

$$y = f(x) = \frac{3000}{x}$$

where $y$ is the demand and $x$ the price.

31. Find $f(1)$.
32. Find $f(3)$.
33. Find $f(10)$.

*Psychology application*

For a certain task, the *learning function* is given by

$$y = f(x) = \frac{100x}{x + 10}$$

where $x$ is the number of trials at the task and $y$ the percent of work done on the task.

34. Find $f(0)$.
35. Find $f(10)$.
36. Find $f(40)$.

*Physical application*

The *distance function* for a falling object is

$$y = f(x) = 16x^2$$

where $x$ is the time in seconds and $y$ the distance in feet.

37. Find $f(2)$.
38. Find $f(10)$.
39. Find the distance at time $x + h$; in other words, find $f(x + h)$.
40. Find the change in distance; this is, $f(x + h) - f(x)$.

41. Find the average velocity (or speed) in the time between $x$ and $x + h$; this is

$$\frac{f(x+h) - f(x)}{h}$$

## 9.2 QUADRATIC FUNCTIONS (PARABOLAS)

In this and the next section, we discuss special types of relations known as **conic sections.** These curves are given this name since they can be cut from cones as seen in Fig. 9.3.

In Section 9.3, we discuss the circle, the ellipse, and the hyperbola, but in this section we deal only with the parabola. The **parabola** can be seen as the path of a thrown object, the shape of a flashlight reflector, or the shape of a radar scanner. (The reflector and the scanner are actually paraboloids, three-dimensional versions of a parabola.)

Parabolas occur in mathematics when we graph a quadratic function which has the general form

$$y = ax^2 + bx + c \qquad (a \neq 0)$$

or

$$f(x) = ax^2 + bx + c \qquad (a \neq 0)$$

Recall from the preceding section that $y$ and $f(x)$ are the same quantity.

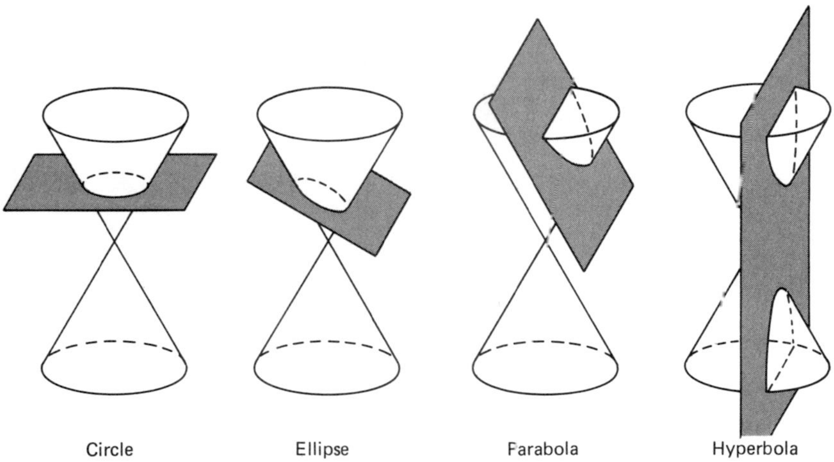


Figure 9.3

> To graph a quadratic function:
>
> 1. Choose an appropriate set of $x$-values.
> 2. Make a table of $(x, y)$ pairs by substituting the $x$-values into the formula for $y$.
> 3. Plot these points on a graph.
> 4. Draw a smooth curve through these points.

**EXAMPLE 6** Graph $y = x^2 + 3x - 4$, for $-3 \le x \le 3$. Then extend the graph.

**SOLUTION** Here we are given the $x$-values to use $-3$ to 3. Now we make a table.

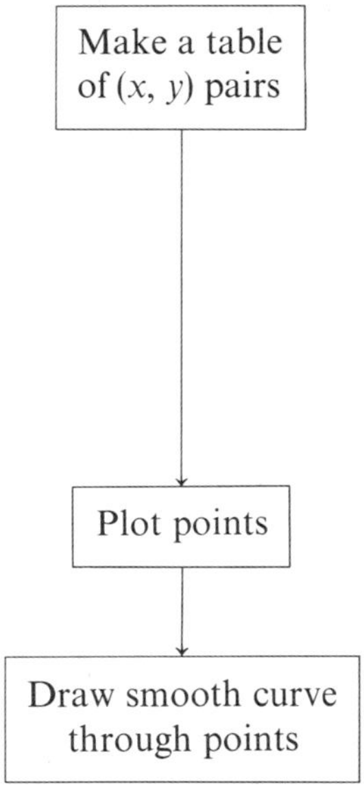


| $x$ | $y=x^2+3x-4$ | $(x, y)$ |
|---|---|---|
| $-3$ | $(-3)^2+3(-3)-4=-4$ | $(-3, -4)$ |
| $-2$ | $(-2)^2+3(-2)-4=-6$ | $(-2, -6)$ |
| $-1$ | $(-1)^2+3(-1)-4=-6$ | $(-1, -6)$ |
| 0 | $(0)^2+3(0)-4=-4$ | $(0, -4)$ |
| 1 | $(1)^2+3(1)-4=0$ | $(1, 0)$ |
| 2 | $(2)^2+3(2)-4=6$ | $(2, 6)$ |
| 3 | $(3)^2+3(3)-4=14$ | $(3, 14)$ |

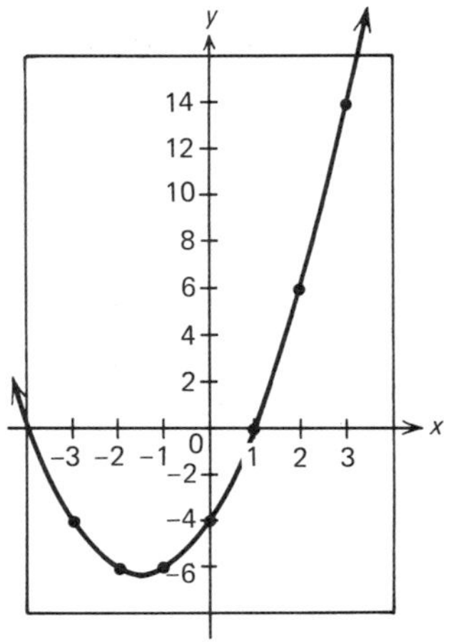


Notice that we extended the curve beyond the interval $-3 \le x \le 3$.

**EXAMPLE 7** Graph $y = -x^2 + 2x + 7$, for $-2 \le x \le 4$. Then extend the graph.

**SOLUTION** We again start with a table of $(x, y)$ pairs.

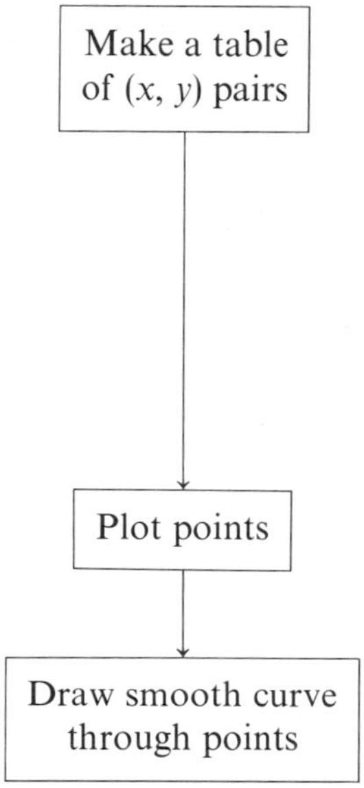


| $x$ | $y=-x^2+2x+7$ | $(x, y)$ |
|---|---|---|
| $-2$ | $-(-2)^2+2(-2)+7=-1$ | $(-2, -1)$ |
| $-1$ | $-(-1)^2+2(-1)+7=4$ | $(-1, 4)$ |
| $0$ | $-(0)^2+2(0)+7=7$ | $(0, 7)$ |
| $1$ | $-(1)^2+2(1)+7=8$ | $(1, 8)$ |
| $2$ | $-(2)^2+2(2)+7=7$ | $(2, 7)$ |
| $3$ | $-(3)^2+2(3)+7=4$ | $(3, 4)$ |
| $4$ | $-(4)^2+2(4)+7=-1$ | $(4, -1)$ |

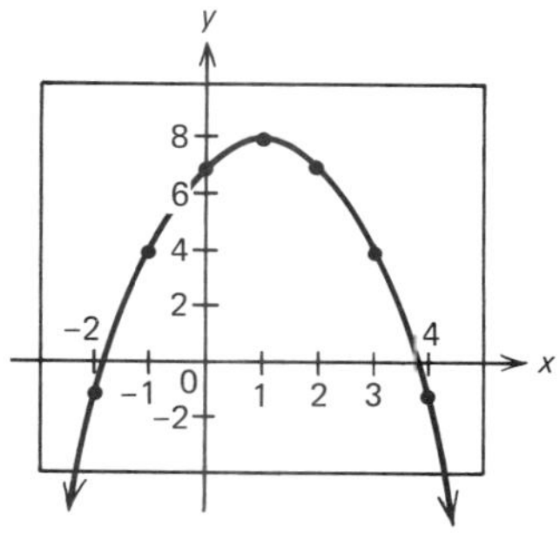


Again, we extended the graph beyond the interval $-2 \leq x \leq 4$.

We now look at another way to graph the quadratic function

$$y = ax^2 + bx + c$$

We do this by finding these key points: the $x$-intercepts, the $y$-intercept, and the vertex.

The ***x*-intercepts** are the points (if any) where the curve crosses the $x$-axis (the line $y = 0$). If we set $y = 0$, we solve $ax^2 + bx + c = 0$ by factoring or using the quadratic formula to find the $x$-intercepts.

The ***y*-intercept** is the point where the curve crosses the $y$-axis (the line $x = 0$). If we set $x = 0$, we get $y = c$. Thus, the $y$-intercept is $(0, c)$.

The **vertex** is the turning point where the curve peaks (as in Example 7) or bottoms out (as in Example 6). It can be shown by calculus that the $x$-coordinate of the vertex is $x = -b/2a$.

The curve **opens upward** (as in Example 6) if the coefficient $a$ is positive, and **opens downward** (as in Example 7) if $a$ is negative.

We can summarize this as follows.

> To graph the curve $y = ax^2 + bx + c$ $(a \neq 0)$:
>
> 1. Solve $ax^2 + bx + c = 0$ by factoring or using the quadratic formula to find the $x$-intercepts.
> 2. The $y$-intercept is at $(0, c)$.
> 3. The vertex is at $\left(\frac{-b}{2a}, f\left(\frac{-b}{2a}\right)\right)$.
> 4. The curve open upward if $a > 0$ and downward if $a < 0$.
> 5. Plot these points and draw a smooth curve through them.

**EXAMPLE 8** Graph $y = x^2 - 4x + 3$.

**SOLUTION** We find the $x$-intercepts, the $y$-intercept, and the vertex.

| | |
|---|---|
| Solve $ax^2 + bx + c = 0$ to find $x$-intercepts | $x - 4x + 3 = (x - 3)(x - 1) = 0$<br>$x - 3 = 0$ or $x - 1 = 0$<br>$x = 3$ or $x = 1$<br>$x$-intercepts are $(3, 0)$ and $(1, 0)$ |
| ↓ | |
| $y$-intercept is $(0, c)$ | $y$-intercept is $(0, 3)$ |
| ↓ | |
| Vertex is at $x = \frac{-b}{2a}$ | Vertex is at $x = \frac{-(-4)}{2(1)} = 2$ |
| Find $y$-value | $y = f(2) = (2)^2 - 4(2) + 3 = -1$<br>Vertex $= (2, -1)$ |
| ↓ | |
| $a = 1$, so curve opens up | |
| ↓ | |
| Plot points | |
| ↓ | |
| Draw smooth curve through points | 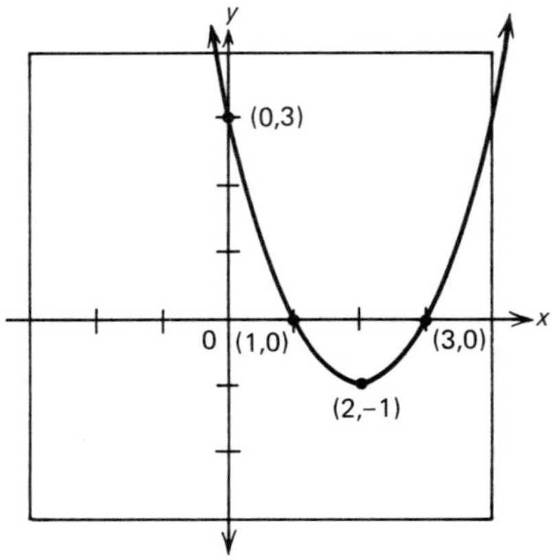 |

**EXAMPLE 9** Graph $y = -2x^2 + 2x + 12$.

**SOLUTION** We locate the $x$-intercepts, the $y$-intercept, and the vertex.

Solve $ax^2 + bx + c = 0$ to find $x$-intercepts

$$-2x^2 + 2x + 12 = -2(x^2 - x - 6) = 0$$
$$= -2(x+2)(x-3) = 0$$
$$x + 2 = 0 \quad \text{or} \quad x - 3 = 0$$
$$x = -2 \quad \text{or} \quad x = 3$$

$x$-intercepts are $(-2, 0)$ and $(3, 0)$

$y$-intercept is $(0, c)$

$y$-intercept is $(0, 12)$

Vertex is at $x = \dfrac{-b}{2a}$

Find $y$-value

$$\text{Vertex is at } x = \frac{-(2)}{2(-2)} = \frac{1}{2}$$

$$y = f(\tfrac{1}{2}) = -2(\tfrac{1}{2})^2 + 2(\tfrac{1}{2}) + 12 = 12\tfrac{1}{2}$$

$$\text{Vertex} = (\tfrac{1}{2}, 12\tfrac{1}{2})$$

$a = -2$, so curve opens down

Plot points

Draw smooth curve through points

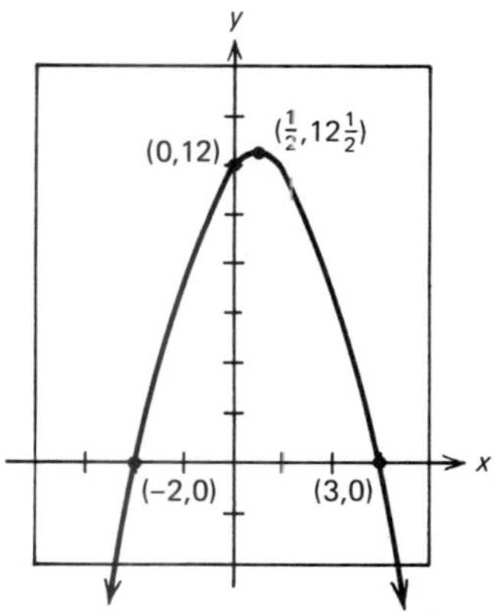


## PROBLEM SET 9.2

For the following quadratic functions, make a table of $(x, y)$ pairs for the indicated $x$-values. Then graph and extend the curve.

*See Examples 6 and 7*

1. $y = x^2; \quad -3 \le x \le 3$
2. $y = 2x^2; \quad -2 \le x \le 4$
3. $y = -x^2 + 3; \quad -1 \le x \le 5$
4. $y = x^2 - 5; \quad -3 \le x \le 3$
5. $y = x^2 + x + 1; \quad -2 \le x \le 4$
6. $y = x^2 - 2x - 3; \quad -1 \le x \le 5$

7. $y = 2x^2 - x + 2; \quad 0 \le x \le 6$
8. $y = 3x^2 + x - 10; \quad -3 \le x \le 3$
9. $y = -x^2 - 5x + 8; \quad -1 \le x \le 5$
10. $y = 4x^2 - x + 5; \quad 0 \le x \le 6$
11. $y = 3x^2 + 5x - 4; \quad -3 \le x \le 3$
12. $y = -3x^2 + 2x + 5; \quad -1 \le x \le 5$

For the following quadratic functions, find the (a) $x$-intercepts, (b) $y$-intercept, and (c) vertex. Then graph the curve.

*See Examples 8 and 9*

13. $y = x^2 - 6x + 5$
14. $y = x^2 - 2x - 3$
15. $y = x^2 - 4$
16. $y = -x^2 + 9$
17. $y = -x^2 + 2x + 8$
18. $y = x^2 + 2x - 15$
19. $y = 2x^2 - 2x - 4$
20. $y = -x^2 + 5x - 4$
21. $y = -x^2 + 6x$
22. $y = 2x^2 + 8x$
23. $y = 3x^2 - 21x + 30$
24. $y = -2x^2 + 10x + 28$

*Physical application*

25. An object is thrown into the air, and its height is given by $y = -16t^2 + 80t$. Graph this for $0 \le t \le 5$ and give the vertex.

*Health applications*

26. Blood flows through blood vessels with a velocity $v = 1.28 - 20{,}000x^2$ (see page 241). Graph this equation for $0 \le x \le 0.008$. (Scale $x$-axis as 0, 0.001, 0.002, and so on.)

27. The population of a bacteria culture is given by

$$p = -125t^2 + 1000t + 10{,}000$$

where $t$ is the time in hours. Graph this equation for $0 \le t \le 8$. Give the vertex.

*Life science application*

28. The yield $y$ in bushels per acre in a rice field is given by $y = 10x - x^2/2$, where $x$ is the density of the rice plant. Find the vertex and graph the equation for $0 \le x \le 20$. (Take $x = 0, 5, 10, 15, 20$.)

*Business application*

29. The profit $P$ on a item is given by $P = -2x^2 + 200x - 100$, where $x$ = number produced. Find the vertex (maximum profit) and graph for $0 \le x \le 60$. (Graph $x = 0, 10, 20, 30$, and so on.)

## 9.3 CIRCLES, ELLIPSES, AND HYPERBOLAS

In Section 9.2, we looked at parabolas. In this section, we graph **circles, ellipses,** and **hyperbolas.** These have the general form

$$Ax^2 + By^2 = C \qquad (A, B \neq 0)$$

with $A \neq 0$, $B \neq 0$, and $C > 0$. Table 9.1 summarizes the different situations that we will discuss.

**Table 9.1**

| CASE | EXAMPLE | SKETCH |
|---|---|---|
| *A, B both positive.* <br> $A = B$ <br> *Circle* | $x^2 + y^2 = 25$ | 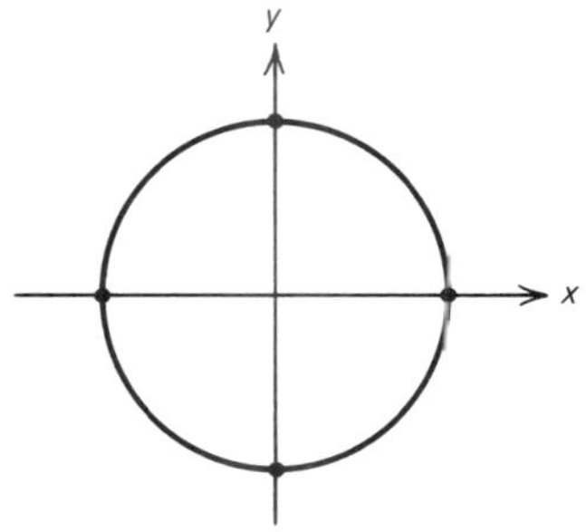  |
| *A, B both positive.* <br> $A \neq B$ <br> *Ellipse* | $\frac{x^2}{4} + \frac{y^2}{25} = 1$ | 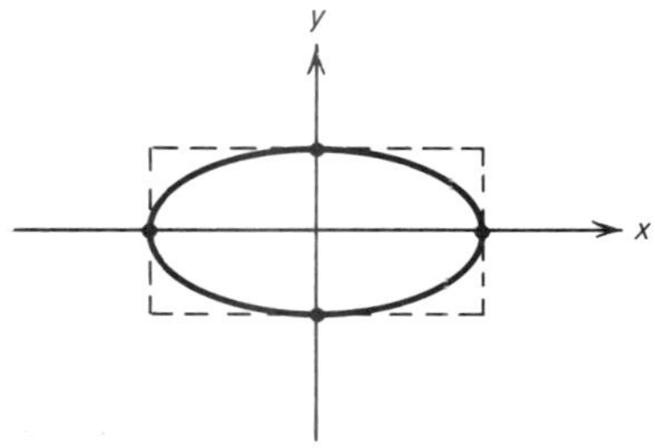  |
| *A, B are of opposite signs.* <br> *Hyperbola* | $x^2 - 4y^2 = 16$ | 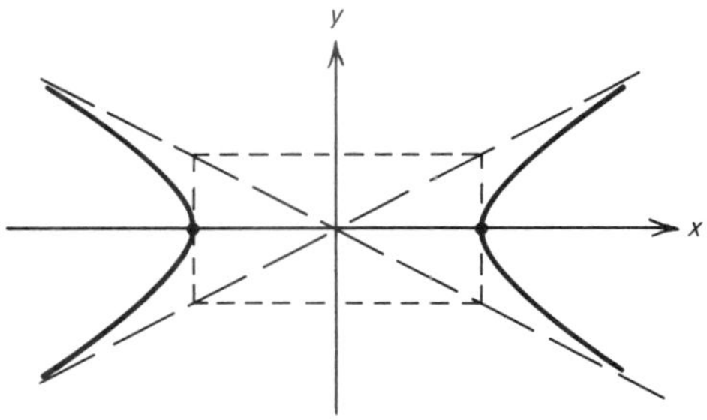  |
| | $-\frac{x^2}{25} + \frac{y^2}{16} = 1$ | 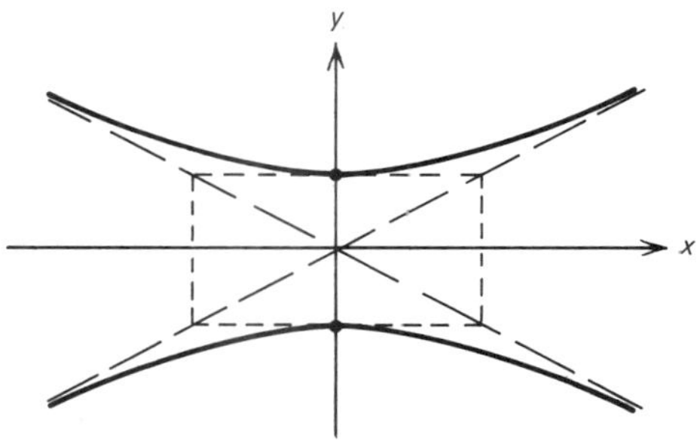  |

A circle centered at the origin is a fixed distance $r$ from (0, 0). Recall from Section 8.2 the distance formula that gives us the relation $x^2 + y^2 = r^2$ for a circle with center at (0, 0) and radius $r$.

> To graph a circle, $Ax^2 + Ay^2 = C$:
>
> 1. Put the equation into the form $x^2 + y^2 = C/A$.
> 2. The circle is centered at (0, 0), and the radius is $\sqrt{C/A}$.

**EXAMPLE 10** Graph the equation $4x^2 + 4y^2 = 36$.

**SOLUTION** Since $A = B = 4$, this is a circle. We divide both sides by 4 to find the radius.

| | |
|---|---|
| Given | $4x^2 + 4y^2 = 36$ |
| ↓ Divide by 4 | $x^2 + y^2 = 9$ |
| ↓ Find radius | Radius $= \sqrt{9} = 3$ |
| ↓ Graph circle: Center = (0, 0) Radius = 3 | 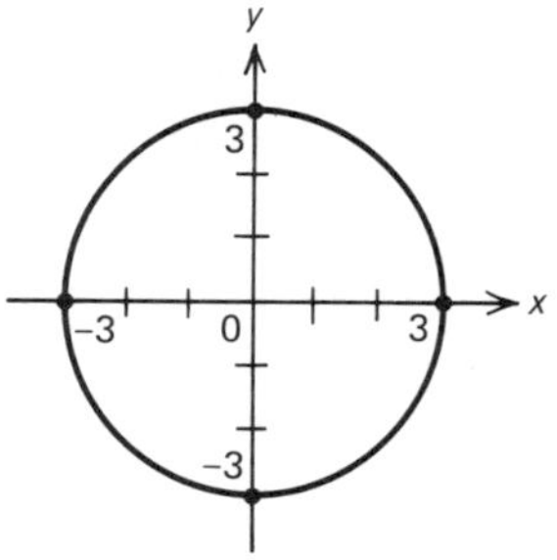 |



> To graph an ellipse, $Ax^2 + By^2 = C$:
>
> 1. Set $x = 0$ and find the $y$-intercepts.
> 2. Set $y = 0$ and find the $x$-intercepts.
> 3. Plot these intercepts.
> 4. Sketch a "helping rectangle" through these points.
> 5. Draw a curve through the intercepts and within the rectangle.

**EXAMPLE 11** Graph $4x^2 + 9y^2 = 36$.

**SOLUTION** Since $4 \neq 9$, this is an ellipse. We find the intercepts.

| | |
|---|---|
| Set $x = 0$ | $4(0) + 9y^2 = 36$ |
| ↓ Solve for $y$ | $y^2 = 4$ <br> $y = \pm 2$ |

The $y$-intercepts are therefore $(0, -2)$ and $(0, 2)$.

| | |
|---|---|
| Set $y = 0$ | $4x^2 + 9(0) = 36$ |
| ↓ Solve for $x$ | $x^2 = 9$ <br> $x = \pm 3$ |

The $x$-intercepts are therefore $(-3, 0)$ and $(3, 0)$.

Plot intercepts and sketch "helping rectangle"

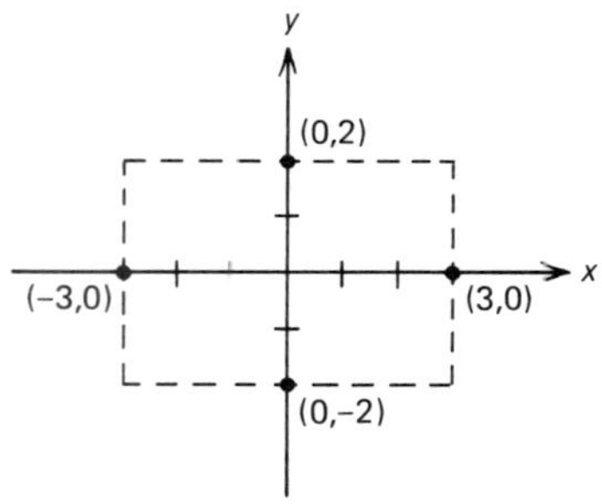


↓

Draw ellipse within rectangle through intercepts

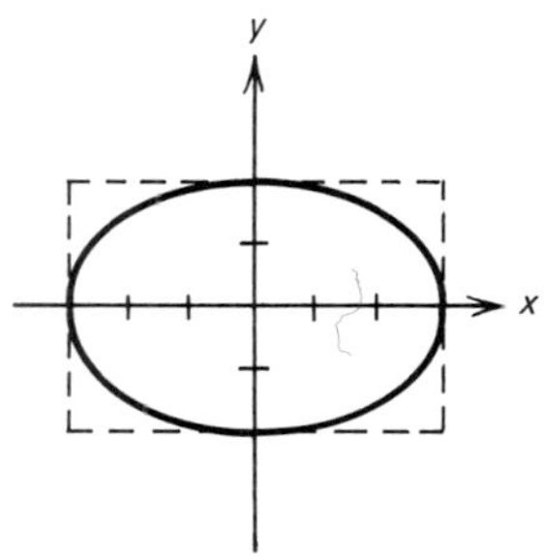

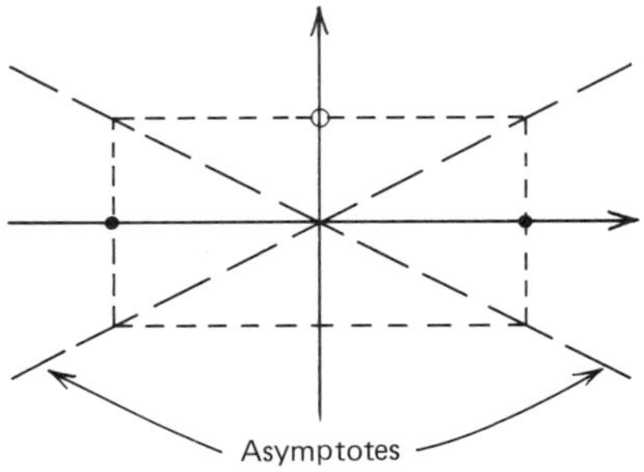


**Figure 9.4**

The hyperbola is different from the ellipse in that it has only one pair of intercepts: either on the $x$-axis or the $y$-axis, but *not* both. As with the ellipse, we will use a "helping rectangle," shown in Fig. 9.4. The diagonals drawn through the corners are called **asymptotes.** These are lines that the hyperbola approaches, but never hits. The asymptotes are *not* part of the graph, just helpers.

**EXAMPLE 12** Graph $x^2 - 4y^2 = 16$.

**SOLUTION** Since the coefficients of $x^2$ and $y^2$ have opposite signs, this is a hyperbola. We begin by finding the intercepts.

Set $y = 0$ $\qquad x^2 - 4(0) = 16$

↓

Solve for $x$ $\qquad x = \pm 4$

Therefore, the $x$-intercepts are $(-4, 0)$ and $(4, 0)$.

Set $x = 0$ $\qquad 0 - 4y^2 = 16$

↓

Solve for $y$ $\qquad y^2 = -4$

$$y = \pm 2i$$

These are *not* real numbers, and so there are *no* $y$-intercepts. Nevertheless, we will use the points $(0, -2)$ and $(0, 2)$ to construct our "helping rectangle." (We will call these points **imaginary intercepts** and plot them with ○.)

Plot real intercepts (●) and imaginary intercepts (○)

↓

Draw "helping rectangle" and asymptotes through corners

↓

Draw hyperbola through $x$-intercepts approaching asymptotes

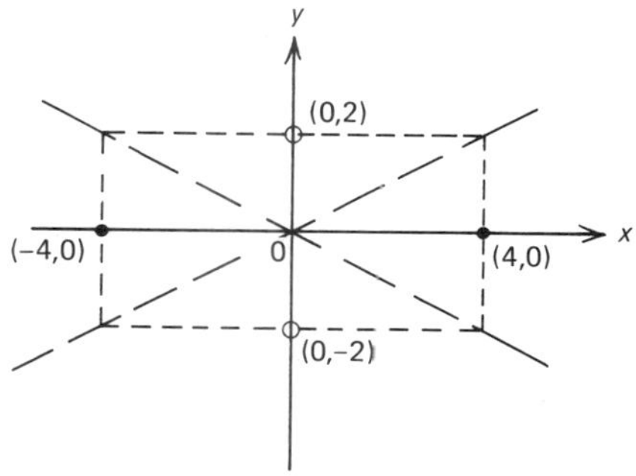


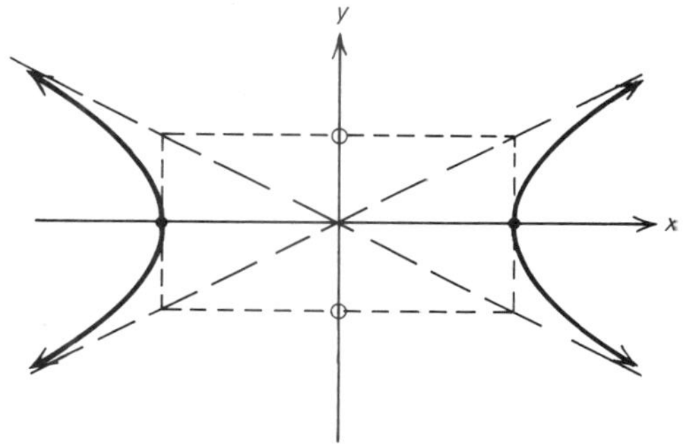


**EXAMPLE 13** Graph $-\dfrac{x^2}{25} + \dfrac{y^2}{4} = 1.$

**SOLUTION** Since the coefficients have opposite signs, this also is a hyperbola.

Set $x = 0$ → Solve for $y$

$$\frac{0}{25} + \frac{y^2}{4} = 1$$

$$y^2 = 4$$

$$y = \pm 2$$

Thus, the $y$-intercepts are $(0, -2)$ and $(0, 2)$.

Set $y = 0$ → Solve for $x$

$$-\frac{x^2}{25} + \frac{0}{4} = 1$$

$$x^2 = -25$$

$$x = \pm 5i$$

These are *not* real numbers, so that there are *no* $x$-intercepts. However, we do use the points $(5, 0)$ and $(-5, 0)$ to construct the "helping rectangle."

Plot real intercepts (●) and imaginary intercepts (○)

↓

Draw "helping rectangle" and asymptotes through corners

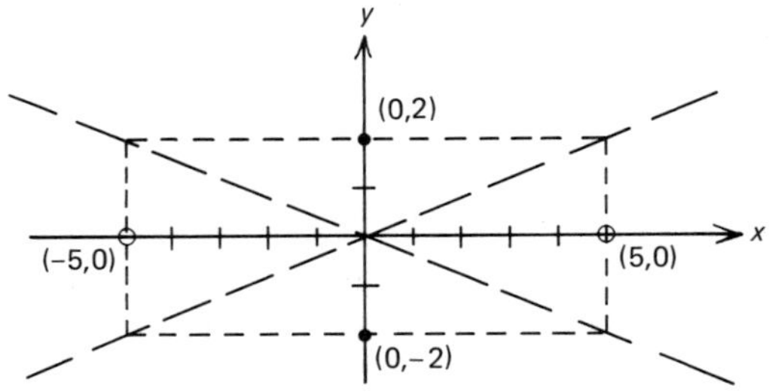


↓

Draw hyperbola through $y$-intercepts approaching asymptotes

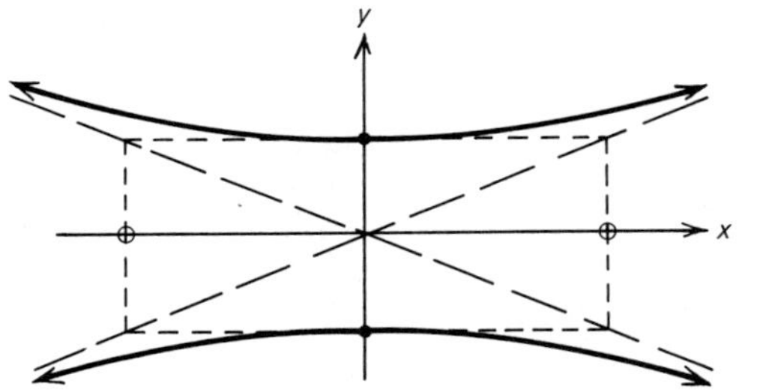


## PROBLEM SET 9.3

Graph the following circles, ellipses, and hyperbolas.

*See Example 10*

1. $\dfrac{x^2}{4^2} + \dfrac{y^2}{4^2} = 1$
2. $\dfrac{x^2}{2^2} + \dfrac{y^2}{2^2} = 1$
3. $\dfrac{x^2}{100} + \dfrac{y^2}{100} = 1$
4. $\dfrac{x^2}{64} + \dfrac{y^2}{64} = 1$
5. $x^2 + y^2 = 49$
6. $x^2 + y^2 = 36$
7. $2x^2 + 2y^2 = 50$
8. $5x^2 + 5y^2 = 5$

*See Example 11*

9. $\dfrac{x^2}{2^2} + \dfrac{y^2}{3^2} = 1$
10. $\dfrac{x^2}{5^2} + \dfrac{y^2}{1^2} = 1$
11. $\dfrac{x^2}{16} + \dfrac{y^2}{9} = 1$
12. $\dfrac{x^2}{64} + \dfrac{y^2}{25} = 1$
13. $16x^2 + y^2 = 64$
14. $x^2 + 9y^2 = 36$
15. $4x^2 + 25y^2 = 100$
16. $9x^2 + 4y^2 = 144$

*See Examples 12 and 13*

17. $\dfrac{x^2}{2^2} - \dfrac{y^2}{5^2} = 1$
18. $\dfrac{-x^2}{3^2} + \dfrac{y^2}{4^2} = 1$

19. $\dfrac{-x^2}{16} + \dfrac{y^2}{4} = 1$

20. $\dfrac{x^2}{9} - \dfrac{y^2}{25} = 1$

21. $\dfrac{x^2}{144} - \dfrac{y^2}{64} = 1$

22. $\dfrac{-x^2}{81} + \dfrac{y^2}{4} = 1$

23. $-4x^2 + 9y^2 = 36$

24. $x^2 - 16y^2 = 16$

25. $-4x^2 + y^2 = 16$

26. $-25x^2 + 4y^2 = 100$

27. $2x^2 - 18y^2 = 72$

28. $20x^2 - 5y^2 = 20$

*Physical applications*

29. The paths of orbiting planets are ellipses, but they are almost circles. An orbit might have a relation such as

$$100x^2 + 81y^2 = 8100$$

Graph this ellipse.

30. A whispering gallery is usually a room with a floor shaped like an ellipse. The ellipse has two focal points, and a whispering person standing at one focal point can be heard by someone standing at the other. The floor might be defined by the equation

$$144x^2 + 400y^2 = 57{,}600$$

with focal points at $(-16, 0)$ and $(16, 0)$. Graph the ellipse and the focal points.

## 9.4 SECOND-DEGREE INEQUALITIES

In this section, we look at the regions determined by **second-degree inequalities** such as

$$y \le x^2 - 4x + 3 \qquad 4x^2 + 9y^2 > 36 \qquad x^2 - 4y^2 < 16$$

Recall that with linear inequalities (Section 8.4), the corresponding equation cuts the plane into two regions. Similarly with these second-degree inequalities, the corresponding equation will cut the plane into two regions, as shown in Fig. 9.5. Notice with the hyperbola on the right that we consider the two outer

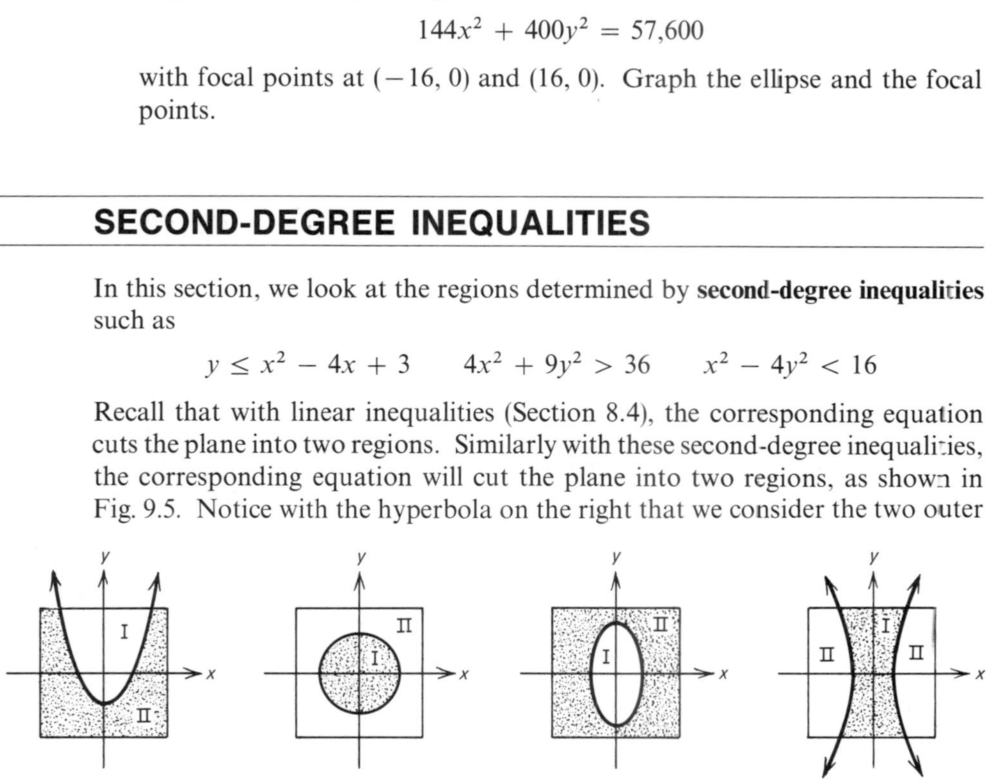


**Figure 9.5**

regions as one region. The procedure for graphing these inequalities is almost exactly as that for linear inequalities (see page 276).

> To graph a second-degree inequality:
>
> 1. Graph the corresponding equation.
> 2. Determine if a test point $P$ (not on the curve) satisfies the inequality.
>
> 3a. If $P$ satisfies the inequality, shade the region *containing P*.
>
> 3b. If $P$ does not satisfy the inequality, shade the region *not containing P*.
>
> 4a. If the inequality is $\leq$ or $\geq$, include the curve.
>
> 4b. If the inequality is $<$ or $>$, exclude the curve (use a dashed line).

**EXAMPLE 14** Graph $y \leq x^2 - 4x + 3$.

**SOLUTION** We begin by graphing the equation $y = x^2 - 4x + 3$. This was done in Example 8. (See page 302 for details.)

Graph $y = x^2 - 4x + 3$

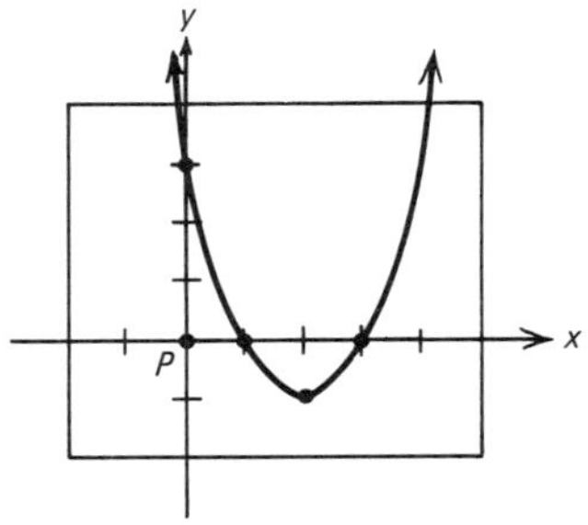


Check $P = (0, 0)$ in inequality

Is $0 \leq (0)^2 - 4(0) + 3$? YES

"YES" means shade region containing $P$

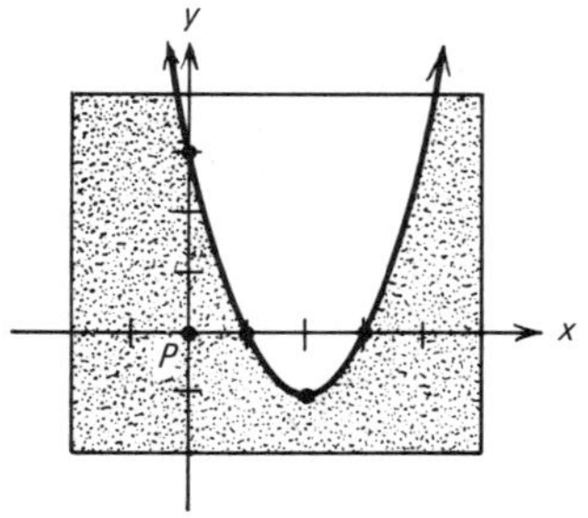


$\leq$ means include curve

**EXAMPLE 15** Graph $4x^2 + 9y^2 > 36$.

**SOLUTION** We first graph the ellipse $4x^2 + 9y^2 = 36$. This was done in Example 11. (See page 307 for details.)

Graph the equation
$4x^2 + 9y^2 = 36$

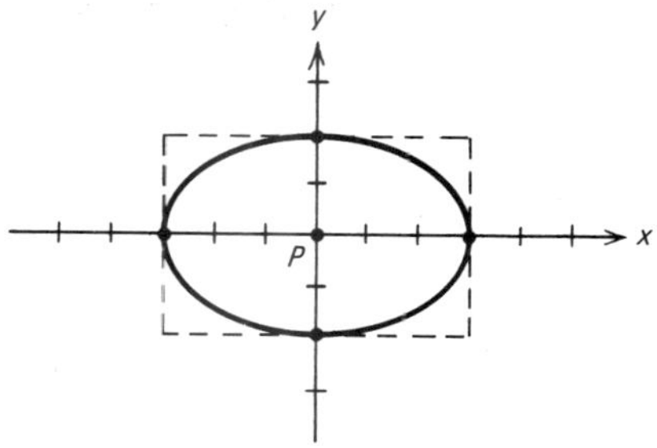


Check $P = (0, 0)$
in inequality

Is $4(0)^2 + 9(0)^2 > 36$? NO

"NO" means shade
region not containing $P$

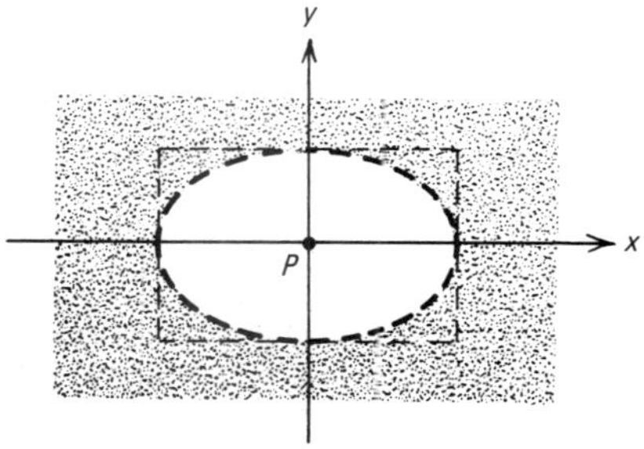


$>$ means
exclude curve

**EXAMPLE 16** Graph $x^2 - 4y^2 < 16$.

**SOLUTION** We first graph the hyperbola $x^2 - 4y^2 = 16$. This was done in Example 12. (See page 308 for details.)

Graph equation
$x^2 - 4y^2 = 16$

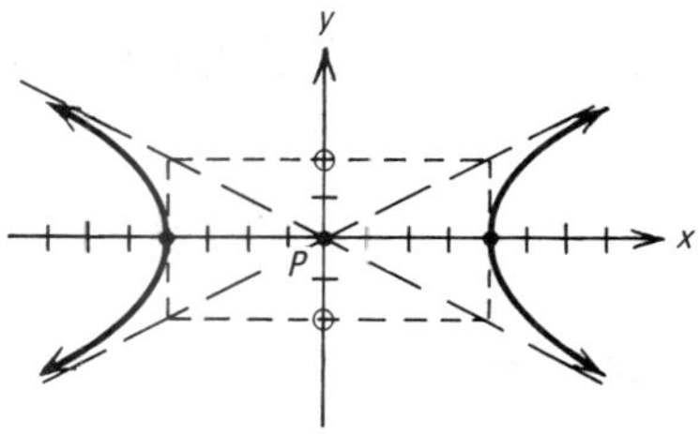


Check $P = (0, 0)$
in inequality

Is $(0)^2 - 4(0)^2 < 16$? YES

"YES" means shade
region containing $P$

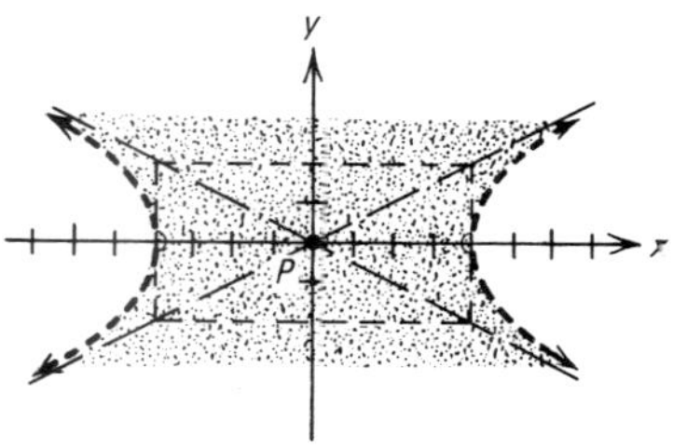


$<$ means
exclude curve

## PROBLEM SET 9.4

Graph the following second-degree inequalities.

See Example 14

1. $y < x^2 - 6x + 5$
2. $y > x^2 - 2x - 3$
3. $y \geq x^2 - 4$
4. $y \leq -x^2 + 9$
5. $y > -x^2 + 2x + 8$
6. $y < x^2 + 2x - 15$
7. $y \leq 2x^2 - 2x - 4$
8. $y \geq -x^2 + 5x - 4$
9. $y < -x^2 + 6x$
10. $y > 2x^2 + 8x$

See Example 15

11. $x^2 + y^2 \geq 100$
12. $x^2 + y^2 \leq 4$
13. $x^2 + y^2 > 49$
14. $2x^2 + 2y^2 < 50$
15. $x^2 + 25y^2 \leq 25$
16. $9x^2 + 16y^2 \geq 144$
17. $16x^2 + y^2 < 64$
18. $x^2 + 9y^2 > 36$
19. $4x^2 + 25y^2 \geq 100$
20. $9x^2 + 4y^2 \leq 144$

See Example 16

21. $25x^2 - 4y^2 > 100$
22. $-16x^2 + 9y^2 < 144$
23. $-4x^2 + 16y^2 \leq 64$
24. $25x^2 - 9y^2 \geq 225$
25. $4x^2 - y^2 < 16$
26. $-4x^2 + 9y^2 > 36$
27. $-2x^2 + 18y^2 \geq 72$
28. $20x^2 - 5y^2 \leq 20$
29. $x^2 - 25y^2 > 100$
30. $-x^2 + y^2 < 4$

## 9.5 INVERSE RELATIONS

If $f$ is a function or a relation, we get the **inverse of $f$** (also called **$f$-inverse**), written $f^{-1}$, by interchanging all the $x$- and $y$-coordinates.

**EXAMPLE 17** The following are a function $f$ and its inverse $f^{-1}$.

Interchange $x$ and $y$ to get $f^{-1}$

$$f = \{(1, 7), (2, 10), (3, 14), (4, 19)\}$$

$$f^{-1} = \{(7, 1), (10, 2), (14, 3), (19, 4)\}$$

| YES | NO |
|---|---|
| $f^{-1}$ is the inverse of $f$ | ~~$f^{-1}(x) = \frac{1}{f(x)}$~~ |

The $-1$ is *not* an exponent here, but a symbol for the inverse.

To find and graph the inverse of a function or a relation (defined by an equation):

1. Interchange the $x$- and $y$-variables in the equation.
2. Make a table of $(x, y)$ pairs for the original equation.
3. Reverse the $x$- and $y$-columns to get points for the inverse graph.
4. As a check, the graphs of $f$ and $f^{-1}$ will be symmetric about the line $y = x$.

**EXAMPLE 18** Find and graph the inverse of $f$: $y = 2x + 3$.

**SOLUTION** We first find the inverse.

Interchange $x$ and $y$

$$f^{-1}: \quad x = 2y + 3$$

Now, to graph $f$ and $f^{-1}$, we make a table for $f$ and then reverse the columns to get $f^{-1}$.

Make table for $f$

↓

Reverse columns for $f^{-1}$

↓

Graph both curves from tables

| $f$: $x$ | $f$: $y$ | $f^{-1}$: $x$ | $f^{-1}$: $y$ |
|---|---|---|---|
| 0 | 3 | 3 | 0 |
| 1 | 5 | 5 | 1 |
| 2 | 7 | 7 | 2 |

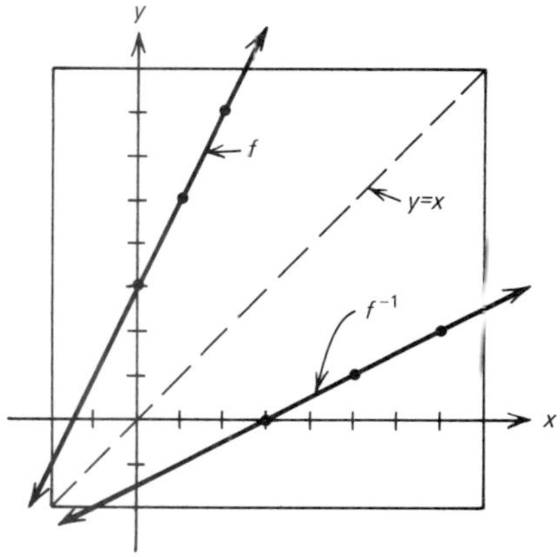


Notice that the lines for $f$ and $f^{-1}$ are symmetric about the line $y = x$. In other words, the line $y = x$ acts like a mirror, and $f$ and $f^{-1}$ are mirror images of each other. In this case both $f$ and $f^{-1}$ are functions since their graphs satisfy the vertical-line test (see page 295).

**EXAMPLE 19** Find and graph the inverse of $f$: $y = x^2 + 1$.

**SOLUTION** We first find the inverse.

Interchange $x$ and $y$

$$f^{-1}: \quad x = y^2 + 1$$

We now make our tables.

Make table for $f$

Reverse column for $f^{-1}$

| $f$: $x$ | $y$ | $f^{-1}$: $x$ | $y$ |
|---|---|---|---|
| $-2$ | 5 | 5 | $-2$ |
| $-1$ | 2 | 2 | $-1$ |
| 0 | 1 | 1 | 0 |
| 1 | 2 | 2 | 1 |
| 2 | 5 | 5 | 2 |

Graph both curves from tables

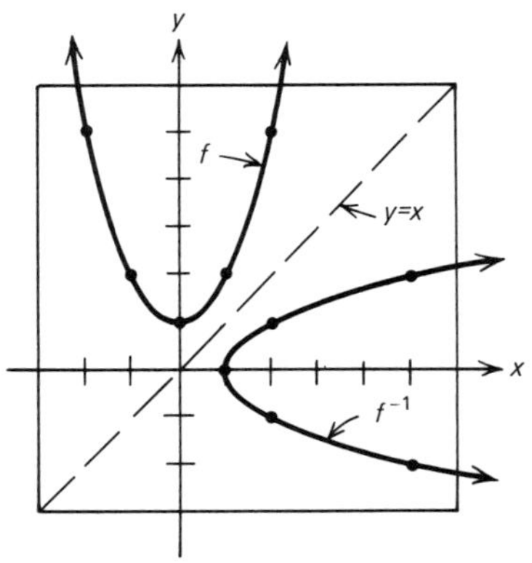


Again, notice the symmetry of $f$ and $f^{-1}$ about the line $y = x$. Although $f$ is a function, $f^{-1}$ is *not* a function, as can be seen using the vertical-line test.

**EXAMPLE 20** Find and graph the inverse of $f$: $\dfrac{x^2}{4} - \dfrac{y^2}{9} = 1$.

**SOLUTION** We first find the inverse.

| Interchange $x$ and $y$ | $f^{-1}: \frac{y^2}{4} - \frac{x^2}{9} = 1$ |
|---|---|

Both curves are hyperbolas. Recall that hyperbolas have real and imaginary intercepts. Here, we reverse roles for $f$ and $f^{-1}$.

Make table for $f$

$f$

| $x$ | $y$ | |
|---|---|---|
| 2 | 0 | real |
| −2 | 0 | |
| 0 | 3 | imaginary |
| 0 | −3 | |

Reverse columns for $f^{-1}$

$f^{-1}$

| $x$ | $y$ | |
|---|---|---|
| 0 | 2 | imaginary |
| 0 | −2 | |
| 3 | 0 | real |
| −3 | 0 | |

Graph both hyperbolas

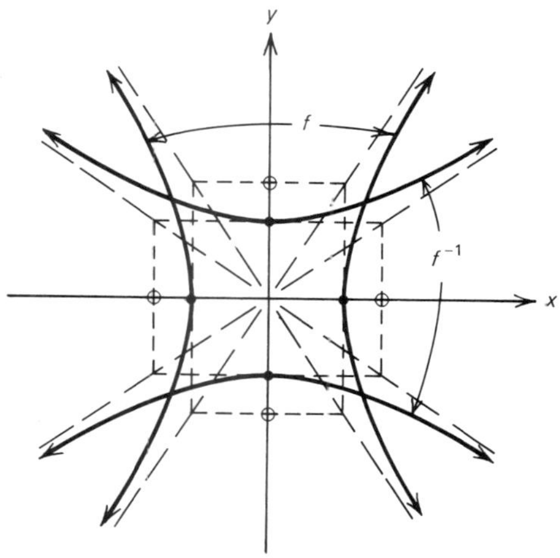


Here, neither $f$ nor $f^{-1}$ is a function, since they both fail the vertical-line test.

## PROBLEM SET 9.5

For each of the following relations, find the inverse. Then state whether $f$ and/or $f^{-1}$ is a function.

*See Example 17*

1. $f = \{(1, 5), (2, 7), (3, 10), (4, 14), (5, 19), (6, 25)\}$
2. $f = \{(10, 3), (20, 3), (30, 3), (40, 3)\}$
3. $f = \{(1, 5), (1, 6), (1, 7), (1, 8), (1, 9)\}$
4. $f = \{(2, 3), (4, 3), (3, 7), (1, 9), (5, 4), (2, 6)\}$

For each of the following relations $f$, find the relation for $f^{-1}$; then graph $f$ and $f^{-1}$ together and state whether $f$ and/or $f^{-1}$ is a function.

*See Example 18*

5. $f\colon\ y = 3x - 1$
6. $f\colon\ y = 2x + 3$
7. $f\colon\ y = 5x - 2$
8. $f\colon\ y = x + 7$
9. $f\colon\ 2x + 5y = 10$
10. $f\colon\ 3x - 4y = 12$

*See Example 19*

11. $f\colon\ y = x^2$
12. $f\colon\ y = -x^2$
13. $f\colon\ y = x^2 + 4$
14. $f\colon\ y = 2x^2 - 7$
15. $f\colon\ y = x^2 - 5x + 4$
16. $f\colon\ y = x^2 - x - 2$

*See Example 20*

17. $f\colon\ \dfrac{x^2}{4} + \dfrac{y^2}{25} = 1$
18. $f\colon\ \dfrac{x^2}{9} + \dfrac{y^2}{4} = 1$
19. $f\colon\ 16x^2 + 16y^2 = 64$
20. $f\colon\ 9x^2 + y^2 = 36$
21. $f\colon\ \dfrac{x^2}{9} - \dfrac{y^2}{1} = 1$
22. $f\colon\ \dfrac{-x^2}{16} + \dfrac{y^2}{4} = 1$
23. $f\colon\ -9x^2 + 25y^2 = 225$
24. $f\colon\ 4x^2 - 36y^2 = 36$

# CHAPTER 9 SUMMARY

## Important Words and Phrases

asymptotes (9.3)
circle (9.3)
conic sections (9.2)
domain (9.1)
ellipse (9.3)
function (9.1)
helping rectangle (9.3)
hyperbola (9.3)
imaginary intercepts (9.3)
intercept form (9.3)

inverse (9.5)
line $y = x$ (9.5)
opens downward (9.2)
opens upward (9.2)
parabola (9.2)
quadratic function (9.2)
range (9.1)
relation (9.1)
second-degree inequality (9.4)
vertex (9.2)
vertical-line test (9.1)
$x$-intercept (9.2)
$y$-intercept (9.2)

## Important Procedures

*To graph the quadratic function, $y = f(x) = ax^2 + bx + c$:*

Method I:
1. Make a table of $(x, y)$ pairs.
2. Plot these pairs.
3. Draw a smooth curve through these points.

Method II:
1. Factor and solve to find the $x$-intercepts.
2. The $y$-intercepts is at $(0, c)$.
3. The vertex is at $(-b/2a, f(-b/2a))$;
4. Plot these points and draw a smooth curve through them.

*To graph an ellipse, $Ax^2 + By^2 = C$:*
1. Find the intercepts by setting $x = 0$ and $y = 0$.
2. Sketch a "helping rectangle" through these points.
3. Draw the ellipse within the rectangle through the intercepts.

*To graph a hyperbola, $Ax^2 - By^2 = C$ or $-Ax^2 + By^2 = C$:*
1. Find the real and imaginary intercepts.
2. Sketch a "helping rectangle."
3. Draw asymptotes through corners.
4. Draw two parts of hyperbola through real intercepts and approaching asymptotes.

*To graph a second-degree inequality:*

1. Graph the corresponding equality.
2. Determine if a test point $P$ satisfies the inequality.

3a. If $P$ satisfies inequality, shade region containing $P$.
3b. If not, shade region not containing $P$.
4a. If inequality is $\leq$ or $\geq$, include curve.
4b. If inequality is $<$ or $>$, exclude curve.

*To find and graph an inverse:*
1. Interchange $x$- and $y$-variables.
2. Make a table of $(x, y)$ pairs for equation of $f$.
3. Reverse columns for $f^{-1}$.
4. Graph both tables.

## CHAPTER 9 REVIEW EXERCISES

Let $f = \{(1, 3), (3, 5), (8, 10), (11, 5), (13, 2)\}$.

1. Find the domain of $f$.
2. Find the range of $f$.
3. Find the inverse of $f$.
4. Graph $f$ and $f^{-1}$.
5. Is $f$ a function?
6. If $f^{-1}$ a function?

Let $f(x) = x^2 + 1$.

7. Find $f(0)$.
8. Find $f(3)$.
9. Find $f[f(3)]$.
10. Find $f(x + h)$.
11. Find $f(x + h) - f(x)$.
12. Find $\dfrac{f(x + h) - f(x)}{h}$.

Graph the following equations and inequalities.

13. $y = 3x^2 + x - 7; \quad -3 \leq x \leq 3$
14. $y = x^2 - 3x - 10$
15. $x^2 + y^2 = 16$
16. $\dfrac{x^2}{9} + \dfrac{y^2}{9} = 1$
17. $4x^2 + 16y^2 = 16$
18. $\dfrac{x^2}{9} + \dfrac{y^2}{25} = 1$
19. $9x^2 - y^2 = 36$
20. $\dfrac{-x^2}{4} + \dfrac{y^2}{25} = 1$
21. $x^2 - 25y^2 = 100$
22. $y < x^2 - 9$
23. $x^2 + 4y^2 \leq 4$
24. $9x^2 - 4y^2 > 36$

For the following relations $f$, find the relation for $f^{-1}$; graph both $f$ and $f^{-1}$; and state whether $f$ and/or $f^{-1}$ is a function.

25. $f$: $y = 2x - 5$
26. $f$: $y = x^2 - 1$
27. $f$: $4x^2 + y^2 = 36$
28. $f$: $9x^2 - 4y^2 = 36$

# 10

# SYSTEMS OF EQUATIONS AND INEQUALITIES

Consider the two equations

$$A: \quad x + y = 8$$
$$B: \quad x - y = 2$$

The pair (6, 2) satisfies $A$, but not $B$. The pair (9, 7) satisfies $B$, but not $A$. However, the **ordered pair** (5, 3) satisfies *both* $A$ and $B$.

A set of two or more equations for which we seek a common solution is called a **system of equations.** We seek a solution set that will satisfy all the equations at the same time.

## 10.1 SYSTEMS OF LINEAR EQUATIONS IN TWO VARIABLES

The simplest system to study is that of linear equations. Such a system is called a **linear system.** In this section, we consider linear systems with just two variables, $x$ and $y$. These are the equations of two straight lines, $l_1$ and $l_2$, and there are three possible cases, as shown in Fig. 10.1.

*Case* (a). The two lines intersect in a single point. This point $(a, b)$ is the solution set of the system. This is the most common case that we will see.

*Case* (b). The two lines are parallel and, of course, do not intersect. The solution set is empty. We call this system **inconsistent.**

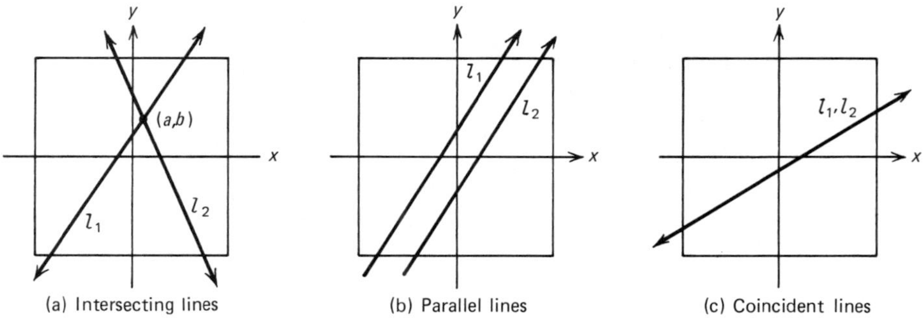


(a) Intersecting lines (b) Parallel lines (c) Coincident lines

**Figure 10.1**

*Case* (c). The two lines turn out to really be the same line. The solution set is either line. This system is called **dependent.**

One way to find the solution (if any) is to graph the two equations. But this can be difficult and inaccurate. Instead, we find the solution algebraically, using the **elimination method.** The following theorem allows us to eliminate $x$ or $y$ and then solve a simpler equation.

***THEOREM 10.1*** If the ordered pair $(x, y)$ satisfies

$$A: \quad ax + by = e$$

$$B: \quad cx + dy = f$$

then $(x, y)$ also satisfies $M(ax + by) + N(cx + dy) = Me + Nf$, for any real numbers $M$ and $N$. This theorem is best seen in the examples that follow.

**EXAMPLE 1** Solve the system

$$A: \quad x + y = 8$$

$$B: \quad x - y = 2$$

**SOLUTION** Theorem 10.1 allows us to add these equations together. This will eliminate $y$.

| | |
|---|---|
| Given | $A: \quad x + y = 8$<br>$B: \quad x - y = 2$ |
| Add equations to eliminate $y$ | $A + B: \quad 2x = 10$ |
| Solve for $x$ | $x = 5$ |
| Substitute $x = 5$ into $A$ | $A: \quad 5 + y = 8$ |
| Solve for $y$ | $y = 3$ |

Therefore, the solution is (5, 3). We can check by substituting this into $B$.

$$B: \quad 5 - 3 = 2 \qquad \text{Checks}$$

> To solve a system of two linear equations:
>
> 1. Write both equations in the form $ax + by = c$.
> 2. Choose a variable ($x$ or $y$) to be eliminated.
> 3. Multiply each equation to make the coefficients of the variable to be eliminated the opposites of each other (such as $5y$ and $-5y$).
> 4. Add the equations.
> 5. Solve for the remaining variable.
> 6. Substitute this value back into one of the original equations and solve for the other variable.
> 7. Check the solution by substituting into the other equation.
>
> *Note:* Step 4 may produce two unusual results:
>
> a. A false result, such as $0 = 4$, means that the lines are parallel, and there is no solution.
>
> b. A true result, such as $0 = 0$, means that the lines or equations are the same.

**EXAMPLE 2** Solve the system

$$\begin{aligned} A: &\quad x + 2y = 1 \\ B: &\quad 3x - y = 17 \end{aligned}$$

**SOLUTION** If we multiply both sides of equation $B$ by 2, we will have $2y$ in equation $A$ and $-2y$ in equation $B$. These will cancel when we add. We purposely chose to multiply both sides of equation $B$ by 2 to cancel the $y$-terms.

Given ↓

$$\begin{aligned} A: &\quad x + 2y = 1 \\ B: &\quad 3x - y = 17 \end{aligned}$$

| | |
|---|---|
| ↓ | |
| Multiply both sides of $B$ by 2 to eliminate $y$ | $A: \quad x + 2y = 1$<br>$2B: \quad 6x - 2y = 34$ |
| ↓ | |
| Add equations | $A + 2B: \quad 7x = 35$ |
| ↓ | |
| Solve for $x$ | $x = 5$ |
| ↓ | |
| Substitute $x = 5$ into $A$ | $A: \quad 5 + 2y = 1$ |
| ↓ | |
| Solve for $y$ | $y = -2$ |

The solution is $(5, -2)$. Let us check this in equation $B$:

$$B: \quad 3(5) - (-2) = 15 + 2 = 17 \qquad \text{Checks}$$

**EXAMPLE 3** Solve the system

$$\begin{aligned} A&: \quad 3x - 2y = -12 \\ B&: \quad 2x + 5y = 11 \end{aligned}$$

**SOLUTION** Here, we have to multiply both equations. If we multiply both sides of equation $A$ by 5 and both sides of equation $B$ by 2, we get $-10y$ in equation $A$ and $10y$ in equation $B$. These will cancel when we add.

| | |
|---|---|
| Given | $A: \quad 3x - 2y = -12$<br>$B: \quad 2x + 5y = 11$ |
| ↓ | |

| | |
|---|---|
| Multiply both sides of $A$ by 5, both sides of $B$ by 2 to eliminate $y$ | $5A\colon\ 15x - 10y = -60$<br>$2B\colon\ 4x + 10y = 22$ |
| Add equations | $5A + 2B\colon\ 19x = -38$ |
| Solve for $x$ | $x = -2$ |
| Substitute $x = -2$ into $A$ | $A\colon\ 3(-2) - 2y = -12$ |
| Simplify | $-6 - 2y = -12$ |
| Solve for $y$ | $y = 3$ |

The solution is $(-2, 3)$. We check this in equation $B$:

$$B\colon\ 2(-2) + 5(3) = -4 + 15 = 11 \qquad \text{Checks}$$

**EXAMPLE 4** Solve the system

$$A\colon\quad \frac{x}{4} - \frac{y}{2} = \frac{1}{2}$$

$$B\colon\quad \frac{-2x}{7} + \frac{4y}{7} = 1$$

**SOLUTION** Before we try to eliminate a variable, we first clear the fractions by multiplying each equation by its LCD (4 for equation $A$ and 7 for equation $B$).

| | |
|---|---|
| Multiply by LCDs to clear fractions | $4A\colon\ x - 2y = 2$<br>$7B\colon\ -2x + 4y = 7$ |

| | |
|---|---|
| Multiply both sides of (4$A$) by 2 to eliminate $x$ | $2(4A)$: $2x - 4y = 4$<br>$7B$: $-2x + 4y = 7$ |
| Add equations | $8A + 7B$: $0 = 11$ |
| Result is false | ????? |

We worked the problem correctly but got $0 = 11$, which is false. This tells us that there is *no solution* (the lines are parallel). The solution set is $\varnothing$. (The student should check that both lines have the slope $\frac{1}{2}$.)

**EXAMPLE 5** Solve the system

$$A: \quad 2x + 5y = 11$$
$$B: \quad 4x + 10y = 22$$

**SOLUTION** We multiply both sides of equation $A$ by $-2$ to eliminate $y$.

| | |
|---|---|
| Multiply both sides of $A$ by 2 to eliminate $y$ | $-2A$: $-4x - 10y = -22$<br>$B$: $4x + 10y = 22$ |
| Add equations | $-2A + B$: $0 = 0$ |
| Result has no unknowns! | ????? |

Unlike Example 4, $0 = 0$ is true, but it does not enable us to find an intersection point. In this case, $0 = 0$ means that both lines are really the *same* line. Therefore, the solution is the entire equation $A$ (or $B$).

**EXAMPLE 6** Solve the following linear system.

$$A: \quad 2x + 3y = 25$$
$$B: \quad y = 4x - 1$$

**SOLUTION** We solve this in a different way. We take the expression for $y$ in equation $B$ and substitute it into equation $A$ for $y$. This is called the **substitution method.**

| | |
|---|---|
| Given | $A$: $2x + 3y = 25$<br>$B$: $y = 4x - 1$ |
| Substitute $y = 4x - 1$ into $A$ | $2x + 3(4x - 1) = 25$ |
| Distributive law | $2x + 12x - 3 = 25$ |
| Combine terms | $14x - 3 = 25$ |
| Add 3 to both sides | $14x = 28$ |
| Divide both sides by 2 | $x = 2$ |
| Substitute $x = 2$ into $B$ | $y = 4(2) - 1$ |
| Simplify | $y = 7$ |

Thus, the solution is (2, 7). We can check this in equation $A$.

$$A: \quad 2(2) + 3(7) = 4 + 21 = 25 \qquad \text{Checks}$$

## PROBLEM SET 10.1

Solve the following linear systems. (*Be careful:* Some may be inconsistent or dependent—see Examples 4 and 5.)

*See Examples 1 and 2*

1. $A$: $x + y = 6$
   $B$: $x - y = 4$

2. $A$: $x + y = 1$
   $B$: $x - y = 7$

3. $A$: $-x - y = 2$
   $B$: $x - y = 2$

4. $A$: $-x + y = 2$
   $B$: $x + y = 10$

5. $A$: $x - y = -1$
   $B$: $2x + 3y = 8$

6. $A$: $x + y = 4$
   $B$: $5x - 2y = 13$

7. $A$: $x + 2y = 5$
   $B$: $3x - y = -6$

8. $A$: $3x + y = 2$
   $B$: $6x + 2y = 4$

9. $A$: $3x + 4y = 14$
   $B$: $x + y = 3$

10. $A$: $2x + 3y = 10$
    $B$: $x - 4y = -6$

11. $A$: $x + 5y = 3$
    $B$: $3x + 7y = 9$

12. $A$: $-4x + 3y = 23$
    $B$: $x - 5y = -10$

*See Example 3*

13. $A$: $2x + 3y = 7$
    $B$: $3x - 4y = 2$

14. $A$: $3x - 2y = 14$
    $B$: $4x + 5y = 11$

15. $A$: $2x + 5y = 15$
    $B$: $-3x + 2y = 6$

16. $A$: $6x - 5y = -32$
    $B$: $3x + 2y = 2$

17. $A$: $5x - 2y = 3$
    $B$: $-10x + 4y = -6$

18. $A$: $4x + 5y = -1$
    $B$: $6x - 7y = 13$

19. $A$: $-3x + 4y = 6$
    $B$: $5x - 7y = 2$

20. $A$: $2x - 9y = 5$
    $B$: $3x + 4y = 2$

21. $A$: $10x + 3y = 8$
    $B$: $3x - 4y = 1$

22. $A$: $3x - 5y = 2$
    $B$: $4x - 3y = 3$

*See Example 4*

23. $A$: $\dfrac{-x}{5} + \dfrac{3y}{10} = 1$
    $B$: $\dfrac{x}{2} + \dfrac{y}{2} = \dfrac{5}{2}$

24. $A$: $\dfrac{4x}{3} - y = \dfrac{-17}{3}$
    $B$: $x + \dfrac{5y}{3} = 3$

25. $A: \quad \dfrac{-x}{2} - \dfrac{5y}{4} = 1$

$B: \quad -x + \dfrac{2y}{3} = 2$

26. $A: \quad \dfrac{x}{7} + y = \dfrac{13}{7}$

$B: \quad x - \dfrac{y}{3} = \dfrac{17}{3}$

27. $A: \quad \dfrac{x}{3} - \dfrac{4}{3}y = 5$

$B: \quad \dfrac{-x}{2} + 2y = 3$

28. $A: \quad \dfrac{x}{2} - \dfrac{y}{4} = 2$

$B: \quad \dfrac{x}{2} - \dfrac{y}{3} = \dfrac{-5}{6}$

29. $A: \quad \dfrac{x}{5} - \dfrac{y}{3} = \dfrac{4}{15}$

$B: \quad \dfrac{x}{4} + \dfrac{y}{3} = \dfrac{7}{12}$

30. $A: \quad \dfrac{x}{6} - \dfrac{y}{3} = \dfrac{5}{9}$

$B: \quad \dfrac{x}{5} + \dfrac{y}{10} = \dfrac{8}{15}$

*See Example 6*

31. $A: \quad 3x + 2y = 13$
$B: \quad y = 5x$

32. $A: \quad 4x - 3y = -4$
$B: \quad y = 2x$

33. $A: \quad 4x - 3y = 30$
$B: \quad y = x - 9$

34. $A: \quad 5x + 2y = -9$
$B: \quad y = 3x + 1$

35. $A: \quad 6x + 5y = -3$
$B: \quad y = 2x - 7$

36. $A: \quad 4x - 3y = 26$
$B: \quad y = -2x + 8$

37. $A: \quad y = 4x + 2$
$B: \quad y = 7x - 1$

38. $A: \quad y = 4x - 5$
$B: \quad y = x + 1$

39. $A: \quad y = -2x - 8$
$B: \quad y = x + 7$

40. $A: \quad y = 2x + 10$
$B: \quad y = -x - 5$

*Consumer application*

41. The annual cost to run a certain used car is given by $C = 800 + 0.08m$, where $m$ is miles driven. The annual cost to run a certain new car is $C = 1200 + 0.06m$. Find the intersection, the number of miles at which the cost is the same.

*Business applications*

42. Susan has two job offers. One pays \$12,000 to start plus a \$500-per-year raise. The second starts at \$15,000 with a \$350-per-year raise. If $x$ is the number of years and $y$ is pay, solve the following system to find when the salaries will be the same (and what the pay will be).

$$A: \quad y = 12{,}000 + 500x$$

$$B: \quad y = 15{,}000 + 350x$$

43. The cost to produce $x$ number of an item is $y = 20{,}000 + 60x$. The total revenue from their sales is given by $y = 100x$. Solve these two equations to find the *break-even* point, where the cost equals the revenue.

*Social application*

44. The population of Beaverville is 4000 and is growing by 600 people per year. Its neighbor, Squirrel Junction, has 8000 people but is losing 400 per year. If $x$ is years and $y$ is population, solve the following system to find when the populations will be the same.

$$A: \quad y = 4000 + 600x$$
$$B: \quad y = 8000 - 400x$$

*Engineering application*

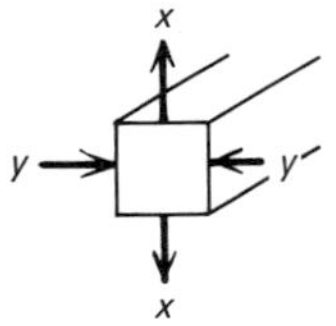


45. The stresses on a certain column satisfy the following system:

$$A: \quad x + y = 49{,}285$$
$$B: \quad x - y = 49{,}615$$

Solve this system for the stresses $x$ and $y$.

## 10.2 WORD PROBLEMS

Thus far, we have seen word problems with only one unknown, usually $x$. In this section, we study problems with two unknowns, $x$ and $y$. Just as with the previous word problems, the key will be identifying the unknowns and setting up the equations.

### Translation Problems

As in Chapters 2, 5, and 7, some word problems are straight translations from English into mathematical equations.

**EXAMPLE 7** The larger of two numbers is one more than three times the smaller. Their difference is 7. Find the numbers.

**SOLUTION** We identify the unknowns and translate.

|  | |
|---|---|
| Identify unknowns | Let $x$ = larger number<br>$y$ = smaller number |
| Translate first sentence | Larger is one more than 3 times smaller<br>$x = 1 + 3 \cdot y$ |
| Rewrite | $A$: $x - 3y = 1$ |
| Translate second sentence | Difference is 7<br>$B$: $x - y = 7$ |
| Multiply both sides of $A$ by $-1$ to eliminate $x$ | $-A$: $-x + 3y = -1$<br>$B$: $x - y = 7$ |
| Add equations | $-A + B$: $2y = 6$ |
| Solve for $y$ | $y = 3$ |
| Substitute $y = 3$ into $B$ | $B$: $x - 3 = 7$ |
| Solve for $x$ | $x = 10$ |

Therefore, the numbers are 10 and 3.

## Mixture Problems

As we saw in Chapter 2, mixture and interest problems are very similar.

> To solve mixture or interest problems:
>
> 1. Let $x$ and $y$ be the two unknown amounts mixed or invested.
> 2. Identify the amounts of pure substance or interest.

3. Write an equation for the total amount mixed or invested.
4. Write an equation for the amount of pure substance or interest.
5. Solve this system.

**EXAMPLE 8** How many milliliters of a 30% acid solution and a 50% acid solution must be mixed to get 1000 ml of a 45% acid solution?

**SOLUTION** We have two amounts: the total solution of 1000 ml, and 450 ml (= 45% of 1000) of pure acid. We now identify the unknown amounts of the solutions and the acid they contribute.

| | |
|---|---|
| Identify unknowns | Let $x$ = amount of 30% solution<br>$y$ = amount of 50% solution |
| Identify acid amounts | $0.30x$ = amount of acid in 30% solution<br>$0.50y$ = amount of acid in 50% solution |
| Equation for total | $A: \quad x + y = 1000$ |
| Equation for acid | $B: \quad 0.30x + 0.50y = 450$ |
| Multiply both sides of $A$ $A$ by $-0.30$ to eliminate $x$ | $-0.30A: \quad -0.30x - 0.30y = -300$<br>$B: \quad 0.30x + 0.50y = 450$ |
| Add equations | $-0.30A + B: \quad 0.20y = 150$ |
| Solve for $y$ | $y = 750$ |
| Substitute $y = 750$ into $A$ | $A: \quad x + 750 = 1000$ |
| Solve for $x$ | $x = 250$ |

Therefore, the solution is 250 ml of 30% solution (= 75 ml of acid) and 750 ml of 50% solution (= 375 ml of acid). Together, this is 450 ml of acid.

## Value Problems

These problems involve coin values, nutrition values, and so on.

> To solve value problems:
>
> 1. Let $x$ and $y$ be the unknown amounts.
> 2. Find the coin (or other) value of these amounts.
> 3. Write an equation for the total amount.
> 4. Write an equation for the total value.
> 5. Solve this system.

**EXAMPLE 9** Jamie has 19 coins (in nickels and dimes). Their total value is \$1.30. How many of each coin does she have?

**SOLUTION** If she has $x$ nickels, they are worth $5x$ cents; similarly, $y$ dimes are worth $10y$ cents. We write \$1.30 as 130¢.

| Step | |
|---|---|
| Identify unknowns ↓ | Let $x$ = number of nickels<br>$y$ = number of dimes |
| Identify value of coins ↓ | $5x$ = value of nickels<br>$10y$ = value of dimes |
| Equation for number of coins ↓ | $A$: $x + y = 19$ |
| Equation for value of coins ↓ | $B$: $5x + 10y = 130$ |

| | |
|---|---|
| Multiply both sides of $A$ by $-5$ to eliminate $x$ | $-5A$: $-5x - 5y = -95$ <br> $B$: $5x + 10y = 130$ |
| Add equations | $-5A + B$: $5y = 35$ |
| Solve for $y$ | $y = 7$ |
| Substitute $y = 7$ into $A$ | $A$: $x + 7 = 19$ |
| Solve for $x$ | $x = 12$ |

Therefore, she has 12 nickels (60¢) and 7 dimes (70¢). This is a total of 130¢, or \$1.30.

## PROBLEM SET 10.2

*See Example 7*

1. The sum of two numbers is 15, and their difference is 7. Find the numbers.
2. The sum of two numbers is 21, and their difference is 15. Find the numbers.
3. The larger of two numbers is two more than twice the smaller. Their sum is 17. Find the numbers.
4. The larger of two numbers is one more than twice the smaller. Their difference is 8. Find the numbers.
5. The larger of two numbers is three more than twice the smaller. The larger is two less than three times the smaller. Find both numbers.
6. The larger of two numbers is one more than three times the smaller. The larger is five less than four times the smaller. Find the numbers.

*See Example 8*

7. How many ounces of a 20% iodine solution and a 40% iodine solution must be mixed to produce 60 ounces of a 25% iodine solution?

8. How many pounds of 70% lean meat and 85% lean meat must be mixed to produce 200 pounds of 80% lean meat?

9. How many ounces of a 70% gold alloy and a 90% gold alloy must be mixed to produce 20 ounces of a 85% gold alloy?

10. How many liters of a 40% alcohol solution and a 90% alcohol solution must be mixed to produce 50 liters of a 50% alcohol solution?

11. Mr. Marx has \$8000. He invests some at 5% and the rest at 7%. Together, he earns \$440 interest. How much did he invest at each rate?

12. Mrs. Hadley invests \$12,000, some at 6%, the rest at 10%. Together, she earns \$840 interest. How much did she invest at each rate?

*See Example 9*

13. Eddie has 18 coins (nickels and dimes). Their value is \$1.40 (140¢). How many coins of each does he have?

14. Judy has 15 coins (dimes and quarters). Their value is \$2.70 (270¢). How many coins of each does she have?

15. A theater sells 1000 tickets to a show, some at \$3, the rest at \$5. If their revenue is \$3600, how much of each ticket did they sell?

16. A bazaar sells 350 tickets, \$0.50 for children, \$1.50 for adults. If they collect \$325 for the tickets, how much of each did they sell?

17. A confectioner has \$2-per-pound candy and \$3-per-pound candy. How much of each must she mix to have 20 pounds of \$2.70-per-pound candy?

18. A patient in a hospital drank 12 ounces of juice, some prune and the rest orange. An ounce of prune juice provides 5 mg of thiamine, while orange juice provides 8 mg of thiamine. If the patient got 84 mg of thiamine, how much of each juice did he drink?

## 10.3 SYSTEMS OF LINEAR EQUATIONS IN THREE VARIABLES

Consider the system

$$\begin{aligned} A&: \quad x + y + z = 6 \\ B&: \quad x + y - z = 2 \\ C&: \quad x - 2y - z = -1 \end{aligned}$$

Here we have three linear equations ($A$, $B$, $C$) and three unknown variables

$(x, y, z)$. The solution is the **ordered triple** $(3, 1, 2)$ which satisfies all three equations at the same time.

Naturally, solving a linear system of three equations is more difficult than a system of two equations, but the principle is the same: *eliminate variables until there is one equation with one variable.*

> To solve a linear system of three equations $A$, $B$, $C$:
>
> 1. Using two of the equations, eliminate one of the variables; call the resulting equation $D$.
> 2. Using another pair of the equations, eliminate the *same* variable; call the resulting equation $E$.
> 3. Taking these new equations, $D$ and $E$, eliminate a second variable.
> 4. Solve for the last remaining variable.
> 5. Substitute this variable into $D$ or $E$ and solve for the second variable.
> 6. Substitute these two variables into $A$, $B$, or $C$ and solve for the first variable.

**EXAMPLE 10** Solve the system

$$\begin{aligned} A&: \ x + y + z = 6 \\ B&: \ x + y - z = 2 \\ C&: \ x - 2y - z = -1 \end{aligned}$$

**SOLUTION** We first eliminate $z$, then eliminate $x$, and finally solve for $y$. Once we have $y$, we then find $x$, and finally $z$.

Add $A + B$ to eliminate $z$; call new equation $D$

$$\begin{array}{rl} A: & x + y + z = 6 \\ B: & x + y - z = 2 \\ \hline D = A + B: & 2x + 2y = 8 \end{array}$$

↓

Add $A + C$ to eliminate $z$; call new equation $E$

$$\begin{array}{rl} A: & x + y - z = 6 \\ C: & x - 2y - z = -1 \\ \hline E = A + C: & 2x - y = 5 \end{array}$$

↓

| Step | |
|---|---|
| Multiply both sides of $E$ by $(-1)$ and add to $D$ to eliminate $x$ | $D$: $2x + 2y = 8$ <br> $(-1)E$: $-2x + y = -5$ <br> $D + (-1)E$: $3y = 3$ |
| ↓ Solve for $y$ | $y = 1$ |
| ↓ Substitute $y = 1$ into $E$ | $E$: $2x - 1 = 5$ |
| ↓ Solve for $x$ | $x = 3$ |
| ↓ Substitute $x = 3$ and $y = 1$ into $A$ | $A$: $3 + 1 + z = 6$ |
| ↓ Solve for $z$ | $z = 2$ |

Therefore, the solution is (3, 1, 2). We can check this in equation $B$ or $C$.

$B$: $3 + 1 - 2 = 2$ Checks

$C$: $3 - 2 - 2 = -1$ Checks

**EXAMPLE 11** Solve the system

$$\begin{aligned} A&: \ 2x - 3y + z = 13 \\ B&: \ 3x - 2y + 4z = 22 \\ C&: \ x - 5y - 2z = 6 \end{aligned}$$

**SOLUTION** We first eliminate $z$.

| | |
|---|---|
| Add $-4A + B$ to eliminate $z$; call new equation $D$ | $-4A$: $-8x + 12y - 4z = -52$<br>$B$: $3x - 2y + 4z = 22$<br>$D = -4A + B$: $-5x + 10y = -30$ |
| ↓ Add $2A + C$ to eliminate $z$; call new equation $E$ | $2A$: $4x - 6y + 2z = 26$<br>$C$: $x - 5y - 2z = 6$<br>$E = 2A + C$: $5x - 11y = 32$ |
| ↓ Add $D + E$ to eliminate $x$ | $D$: $-5x + 10y = -30$<br>$E$: $5x - 11y = 32$<br>$D + E$: $-y = 2$ |
| ↓ Solve for $y$ | $y = -2$ |
| ↓ Substitute $y = -2$ into $E$ | $E$: $5x - 11(-2) = 32$ |
| ↓ Solve for $x$ | $x = 2$ |
| ↓ Substitute $x = 2$ and $y = -2$ into $A$ | $A$: $2(2) - 3(-2) + z = 13$<br>$4 + 6 + z = 13$ |
| ↓ Solve for $z$ | $z = 3$ |

Therefore, the solution is $(2, -2, 3)$. We check this in $B$ and $C$.

$B$: $3(2) - 2(-2) + 4(3) = 6 + 4 + 12 = 22$ Checks

$C$: $2 - 5(-2) - 2(3) = 2 + 10 - 6 = 6$ Checks

**EXAMPLE 12** Solve the system

$$\begin{aligned} A:&\quad x - 2z = 4 \\ B:&\quad x + y = 11 \\ C:&\quad y + 2z = 5 \end{aligned}$$

**SOLUTION** It is easier to work with these equations if we rewrite them as follows. Notice that equation $B$ does not have a $z$ term. Thus, we add equations $A$ and $C$ to also eliminate $z$.

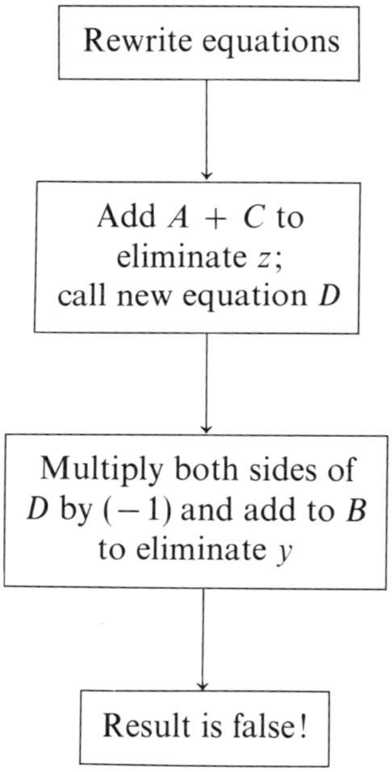


$$\begin{array}{rrrr} A: & x & -2z = & 4 \\ B: & x + y & = & 11 \\ C: & & y + 2z = & 5 \end{array}$$

$$\begin{array}{rrrr} A: & x & -2z = & 4 \\ C: & & y + 2z = & 5 \\ \hline D = A + C: & x + y & = & 9 \end{array}$$

$$\begin{array}{rrr} B: & x + y & = 11 \\ (-1)D: & -x - y & = -9 \\ \hline & 0 & = 2 \end{array}$$

?????

As in Section 10.1, $0 = 2$ means that we have an inconsistent system, or *no solution.*

## PROBLEM SET 10.3

Solve the following linear systems. (*Be careful:* Some may not have solutions—see Example 12.)

*See Examples 10 and 11*

1. $A$: $x + y + z = 5$
   $B$: $x + y - z = 9$
   $C$: $x - y - z = 1$

2. $A$: $x + y + z = 8$
   $B$: $x - y + z = 4$
   $C$: $x - y - z = 2$

3. $A$: $x + y - z = 6$
   $B$: $2x + y + z = 3$
   $C$: $3x - y - z = 2$

4. $A$: $x - y + z = -8$
   $B$: $2x - y + 3z = -10$
   $C$: $4x + y - z = -12$

5. $A$: $x + 2y + 3z = 5$
   $B$: $2x + y + 4z = 4$
   $C$: $x - y + z = 6$

6. $A$: $x - 4y + 2z = 20$
   $B$: $3x - y - 3z = -7$
   $C$: $4x - 5y + 4z = 38$

7. $A$: $2x - y + 5z = 31$
   $B$: $3x + 4y + 2z = 19$
   $C$: $4x - 2y + 3z = 27$

8. $A$: $3x + 4y - 5z = 26$
   $B$: $2x - 5y - 3z = 1$
   $C$: $6x - 2y - z = 5$

9. $A$: $5x - 2y + 3z = 18$
   $B$: $2x - 4y - 3z = 1$
   $C$: $4x - 3y - 2z = 4$

10. $A$: $2x + 3y - 4z = -6$
    $B$: $3x + 4y - 5z = -7$
    $C$: $4x - 5y + 6z = 8$

*See Examples 10, 11, and 12*

11. $A$: $x + y = 1$
    $B$: $x - z = 4$
    $C$: $y - z = 3$

12. $A$: $x + y = 6$
    $B$: $y + z = 2$
    $C$: $x + z = 4$

13. $A$: $x + 2y = 10$
    $B$: $x - 3z = 23$
    $C$: $4y - z = 9$

14. $A$: $y - 2z = -1$
    $B$: $x + 5y = -10$
    $C$: $x - 3z = 8$

15. $A$: $3x + 2y = -3$
    $B$: $2x - 5z = -38$
    $C$: $4y + 3z = 12$

16. $A$: $4x - 3z = -5$
    $B$: $3y + 4z = 12$
    $C$: $2x - 5y = 2$

*See Examples 4, 10, and 11*

17. $A$: $\frac{x}{2} - \frac{3y}{4} + z = \frac{23}{4}$
    $B$: $\frac{x}{5} - \frac{y}{40} + \frac{z}{8} = \frac{37}{40}$
    $C$: $\frac{x}{4} + \frac{y}{3} - \frac{z}{6} = \frac{-1}{2}$

18. $A$: $\frac{2x}{5} - \frac{2y}{15} + \frac{z}{5} = 2$
    $B$: $\frac{x}{2} - \frac{3y}{10} + \frac{z}{5} = \frac{12}{5}$
    $C$: $\frac{x}{2} - \frac{y}{8} + \frac{5z}{8} = \frac{13}{4}$

19. $A$: $\frac{x}{3} - \frac{y}{4} + \frac{z}{6} = \frac{-1}{6}$
    $B$: $\frac{x}{5} + \frac{y}{3} + \frac{z}{5} = \frac{32}{15}$
    $C$: $\frac{x}{6} + \frac{y}{2} - \frac{z}{4} = \frac{17}{12}$

20. $A$: $\frac{x}{2} - \frac{y}{5} - \frac{z}{10} = \frac{11}{10}$
    $B$: $\frac{x}{3} - \frac{y}{12} + \frac{z}{4} = \frac{19}{12}$
    $C$: $\frac{x}{8} - \frac{y}{12} - \frac{z}{6} = \frac{-1}{12}$

*Business application*

21. A small furniture company makes chairs, sofas, and tables. Each day, they use 1790 pounds of wood, 123 yards of fabric, and 370 employee-

hours. The different usages of wood, fabric, and labor by the chairs, sofas, and tables lead to the following system:

$$\begin{aligned} A\colon&\quad 50x + 70y + 100z = 1790 \\ B\colon&\quad 6x + 9y = 123 \\ C\colon&\quad 15x + 20y + 10z = 370 \end{aligned}$$

Solve this system to determine how many chairs ($x$), sofas ($y$), and tables ($z$) the company makes each day.

*Electricity application* 22. The circuit shown has currents $x$, $y$, and $z$ through the three resistors of 10 ohms, 20 ohms, and 30 ohms. Kirchhoff's and Ohm's laws lead to the following system.

$$\begin{aligned} A\colon&\quad x + y + z = 0 \\ B\colon&\quad 10x - 20y = 70 \\ C\colon&\quad -20y + 30z = 10 \end{aligned}$$

Solve this system. (A negative result means that the current is opposite to the arrow in the sketch.)

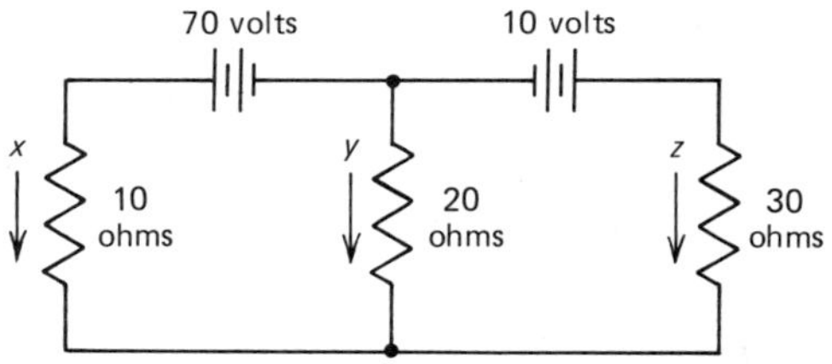


## 10.4 DETERMINANTS

A **determinant** is a real number associated with a square array of real numbers. Determinants help us calculate the solutions to linear systems with up to three equations and variables.

**DEFINITION 10.1** For real numbers $a$, $b$, $c$, and $d$, the value of a **2 by 2 determinant** is given by

$$\begin{vmatrix} a & b \\ c & d \end{vmatrix} = \begin{vmatrix} a & b \\ c & d \end{vmatrix} = ad - bc$$

**EXAMPLE 13** **(a)** $\begin{vmatrix} 3 & 5 \\ 2 & 8 \end{vmatrix} = \begin{vmatrix} 3 & 5 \\ 2 & 8 \end{vmatrix} = 3(8) - 2(5) = 24 - 10 = 14$

**(b)** $\begin{vmatrix} 7 & -4 \\ 2 & 1 \end{vmatrix} = \begin{vmatrix} 7 & -4 \\ 2 & 1 \end{vmatrix} = 7(1) - 2(-4) = 7 + 8 = 15$

| YES | NO |
|---|---|
| $\begin{vmatrix} 2 & 5 \\ 3 & 4 \end{vmatrix} = 8 - 15 = -7$ | $\begin{vmatrix} 2 & 5 \\ 3 & 4 \end{vmatrix} = 8 + 15 = 23$ |

We *subtract* the cross products.

The **3 by 3 determinant** is a bit harder. Although there are several ways to evaluate a 3 by 3 determinant, we only show the following method. (Note the different signs for each of the terms.)

**DEFINITION 10.2** The 3 by 3 determinant is defined as

$$\begin{vmatrix} a_1 & b_1 & c_1 \\ a_2 & b_2 & c_2 \\ a_3 & b_3 & c_3 \end{vmatrix} = \oplus\, a_1 \begin{vmatrix} b_2 & c_2 \\ b_3 & c_3 \end{vmatrix} \ominus\, a_2 \begin{vmatrix} b_1 & c_1 \\ b_3 & c_3 \end{vmatrix} \oplus\, a_3 \begin{vmatrix} b_1 & c_1 \\ b_2 & c_2 \end{vmatrix}$$

The 2 by 2 determinants that are multiplied by the elements of the first column are called the **minors.** We can best remember this by thinking of the minor as being formed by "knocking out" the rest of row and column of the first-column element.

**EXAMPLE 14** Evaluate

$$\begin{vmatrix} 1 & 3 & 4 \\ 2 & 6 & 3 \\ 3 & -5 & 8 \end{vmatrix}$$

**SOLUTION** We show how Definition 10.2 works.

| | |
|---|---|
| "Knock out" row and column to get minor | $\oplus\begin{vmatrix} ① & 3 & 4 \\ 2 & 6 & 3 \\ 3 & -5 & 8 \end{vmatrix} \ominus \begin{vmatrix} 1 & 3 & 4 \\ ② & 6 & 3 \\ 3 & -5 & 8 \end{vmatrix} \oplus \begin{vmatrix} 1 & 3 & 4 \\ 2 & 6 & 3 \\ ③ & -5 & 8 \end{vmatrix}$ |
| ↓ First-column elements times minors (use correct signs) | $= 1 \cdot \begin{vmatrix} 6 & 3 \\ -5 & 8 \end{vmatrix} - 2 \cdot \begin{vmatrix} 3 & 4 \\ -5 & 8 \end{vmatrix} + 3 \cdot \begin{vmatrix} 3 & 4 \\ 6 & 3 \end{vmatrix}$ |
| ↓ Evaluate minors | $= 1 \cdot 63 - 2(44) + 3(-15)$ |
| ↓ Simplify | $= -70$ |

**EXAMPLE 15** Evaluate

$$\begin{vmatrix} 4 & 3 & -2 \\ 0 & 6 & 3 \\ 2 & -4 & 5 \end{vmatrix}$$

**SOLUTION** We expand this with minors.

| | |
|---|---|
| "Knock out" row and column to get minor | $= \oplus\begin{vmatrix} ④ & 3 & -2 \\ 0 & 6 & 3 \\ 2 & -4 & 5 \end{vmatrix} \ominus \begin{vmatrix} 4 & 3 & -2 \\ ⓪ & 6 & 3 \\ 2 & -4 & 5 \end{vmatrix} \oplus \begin{vmatrix} 4 & 3 & -2 \\ 0 & 6 & 3 \\ ② & -4 & 5 \end{vmatrix}$ |
| ↓ Elements × minors | $= 4 \cdot \begin{vmatrix} 6 & 3 \\ -4 & 5 \end{vmatrix} - 0 \cdot \begin{vmatrix} 3 & -2 \\ -4 & 5 \end{vmatrix} + 2 \cdot \begin{vmatrix} 3 & -2 \\ 6 & 3 \end{vmatrix}$ |
| ↓ Evaluate minors | $= 4(42) - 0(7) + 2(21)$ |
| ↓ Simplify | $= 210$ |

## PROBLEM SET 10.4

Evaluate the following determinants.

*See Example 13*

1. $\begin{vmatrix} 1 & 2 \\ 3 & 4 \end{vmatrix}$
2. $\begin{vmatrix} 3 & 1 \\ 4 & 2 \end{vmatrix}$
3. $\begin{vmatrix} 2 & 5 \\ 8 & 3 \end{vmatrix}$
4. $\begin{vmatrix} 8 & 5 \\ 1 & 2 \end{vmatrix}$
5. $\begin{vmatrix} 7 & -1 \\ 2 & 5 \end{vmatrix}$
6. $\begin{vmatrix} -6 & 4 \\ 3 & 1 \end{vmatrix}$
7. $\begin{vmatrix} -2 & 7 \\ 4 & -1 \end{vmatrix}$
8. $\begin{vmatrix} 9 & -4 \\ 5 & 2 \end{vmatrix}$
9. $\begin{vmatrix} 3 & 0 \\ -4 & 2 \end{vmatrix}$
10. $\begin{vmatrix} 0 & 5 \\ -3 & 7 \end{vmatrix}$
11. $\begin{vmatrix} 0 & 6 \\ 0 & 4 \end{vmatrix}$
12. $\begin{vmatrix} 2 & -3 \\ 0 & 0 \end{vmatrix}$

*See Examples 14 and 15*

13. $\begin{vmatrix} 1 & 2 & 3 \\ 2 & 4 & 1 \\ 3 & 5 & 2 \end{vmatrix}$
14. $\begin{vmatrix} 2 & 3 & 2 \\ 3 & 4 & 1 \\ 2 & 1 & 4 \end{vmatrix}$
15. $\begin{vmatrix} 2 & -1 & 2 \\ -5 & 3 & 1 \\ 4 & 3 & -2 \end{vmatrix}$
16. $\begin{vmatrix} 4 & -3 & -2 \\ 1 & -2 & 5 \\ -3 & 1 & -5 \end{vmatrix}$
17. $\begin{vmatrix} 3 & 0 & -5 \\ -4 & 1 & 3 \\ 0 & 2 & -7 \end{vmatrix}$
18. $\begin{vmatrix} -2 & 1 & 0 \\ 7 & 0 & -5 \\ 0 & 6 & -3 \end{vmatrix}$
19. $\begin{vmatrix} 7 & -1 & 3 \\ 2 & 0 & -5 \\ -3 & 0 & 4 \end{vmatrix}$
20. $\begin{vmatrix} 10 & -3 & 7 \\ -2 & 1 & -6 \\ 8 & 0 & 0 \end{vmatrix}$

## 10.5 CRAMER'S RULE

The elimination method for solving linear systems is fine unless the numbers are awkward. In this section, we solve linear systems of equations using **Cramer's rule,** which uses determinants.

If we start with the linear system

$$ax + by = e$$
$$cx + dy = f$$

and use the usual elimination methods, we find the solutions to be

$$x = \frac{de - bf}{ad - bc} \qquad y = \frac{af - ce}{ad - bc}$$

The numerators and denominators of these fractions are special determinants.

***THEOREM 10.2* (*Cramer's Rule*)** For the linear system

$$ax + by = e$$
$$cx + dy = f$$

the solutions are given by

$$x = \frac{D_x}{D} \quad \text{and} \quad y = \frac{D_y}{D}$$

where

$$D = \begin{vmatrix} a & b \\ c & d \end{vmatrix} \qquad D_x = \begin{vmatrix} e & b \\ f & d \end{vmatrix} \qquad D_y = \begin{vmatrix} a & e \\ c & f \end{vmatrix}$$

and $D \neq 0$.

To use Cramer's rule to solve a system of two linear equations:

1. Evaluate $D$, the determinant of the coefficients. (If $D = 0$, stop; there are no unique solutions.)
2. Evaluate $D_x$, the determinant formed by replacing the $x$-coefficients (the first column) in $D$ by the right-hand-side terms.
3. Evaluate $D_y$, the determinant formed by replacing the $y$-coefficients (the second column) in $D$ by the right-hand-side terms.
4. Solve, using $x = D_x/D$ and $y = D_y/D$.

**EXAMPLE 16** Solve the following system using Cramer's rule:

$$A: \quad 6x - 10y = 15$$
$$B: \quad 11x + 3y = 7$$

**SOLUTION** The elimination method would be too messy to use here. After we find $D$, we use the right-hand-side coefficients, 15 and 7, to form $D_x$ and $D_y$.

D = determinant of coefficients ↓

$$D = \begin{vmatrix} 6 & -10 \\ 11 & 3 \end{vmatrix} = 18 + 110 = 128$$

| | |
|---|---|
| Replace first column in $D$ by 15 and 7 to get $D_x$ | $D_x = \begin{vmatrix} 15 & -10 \\ 7 & 3 \end{vmatrix} = 45 + 70 = 115$ |
| Replace second column in $D$ by 15 and 7 to get $D_y$ | $D_y = \begin{vmatrix} 6 & 15 \\ 11 & 7 \end{vmatrix} = 42 - 165 = -123$ |
| Theorem 10.2 | $x = \frac{D_x}{D} = \frac{115}{128} \qquad y = \frac{D_y}{D} = \frac{-123}{128}$ |

Therefore, the solution is $\left(\frac{115}{128}, \frac{-123}{128}\right)$.

**EXAMPLE 17** Check the solutions to Example 16 with a hand calculator.

**SOLUTION** We first convert $x$ and $y$ to decimals; then we substitute into equations $A$ and $B$. We use the memory buttons: [M+] to add, [M−] to subtract, and [MR] to recall the total.

| PUNCH | DISPLAY | MEANING |
|---|---|---|
| [C] | 0. | Clear to evaluate $x$ |
| [1] [1] [5] | 115. | $x = \frac{115}{128}$ |
| [÷] [1] [2] [8] | 128. | |
| [=] | 0.8984375 | $x \approx 0.898$ |
| [C] | 0. | Clear to evaluate $y$ |
| [1] [2] [3] [+/−] | −123. | $y = \frac{-123}{128}$ |
| [÷] [1] [2] [8] | 128. | |
| [=] | −0.9609375 | $y \approx -0.961$ |
| [C] | 0. | Clear to check in $A$: $6x - 10y = 15$ |
| [6] [×] [·] [8] [9] [8] [=] | 5.338 | $6x$ |
| [M+] | 5.338 | Add to memory |

| PUNCH | DISPLAY | MEANING |
|---|---|---|
| 1 0 × · 9 6 1 +/− = | −9.61 | $10y$ |
| M− | −9.61 | Subtract from memory |
| MR | 14.998 | $6x - 10y \approx 15$ Checks |
| C | 0. | Clear to check in $B$: $11x + 3y = 7$ |
| 1 1 × · 8 9 8 = | 9.878 | $11x$ |
| M+ | 9.878 | Add to memory |
| 3 × · 9 6 1 +/− = | −2.883 | $3y$ |
| M+ | −2.883 | Add to memory |
| MR | 6.995 | $11x + 3y \approx 7$ Checks |

Since we rounded off $x$ and $y$, the equations do not come out exactly, but close enough to check.

We also have a Cramer's rule for linear systems of three equations.

***THEOREM 10.3 (Cramer's Rule)*** For the linear system

$$A: \quad a_1x + b_1y + c_1z = d_1$$
$$B: \quad a_2x + b_2y + c_2z = d_2$$
$$C: \quad a_3x + b_3y + c_3z = d_3$$

the solutions are given by

$$x = \frac{D_x}{D} \qquad y = \frac{D_y}{D} \qquad z = \frac{D_z}{D}$$

where

$$D = \begin{vmatrix} a_1 & b_1 & c_1 \\ a_2 & b_2 & c_2 \\ a_3 & b_3 & c_3 \end{vmatrix} \qquad D_x = \begin{vmatrix} d_1 & b_1 & c_1 \\ d_2 & b_2 & c_2 \\ d_3 & b_3 & c_3 \end{vmatrix}$$

$$D_y = \begin{vmatrix} a_1 & d_1 & c_1 \\ a_2 & d_2 & c_2 \\ a_3 & d_3 & c_3 \end{vmatrix} \qquad D_z = \begin{vmatrix} a_1 & b_1 & d_1 \\ a_2 & b_2 & d_2 \\ a_3 & b_3 & d_3 \end{vmatrix}$$

and $D \neq 0$.

To use Cramer's rule to solve a system with three linear equations:

1. Evaluate $D$, the determinant of the coefficients. (If $D = 0$, stop.)
2. Find $D_x$, $D_y$, and $D_z$ by replacing the first, second, and third columns by the right-hand-side terms in $D$ (as in the two-variable case).
3. Solve, using $x = D_x/D$, $y = D_y/D$, and $z = D_z/D$.

While Cramer's rule works nicely for systems of two or three equations, it should be pointed out that it is not very practical for systems with four equations or more.

**EXAMPLE 18** Solve the system

$$\begin{aligned} A&: \quad 3x - 4y + 2z = 11 \\ B&: \quad 7x + 2y - 3z = 4 \\ C&: \quad 5x - 3y + 4z = 5 \end{aligned}$$

**SOLUTION** This would be a very messy elimination-method problem. We start by finding $D$; then we find $D_x$, $D_y$, and $D_z$ using the right-hand-side coefficients, 11, 4, and 5.

$D$ = determinant of coefficients

$$D = \begin{vmatrix} 3 & -4 & 2 \\ 7 & 2 & -3 \\ 5 & -3 & 4 \end{vmatrix}$$

Evaluate with minors

$$= \oplus 3 \cdot \begin{vmatrix} 2 & -3 \\ -3 & 4 \end{vmatrix} \ominus 7 \cdot \begin{vmatrix} -4 & 2 \\ -3 & 4 \end{vmatrix} \oplus 5 \cdot \begin{vmatrix} -4 & 2 \\ 2 & -3 \end{vmatrix}$$

$$= 3(-1) - 7(-10) + 5(8) = 107$$

↓

Replace first column in $D$ by 11, 4, and 5 to get $D_x$

$$D_x = \begin{vmatrix} 11 & -4 & 2 \\ 4 & 2 & -3 \\ 5 & -3 & 4 \end{vmatrix}$$

Evaluate with minors

$$= \oplus 11 \cdot \begin{vmatrix} 2 & -3 \\ -3 & 4 \end{vmatrix} \ominus 4 \cdot \begin{vmatrix} -4 & 2 \\ -3 & 4 \end{vmatrix} \oplus 5 \cdot \begin{vmatrix} -4 & 2 \\ 2 & -3 \end{vmatrix}$$

$$= 11(-1) - 4(-10) + 5(8) = 69$$

↓

Replace second column in $D$ by 11, 4, and 5 to get $D_y$

$$D_y = \begin{vmatrix} \boxed{3} & 11 & 2 \\ \boxed{7} & 4 & -3 \\ \boxed{5} & 5 & 4 \end{vmatrix}$$

Evaluate with minors

$$= \oplus 3 \cdot \begin{vmatrix} 4 & -3 \\ 5 & 4 \end{vmatrix} \ominus 7 \cdot \begin{vmatrix} 11 & 2 \\ 5 & 4 \end{vmatrix} \oplus 5 \cdot \begin{vmatrix} 11 & 2 \\ 4 & -3 \end{vmatrix}$$

$$= 3(31) - 7(34) + 5(-41) = -350$$

Replace third column in $D$ by 11, 4, and 5 to get $D_z$

$$D_z = \begin{vmatrix} \boxed{3} & -4 & 11 \\ \boxed{7} & 2 & 4 \\ \boxed{5} & -3 & 5 \end{vmatrix}$$

Evaluate with minors

$$= \oplus 3 \cdot \begin{vmatrix} 2 & 4 \\ -3 & 5 \end{vmatrix} \ominus 7 \cdot \begin{vmatrix} -4 & 11 \\ -3 & 5 \end{vmatrix} \oplus 5 \cdot \begin{vmatrix} -4 & 11 \\ 2 & 4 \end{vmatrix}$$

$$= 3(22) - 7(13) + 5(-38) = -215$$

Theorem 10.3

$$x = \frac{D_x}{D} = \frac{69}{107} \qquad y = \frac{D_y}{D} = \frac{-350}{107}$$

$$z = \frac{D_z}{D} = \frac{-215}{107}$$

Therefore, the solution is $\left(\frac{69}{107}, \frac{-350}{107}, \frac{-215}{107}\right)$. To check, we convert this answer to decimal form with a hand calculator (as in Example 17). Then we *approximate* the solution as $(0.645, -3.271, -2.009)$. We now check:

$$\begin{aligned} A&: \; 3(0.645) - 4(-3.271) + 2(-2.009) = 11.001 \qquad \text{Checks} \\ B&: \; 7(0.645) + 2(-3.271) - 3(-2.009) = 4.000 \qquad \text{Checks} \\ C&: \; 5(0.645) - 3(-3.271) + 4(-2.009) = 5.002 \qquad \text{Checks} \end{aligned}$$

**EXAMPLE 19** Solve the system

$$\begin{aligned} A&: \; 5x - 7y = 11 \\ B&: \; 2x - 13z = 9 \\ C&: \; 6y + 11z = 0 \end{aligned}$$

**SOLUTION** We first rewrite this in standard form.

**Rewrite in standard form**

$$\begin{aligned} A&: \quad 5x - 7y + 0z = 11 \\ B&: \quad 2x + 0y - 13z = 9 \\ C&: \quad 0x + 6y + 11z = 0 \end{aligned}$$

**D = determinant of coefficients**

$$D = \begin{vmatrix} \boxed{5} & -7 & 0 \\ \boxed{2} & 0 & -13 \\ \boxed{0} & 6 & 11 \end{vmatrix}$$

**Evaluate with minors**

$$= \oplus 5 \cdot \begin{vmatrix} 0 & -13 \\ 6 & 11 \end{vmatrix} \ominus 2 \cdot \begin{vmatrix} -7 & 0 \\ 6 & 11 \end{vmatrix} \oplus 0 \cdot \begin{vmatrix} -7 & 0 \\ 0 & -13 \end{vmatrix}$$

$$= 5(78) - 2(-77) + 0(91) = 544$$

**Replace first column in D by 11, 9, and 0 to get $D_x$**

$$D_x = \begin{vmatrix} \boxed{11} & -7 & 0 \\ \boxed{9} & 0 & -13 \\ \boxed{0} & 6 & 11 \end{vmatrix}$$

**Evaluate with minors**

$$= \oplus 11 \cdot \begin{vmatrix} 0 & -13 \\ 6 & 11 \end{vmatrix} \ominus 9 \cdot \begin{vmatrix} -7 & 0 \\ 6 & 11 \end{vmatrix} \oplus 0 \cdot \begin{vmatrix} -7 & 0 \\ 0 & -13 \end{vmatrix}$$

$$= 11(78) - 9(-77) + 0(91) = 1551$$

**Replace second column in D by 11, 9, and 0 to get $D_y$**

$$D_y = \begin{vmatrix} \boxed{5} & 11 & 0 \\ \boxed{2} & 9 & -13 \\ \boxed{0} & 0 & 11 \end{vmatrix}$$

**Evaluate with minors**

$$= \oplus 5 \cdot \begin{vmatrix} 9 & -13 \\ 0 & 11 \end{vmatrix} \ominus 2 \cdot \begin{vmatrix} 11 & 0 \\ 0 & 11 \end{vmatrix} \oplus 0 \cdot \begin{vmatrix} 11 & 0 \\ 9 & -13 \end{vmatrix}$$

$$= 5(99) - 2(121) + 0(-143) = 253$$

**Replace third column in D by 11, 9, and 0 to get $D_z$**

$$D_z = \begin{vmatrix} \boxed{5} & -7 & 11 \\ \boxed{2} & 0 & 9 \\ \boxed{0} & 6 & 0 \end{vmatrix}$$

**Evaluate with minors**

$$= \oplus 5 \cdot \begin{vmatrix} 0 & 9 \\ 6 & 0 \end{vmatrix} \ominus 2 \cdot \begin{vmatrix} -7 & 11 \\ 6 & 0 \end{vmatrix} \oplus 0 \cdot \begin{vmatrix} -7 & 11 \\ 0 & 9 \end{vmatrix}$$

$$= 5(-54) - 2(-66) + 0(-63) = -138$$

**Theorem 10.3**

$$x = \frac{D_x}{D} = \frac{1551}{544} \qquad y = \frac{D_y}{D} = \frac{253}{544} \qquad z = \frac{D_z}{D} = \frac{-138}{544}$$

Therefore, the solution is $\left(\frac{1551}{544}, \frac{253}{544}, \frac{-138}{544}\right)$, and as before we could check this with a hand calculator.

## PROBLEM SET 10.5

Solve the following linear systems using Cramer's rule. Use a hand calculator to check the solutions—see Example 17.

*See Example 16*

1. $A$: $x + 5y = 7$
   $B$: $2x - 3y = 5$

2. $A$: $2x - 3y = 5$
   $B$: $5x + y = 8$

3. $A$: $3x + 4y = 11$
   $B$: $2x - 5y = 7$

4. $A$: $10x - 7y = 8$
   $B$: $9x + 6y = 13$

5. $A$: $7x - 4y = 15$
   $B$: $3x + 8y = 1$

6. $A$: $4x - 11y = 7$
   $B$: $3x + 10y = 12$

7. $A$: $11x - 2y = 5$
   $B$: $3x + 7y = 9$

8. $A$: $2x + 9y = 10$
   $B$: $7x + 5y = 0$

9. $A$: $3x - 7y = 0$
   $B$: $5x - y = 8$

10. $A$: $7x + 10y = 2$
    $B$: $3x - 4y = 0$

*See Example 18*

11. $A$: $2x - y + z = 2$
    $B$: $4x + 2y + 3z = 4$
    $C$: $6x - y - 5z = 1$

12. $A$: $5x - 3y + 2z = 4$
    $B$: $7x + 2y - 3z = 10$
    $C$: $9x - y + 4z = -2$

13. $A$: $3x - 5y - 9z = 2$
    $B$: $4x + 7y + z = 0$
    $C$: $5x - 9y - 2z = 4$

14. $A$: $5x - 2y + 3z = 10$
    $B$: $-6x - 3y - 4z = 3$
    $C$: $2x + 5y + z = -2$

15. $A$: $-2x + 3y + 4z = 7$
    $B$: $4x - y - z = 8$
    $C$: $3x - 2y + 5z = 2$

16. $A$: $4x - 5y + 6z = 11$
    $B$: $3x + 7y - 2z = -3$
    $C$: $-5x - 3y + z = 0$

*See Example 19*

17. $A$: $3x - 4y = 5$
    $B$: $4y - 5z = 7$
    $C$: $5x - 7z = 2$

18. $A$: $8x - 3z = 1$
    $B$: $9y + 2z = -3$
    $C$: $2x + 3y = 4$

19. $A$: $3y + 5z = 0$
    $B$: $2y - x = 5$
    $C$: $7x - 4z = -2$

20. $A$: $4z - x = 6$
    $B$: $3y + x = 8$
    $C$: $4y + 5z = -1$

21. $A$: $4x - 5y = 2$
    $B$: $3z - 7y = 1$
    $C$: $2z + 9x = 0$

22. $A$: $7x - 3y = 6$
    $B$: $-2y + 4z = -7$
    $C$: $6z - 8x = 5$

*Health application*

23. A dietician is trying to satisfy a patient's need for protein (72 grams) and thiamine or $B_1$ (1.2 milligrams) with milk and bread. The protein and thiamine contents of milk and bread lead to the system

$$\begin{aligned} A&: \quad 0.06x + 0.08y = 1.2 \\ B&: \quad 2x + 8y = 72 \end{aligned}$$

where $x$ is slices of bread and $y$ is glasses of milk. Solve this system using Cramer's rule.

*Electricity application*

24. In the circuit shown, $x$, $y$, and $z$ are currents through resistors of 12 ohms, 25 ohms, and 4 ohms. Kirchhoff's and Ohm's laws lead to the following system:

$$\begin{aligned} A&: \quad x + y + z = 0 \\ B&: \quad 12x - 25y = 40 \\ C&: \quad -25y + 4z = 50 \end{aligned}$$

Solve this system using Cramer's rule.

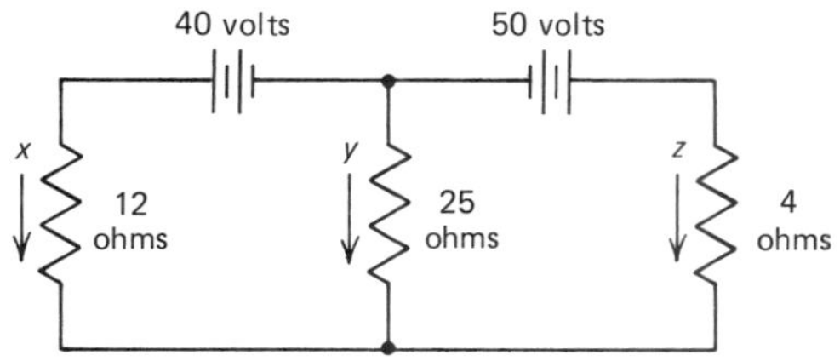


## 10.6 NONLINEAR SYSTEMS OF EQUATIONS

A **nonlinear system of equations** is a system of equations that has at least one term of second degree or higher. As examples,

$$\begin{aligned} A&: \quad -2x + y = -5 \\ B&: \quad x^2 + y^2 = 25 \end{aligned} \qquad \text{and} \qquad \begin{aligned} A&: \quad x^2 - y^2 = 16 \\ B&: \quad x^2 + y^2 = 36 \end{aligned}$$

are nonlinear systems of equations. The first system also involves a linear equation. We solve this type of system using the **substitution method.**

> To solve a nonlinear system with a linear equation $A$ and nonlinear equation $B$:
>
> 1. Solve the linear equation $A$ for one of the variables.
> 2. Substitute this value into the nonlinear equation $B$.
> 3. Solve for the second variable.
> 4. Substitute this value into equation $A$ to find the first variable.
>
> *Note:* There will probably be *two* $x$-answers and *two* $y$-answers. You may have to do step 4 twice.

**EXAMPLE 20** Solve the system

$$\begin{aligned} A&: \quad -2x + y = -5 \\ B&: \quad x^2 + y^2 = 25 \end{aligned}$$

**SOLUTION** We solve the linear equation $A$ for $y$ and then substitute this into equation $B$.

| | |
|---|---|
| Given | $A: \quad -2x + y = -5$ |
| ↓ Solve for $y$ | $y = 2x - 5$ |
| ↓ Given | $B: \quad x^2 + y^2 = 25$ |
| ↓ Substitute $y = 2x - 5$ into $B$ | $x^2 + (2x - 5)^2 = 25$ |
| ↓ Simplify | $x^2 + 4x^2 - 20x + 25 = 25$ |
| ↓ Put into standard form | $5x^2 - 20x = 0$ |
| ↓ | |

| | |
|---|---|
| Factor | $5x(x-4) = 0$ |
| Set each factor = 0 | $5x = 0$ or $x - 4 = 0$ |
| Solve for $x$ | $x = 0$ or $x = 4$ |
| Substitute $x = 0$ and $x = 4$ into $A$ | $A$: $-2(0)+y=-5$ or $A$: $-2(4)+y=-5$ |
| Solve for $y$ | $y = -5$ or $y = 3$ |
| Write $(x, y)$ pairs | $(0, -5)$ or $(4, 3)$ |

The solutions are the *two* points $(0, -5)$ and $(4, 3)$. We can see on the graph in Fig. 10.2 that this is the intersection of a circle and a line.

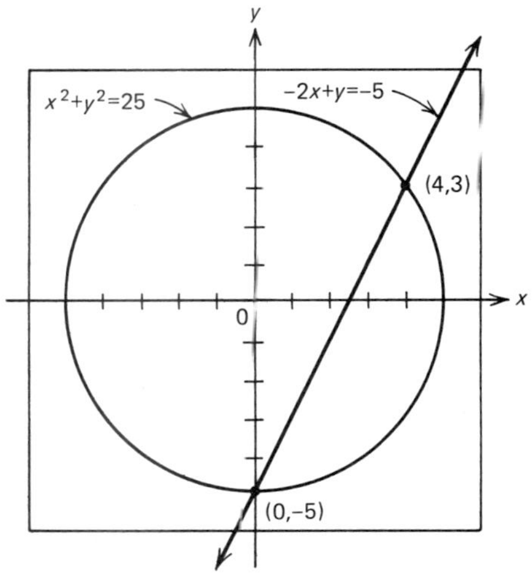


Figure 10.2

**EXAMPLE 21** Solve the system

$$A: \quad 2x + 3y = 25$$

$$B: \quad xy = 14$$

**SOLUTION** Equation $B$ is *not* linear (the degree is the sum of the exponents in $xy$: $1 + 1 = 2$), but we can still solve it by substitution.

| | |
|---|---|
| Given | $B: \quad xy = 14$ |
| Solve for $y$ | $y = \dfrac{14}{x}$ |
| Given | $A: \quad 2x + 3y = 25$ |
| Substitute $y = \dfrac{14}{x}$ into $A$ | $2x + 3\left(\dfrac{14}{x}\right) = 25$ |
| Multiply by LCD, $x$ | $2x^2 + 42 = 25x$ |
| Put into standard form | $2x^2 - 25x + 42 = 0$ |
| Factor | $(2x - 21)(x - 2) = 0$ |
| Set each factor $= 0$ | $2x - 21 = 0$ or $x - 2 = 0$ |
| Solve for $x$ | $x = \dfrac{21}{2}$ or $x = 2$ |
| Substitute $x = \dfrac{21}{2}$ and $x = 2$ into $B$ | $B: \ \dfrac{21}{2} \cdot y = 14$ or $B: \ 2y = 14$ |
| Solve for $y$ | $y = \dfrac{4}{3}$ or $y = 7$ |
| Write $(x, y)$ pairs | $\left(\dfrac{21}{2}, \dfrac{4}{3}\right)$ or $(2, 7)$ |

Again, there are two solution points: $(\frac{21}{2}, \frac{4}{3})$ and $(2, 7)$.

> If both equations are second degree, use the elimination method.

**EXAMPLE 22** Solve the system

$$A: \quad x^2 - y^2 = 16$$
$$B: \quad x^2 + y^2 = 34$$

**SOLUTION** Here, we add equations $A$ and $B$ to eliminate $y^2$.

| | |
|---|---|
| Add $A$ and $B$ to eliminate $y^2$ | $A: \quad x^2 - y^2 = 16$<br>$B: \quad x^2 + y^2 = 34$ |
| Add equations | $A + B: \quad 2x^2 = 50$ |
| Divide by 2 | $x^2 = 25$ |
| Solve for $x$ | $x = -5$ or $x = 5$ |
| Substitute $x = -5$ and $x = 5$ into $B$ | $B: \quad (-5)^2 + y^2 = 34$ or $(5)^2 + y^2 = 34$ |
| Simplify | $25 + y^2 = 34$ or $25 + y^2 = 34$ |
| Subtract 25 | $y^2 = 9$ or $y^2 = 9$ |
| Solve for $y$ | $y = \pm 3$ or $y = \pm 3$ |
| Write $(x, y)$ pairs | $(-5, 3); (-5, -3)$ or $(5, 3); (5, -3)$ |

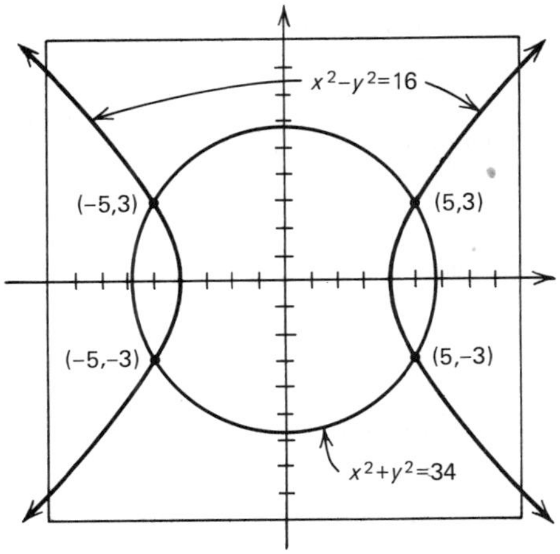


**Figure 10.3**

Now there are *four* solutions: $\{(-5, 3), (-5, -3), (5, 3), (5, -3)\}$. We can see on the graph in Fig. 10.3 that this is the intersection of a circle and a hyperbola.

## PROBLEM SET 10.6

Solve the following nonlinear systems.

*See Example 20*

1. $A$: $x + y = 7$
   $B$: $x^2 + y^2 = 29$

2. $A$: $x - y = 7$
   $B$: $x^2 + y^2 = 109$

3. $A$: $x + y = 8$
   $B$: $2x^2 + y^2 = 51$

4. $A$: $x - y = 2$
   $B$: $x^2 - 3y^2 = 4$

5. $A$: $2x + y = 6$
   $B$: $x^2 + 2y^2 = 12$

6. $A$: $3x - y = 11$
   $B$: $2x^2 + 3y^2 = 35$

7. $A$: $x + 3y = 10$
   $B$: $3x^2 - y^2 = 44$

8. $A$: $4x - y = 1$
   $B$: $2x^2 + 5y^2 = 47$

*See Example 21*

9. $A$: $x + y = 11$
   $B$: $xy = 28$

10. $A$: $x - y = 8$
    $B$: $xy = 20$

11. $A$: $2x - 3y = 7$
    $B$: $xy = 5$

12. $A$: $4x + 5y = 23$
    $B$: $xy = 6$

13. $A$: $y = x^2 + 2x + 7$
    $B$: $y = 2x^2 - 5x + 13$

14. $A$: $y = 2x^2 + 7x - 9$
    $B$: $y = 5x^2 - x - 5$

*See Example 22*

15. $A$: $x^2 - y^2 = 35$
    $B$: $x^2 + y^2 = 37$

16. $A$: $x^2 - 2y^2 = -9$
    $B$: $x^2 + y^2 = 18$

17. $A$: $2x^2 + y^2 = 17$
    $B$: $x^2 + 2y^2 = 22$

18. $A$: $x^2 - 3y^2 = 13$
    $B$: $-3x^2 + y^2 = -71$

19. $A$: $x^2 + 2y^2 = 66$
    $B$: $x^2 + y = 21$

20. $A$: $x^2 - 3y^2 = 24$
    $B$: $x^2 - 5y = 26$

Translate the following problems into nonlinear systems and solve.

*See Examples 7, 20, 21, and 22*

21. The sum of two numbers is 12. Their product is 35. Find the numbers.
22. The product of two numbers is 24. Their difference is 5. Find the numbers.
23. The sum of two numbers is 11. The sum of their squares is 73. Find the numbers.
24. The difference of two numbers is 3. The difference of their squares is 51. Find the numbers.

*Business applications*

25. The total cost to produce $x$ number of a certain item is

$$y = x^2 + 100x + 7500$$

The total revenue from $x$ number of sales is $y = 300x$. Find the intersection points (*break-even points*) and graph these equations.

26. The relation between demand ($x$) and price ($y$) for an item is $xy = 1200$. The relation between supply ($x$) and price ($y$) or the same item is $x - 10y = 20$. Solve this nonlinear system to find the *equilibrium point* where supply = demand.

## 10.7 SYSTEMS OF INEQUALITIES

The **system of inequalities** has as its solution set the intersection of the solutions of each of the inequalities. For example,

$$A: \quad x^2 + y^2 < 25$$
$$B: \quad -2x + y > -5$$

is a system of inequalities.

To graph a system of inequalities:

1. Graph each inequality using the techniques from Sections 8.4 and 9.4.
2. Take the intersection of the regions.

**EXAMPLE 23** Graph the system

$$\begin{aligned} A&: \quad 3x + y \le 9 \\ B&: \quad 2x + 3y \le 12 \\ C&: \quad y \le 3 \end{aligned}$$

**SOLUTION** We graph the three corresponding equations, determine the appropriate regions, and find the intersections.

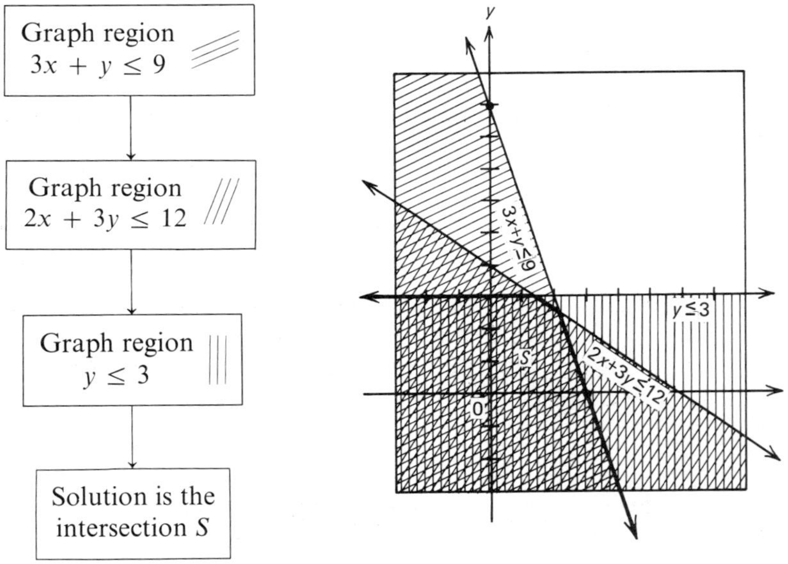


Since all three of these inequalities were $\le$, we included the lines in the solution.

**EXAMPLE 24** Graph the system

$$\begin{aligned} A&: \quad 16x^2 + 9y^2 < 144 \\ B&: \quad x^2 + y^2 \ge 4 \\ C&: \quad x - y > 1 \end{aligned}$$

We begin by graphing each of the corresponding equations: an ellipse, a circle, and a line.

Graph region $16x^2 + 9y^2 < 144$

↓

Graph region $x^2 + y^2 \geq 4$

↓

Graph region $x - y > 1$

↓

Solution is the intersection $S$

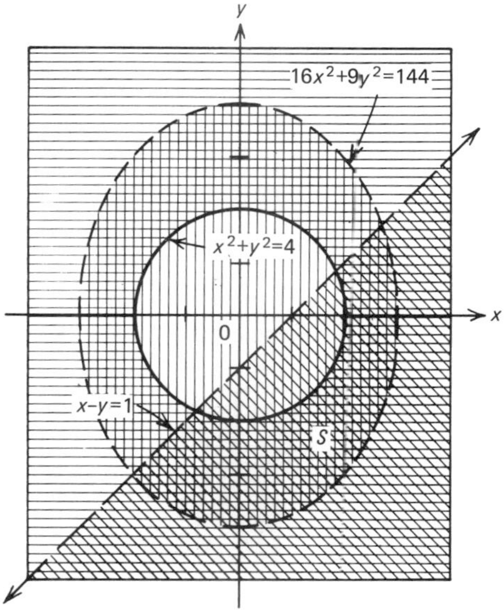


## PROBLEM SET 10.7

Graph the following systems of inequalities.

*See Example 23*

1. $A$: $2x - y > 4$
   $B$: $3x + 5y < 15$

2. $A$: $3x + y \leq 6$
   $B$: $x - 4y > 8$

3. $A$: $3x + 4y \leq 12$
   $B$: $6x + 2y \leq 12$

4. $A$: $4x + 5y > 20$
   $B$: $2x + 10y > 20$

5. $A$: $x + 7y < 7$
   $B$: $5x + y < 10$
   $C$: $x \geq 0$
   $D$: $y \geq 0$

6. $A$: $2x + 5y \leq 10$
   $B$: $3x + 4y \leq 12$
   $C$: $x \geq 0$
   $D$: $y \geq 0$

7. $A$: $2x + y \geq 12$
   $B$: $x + 3y \geq 9$
   $C$: $y < 6$
   $D$: $x < 8$

8. $A$: $x + y > 5$
   $B$: $x + 4y > 8$
   $C$: $y < 7$
   $D$: $x > 0$

*See Example 24*

9. $A$: $x^2 + y^2 < 25$
   $B$: $3x - 2y < 6$

10. $A$: $4x^2 + y^2 \leq 16$
    $B$: $2x + y > 2$

11. $A$: $y > x^2 + 1$
    $B$: $y < x + 5$

12. $A$: $y < -x^2 + 9$
    $B$: $y > 5$

13. $A$: $x^2 + y^2 > 1$
    $B$: $x^2 + y^2 < 4$

14. $A$: $4x^2 + y^2 \geq 4$
    $B$: $4x^2 + y^2 \leq 16$

15. $A: \quad x^2 + y^2 \geq 1$
$B: \quad 9x^2 + y^2 \leq 36$
$C: \quad y \leq 3$

16. $A: \quad x^2 + 4y^2 > 4$
$B: \quad x^2 + y^2 < 25$
$C: \quad x > 1$

17. $A: \quad y \geq x^2 - 1$
$B: \quad x^2 + y^2 \leq 4$
$C: \quad x + y \leq 1$

18. $A: \quad x^2 - y^2 > 1$
$B: \quad 4x^2 + 9y^2 < 36$
$C: \quad -x + 4y < 2$

*Business application*

19. An automobile company is manufacturing two new cars: Squirrels and Beavers. The number of Squirrels ($x$) and Beavers ($y$) that they can produce is related to the amount of labor and steel available with the following system.

$$
\begin{aligned}
A&: \quad 150x + 250y \leq 75{,}000 \\
B&: \quad 1200x + 1200y \leq 480{,}000 \\
C&: \quad x \geq 0 \\
D&: \quad y \geq 0
\end{aligned}
$$

Graph this system of inequalities.

*Health application*

20. A patient needs at least 72 grams of protein and 1.2 milligrams of thiamine. A dietician uses slices of bread ($x$) and glasses of milk ($y$) and arrives at the following system.

$$
\begin{aligned}
A&: \quad 0.06x + 0.08y \geq 1.2 \\
B&: \quad 2x + 8y \geq 72 \\
C&: \quad x \geq 0 \\
D&: \quad y \geq 0
\end{aligned}
$$

Solve this system of inequalities. (Be careful dividing the decimals.)

*Enviromental application*

21. A plant manufacturers two products. Product 1 ($x$) produces 10 grams of $SO_2$ and 50 grams of particulate matter per item. Product 2 ($y$) produces 40 grams of $SO_2$ and 40 grams of particulate matter. The EPA allows them at most 10,000 grams of $SO_2$ and 16,000 grams of particulate matter. Graph the system of inequalities that this gives.

$$
\begin{aligned}
A&: \quad 10x + 40y \leq 10{,}000 \\
B&: \quad 50x + 40y \leq 16{,}000 \\
C&: \quad x \geq 0 \\
D&: \quad y \geq 0
\end{aligned}
$$

# CHAPTER 10 SUMMARY

## Important Words and Phrases

Cramer's rule (10.5)
dependent (10.1)
determinant (10.4)
elimination method (10.1)
inconsistent (10.1)
linear system of equations (10.1)
minor (10.4)
mixture problem (10.2)
nonlinear system of equations (10.6)
ordered pair (introduction)
ordered triple (10.3)
solution set (introduction)
substitution method (10.1)
system of equations (introduction)
system of inequalities (10.7)
3 by 3 determinant (10.4)
translation problem (10.2)
2 by 2 determinant (10.4)
value problem (10.2)

## Important Theorems and Definitions

*Theorem 10.1* If $(x, y)$ satisfies

$$\begin{aligned} A&: \quad ax + by = e \\ B&: \quad cx + dy = f \end{aligned}$$

then $(x, y)$ also satisfies $M(ax + by) + N(cx + dy) = Me + Nf$ for any real numbers $M$ and $N$.

$$\begin{vmatrix} a & b \\ c & d \end{vmatrix} = ad - bc$$

$$\begin{vmatrix} a_1 & b_1 & c_1 \\ a_2 & b_2 & c_2 \\ a_3 & b_3 & c_3 \end{vmatrix} = a_1 \cdot \begin{vmatrix} b_2 & c_2 \\ b_3 & c_3 \end{vmatrix} - a_2 \cdot \begin{vmatrix} b_1 & c_1 \\ b_3 & c_3 \end{vmatrix} + a_3 \cdot \begin{vmatrix} b_1 & c_1 \\ b_2 & c_2 \end{vmatrix}$$

*Cramer's rule* (two variables): The solutions to

$$\begin{aligned} A&: \quad ax + by = e \\ B&: \quad cx + dy = f \end{aligned}$$

are $x = \dfrac{D_x}{D}$ and $y = \dfrac{D_y}{D}$, where

$$D = \begin{vmatrix} a & b \\ c & d \end{vmatrix} \qquad D_x = \begin{vmatrix} e & b \\ f & d \end{vmatrix} \qquad D_y = \begin{vmatrix} a & e \\ c & f \end{vmatrix}$$

*Cramer's rule* (three variables): The solutions to

$$\begin{aligned} A&: \quad a_1x + b_1y + c_1z = d_1 \\ B&: \quad a_2x + b_2y + c_2z = d_2 \\ C&: \quad a_3x + b_3y + c_3z = d_3 \end{aligned}$$

are $x = \dfrac{D_x}{D}$, $y = \dfrac{D_y}{D}$, and $z = \dfrac{D_z}{D}$, where

$$D = \begin{vmatrix} a_1 & b_1 & c_1 \\ a_2 & b_2 & c_2 \\ a_3 & b_3 & c_3 \end{vmatrix} \qquad D_x = \begin{vmatrix} d_1 & b_1 & c_1 \\ d_2 & b_2 & c_2 \\ d_3 & b_3 & c_3 \end{vmatrix}$$

$$D_y = \begin{vmatrix} a_1 & d_1 & c_1 \\ a_2 & d_2 & c_2 \\ a_3 & d_3 & c_3 \end{vmatrix} \qquad D_z = \begin{vmatrix} a_1 & b_1 & d_1 \\ a_2 & b_2 & d_2 \\ a_3 & b_3 & d_3 \end{vmatrix}$$

and $D \neq 0$.

## Important Procedures

*To solve a two-variable linear system of equations:*

1. Multiply each equation by appropriate numbers to eliminate one of the variables.
2. Solve for the second variable.
3. Substitute back to find the first variable.

*To solve a three-variable linear system of equations:*

1. Using two pairs of the equations, eliminate one of the variables.
2. Solve the resulting pair of equations by eliminating a second variable.
3. Solve for the third variable.
4. Substitute back to find the second variable.
5. Substitute the second and third variables to find the first variable.

*To solve a nonlinear system of equations with a linear equation:*

1. Solve the linear equation for one of the variables.
2. Substitute this value into the nonlinear equation.
3. Solve for the second variable.
4. Substitute this into the linear equation to find variable. [*Note:* There may be two $(x, y)$ pairs.]

*To graph a system of inequalities:*

1. Graph each inequality.
2. Take the intersection of these regions.

## CHAPTER 10 REVIEW EXERCISES

Solve the following linear systems of equations using the elimination method.

1. $A$: $x + 2y = 10$
   $B$: $3x - 5y = -3$

2. $A$: $2x + 3y = 4$
   $B$: $3x + 5y = 5$

3. $A$: $\frac{x}{4} + \frac{y}{3} = 0$
   $B$: $\frac{x}{7} - \frac{y}{2} = \frac{29}{14}$

4. $A$: $x + y - z = 13$
   $B$: $x + y + z = 1$
   $C$: $x - y + z = 1$

5. $A$: $2x - 4y - 3z = 20$
   $B$: $4x + 3y + 2z = -8$
   $C$: $3x - 5y - 4z = 27$

6. $A$: $x - 2z = -5$
   $B$: $x - 3y = -5$
   $C$: $y + z = 5$

Evaluate the following determinants.

7. $\begin{vmatrix} 7 & -3 \\ 4 & 5 \end{vmatrix}$

8. $\begin{vmatrix} 1 & 3 & -2 \\ 5 & 0 & 4 \\ -3 & 1 & 6 \end{vmatrix}$

Solve the following linear systems of equations using Cramer's rule.

9. $A$: $5x - 7y = 2$
   $B$: $6x + 11y = 5$

10. $A$: $2x - 3y + 5z = 1$
    $B$: $3x - 4y - 3z = 5$
    $C$: $4x + 5y - 2z = 11$

Solve the following nonlinear systems.

11. $A$: $x + y = 19$
    $B$: $x^2 - y^2 = 19$

12. $A$: $2x + y = 19$
    $B$: $xy = 24$

13. $A$: $x^2 - y^2 = 11$
    $B$: $3x^2 + y^2 = 133$

Graph the following systems of inequalities.

14. $A$: $5x + 4y \leq 20$
    $B$: $6x + 3y \leq 18$
    $C$: $x \geq 0$
    $D$: $y \geq 1$

15. $A$: $x^2 + y^2 > 16$
    $B$: $16x^2 + 25y^2 < 400$
    $C$: $y < x + 1$

16. Twice the smaller of two numbers is 5 more than the larger. Their sum is 16. Find the numbers.

17. How many quarts of a 20% antifreeze solution and a 50% antifreeze solution must be mixed to get 12 quarts of a 30% antifreeze solution?

18. Larry has 18 coins (dimes and quarters) which have a total value of $2.55 (255¢). How many coins of each does he have?

19. The difference of two numbers is 2. The sum of their squares is 130. Find the numbers.

# 11
# LOGARITHMS

## 11.1 LOGARITHMS

A **logarithm** is an exponent. Consider the logarithm concept in contrast to the other exponent concepts that we have seen.

| | | |
|---|---|---|
| *Powers:* | $2^5 = \boxed{?}$ | *Power* is the unknown |
| *Roots:* | $\boxed{?}^5 = 32$ | *Base* is the unknown |
| *Logarithms:* | $2^{\boxed{?}} = 32$ | *Exponent* is the unknown |

In this logarithm problem, we are asking: "2 to what power is 32?" We write this

$$\log_2 32 = 5$$

This is read "log to the base 2 of 32 is 5."

In general, we have the following.

> **DEFINITION 11.1** For real numbers $x$, $N > 0$ and $b > 0$ $(b \neq 1)$,
>
> $$\log_b N = x \quad \text{means} \quad b^x = N$$
>
> We call $b^x = N$ the **exponent form** and $\log_b N = x$ the **logarithm form.** But they both have the same meaning.

**EXAMPLE 1** Here we change exponent forms into their equivalent logarithm forms.

(a) $2^4 = 16$ means $\log_2 16 = 4$

(b) $3^2 = 9$ means $\log_3 9 = 2$

(c) $25^{1/2} = 5$ means $\log_{25} 5 = \frac{1}{2}$

(d) $6^{-2} = \frac{1}{36}$ means $\log_6 \frac{1}{36} = -2$

**(e)** $8^{-1/3} = \frac{1}{2}$ means $\log_8 \frac{1}{2} = -\frac{1}{3}$

base

exponent

**EXAMPLE 2** Here we change logarithm forms into exponent forms.

**(a)** $\log_4 64 = 3$ means $4^3 = 64$

**(b)** $\log_{10} 10{,}000 = 4$ means $10^4 = 10{,}000$

**(c)** $\log_{49} 7 = \frac{1}{2}$ means $49^{1/2} = 7$

**(d)** $\log_{10} 0.1 = -1$ means $10^{-1} = 0.1$

**(e)** $\log_9 \frac{1}{27} = -\frac{3}{2}$ means $9^{-3/2} = \frac{1}{27}$

exponent

base

In working logarithm problems, it is usually easiest to convert to the exponent form. We will use the following theorem to help us.

> ***THEOREM 11.1*** Let $x$, $y$, and $b$ be real numbers ($b > 0, b \neq 1$). If $b^x = b^y$, then $x = y$.

Notice that we must have the *same* base to use Theorem 11.1.

**EXAMPLE 3** Solve $\log_9 x = 2$.

**SOLUTION** We first put this into exponent form.

**Exponent form** $\log_9 x = 2$ means $9^2 = x$

↓

**Compute $x$** $81 = x$

**EXAMPLE 4** Solve $\log_x 16 = 4$.

**SOLUTION** We again put this into exponent form.

| | |
|---|---|
| Exponent form | $\log_x 16 = 4$ means $x^4 = 16$ |
| ↓ Take fourth root | $x = \sqrt[4]{16} = 2$ |

**EXAMPLE 5** Solve $\log_{10} \dfrac{1}{100} = x$.

**SOLUTION** Here we use Theorem 11.1 after we make the bases the same, using the fact that $\frac{1}{100} = 10^{-2}$.

| | |
|---|---|
| Exponent form | $\log_{10} \dfrac{1}{100} = x$ means $10^x = \dfrac{1}{100}$ |
| ↓ Since $\dfrac{1}{100} = 10^{-2}$ | $10^x = 10^{-2}$ |
| ↓ Theorem 11.1 | $x = -2$ |

**EXAMPLE 6** Simplify $\log_8 16$.

**SOLUTION** We write this as $x = \log_8 16$. We use the fact that 8 and 16 are both powers of 2.

| | |
|---|---|
| Write in exponent form | $8^x = 16$ |
| ↓ Since $8 = 2^3$ and $16 = 2^4$ | $(2^3)^x = 2^4$ |
| ↓ Since $(b^m)^n = b^{mn}$ | $2^{3x} = 2^4$ |
| ↓ Exponents must be equal | $3x = 4$ |
| ↓ Solve for $x$ | $\log_8 16 = x = \dfrac{4}{3}$ |

**EXAMPLE 7** Simplify $\log_5 1$.

**SOLUTION** We write this as $x = \log_5 1$. We use the fact that $5^0 = 1$.

| | |
|---|---|
| Exponent form | $5^x = 1$ |
| Since $5^0 = 1$ | $5^x = 5^0$ |
| Exponents must be equal | $\log_5 1 = x = 0$ |

In fact, $\log_b 1 = 0$ for all bases $b > 0$. This is true since $b^0 = 1$ for all $b > 0$.

**EXAMPLE 8** Simplify $\log_2 [\log_{10} 10{,}000]$.

**SOLUTION** We do this in two steps: first the inside logarithm and then the outside logarithm.

| | |
|---|---|
| Given | $\log_2 [\log_{10} 10{,}000]$ |
| Since $10^4 = 10000$ | $= \log_2 [4]$ |
| Since $2^2 = 4$ | $= 2$ |

## PROBLEM SET 11.1

Write the following equations in logarithm form.

*See Example 1*

1. $2^7 = 128$
2. $3^4 = 81$
3. $10^3 = 1000$
4. $8^2 = 64$
5. $100^{1/2} = 10$
6. $8^{1/3} = 2$
7. $5^{-2} = \dfrac{1}{25}$
8. $2^{-5} = \dfrac{1}{32}$

9. $16^{-1/2} = \frac{1}{4}$

10. $27^{-2/3} = \frac{1}{9}$

Write the following in exponent form.

*See Example 2*

11. $\log_2 8 = 3$

12. $\log_3 9 = 2$

13. $\log_5 625 = 4$

14. $\log_{10} 100{,}000 = 5$

15. $\log_{64} 4 = \frac{1}{3}$

16. $\log_{25} 125 = \frac{3}{2}$

17. $\log_4 \frac{1}{16} = -2$

18. $\log_{10} 0.01 = -2$

19. $\log_4 \frac{1}{8} = \frac{-3}{2}$

20. $\log_8 \frac{1}{32} = \frac{-5}{3}$

Solve the following equations for $x$.

*See Example 3*

21. $\log_7 x = 2$

22. $\log_2 x = 6$

23. $\log_3 x = -2$

24. $\log_5 x = -1$

25. $\log_{16} x = \frac{1}{4}$

26. $\log_{25} x = \frac{-3}{2}$

*See Example 4*

27. $\log_x 25 = 2$

28. $\log_x 8 = 3$

29. $\log_x \frac{1}{9} = -2$

30. $\log_x 0.001 = -3$

31. $\log_x 3 = \frac{1}{4}$

32. $\log_x \frac{1}{4} = \frac{-2}{5}$

*See Example 5*

33. $\log_3 27 = x$

34. $\log_4 4 = x$

35. $\log_5 \frac{1}{5} = x$

36. $\log_{10} 10^{-4} = x$

37. $\log_9 \frac{1}{3} = x$

38. $\log_4 \frac{1}{8} = x$

Simplify the following.

*See Examples 6 and 7*

39. $\log_{10} 100$

40. $\log_2 16$

41. $\log_5 5^{10}$

42. $\log_7 7$

43. $\log_{12} 1$

44. $\log_{16} 32$

45. $\log_9 27$

46. $\log_{25} 125$

47. $\log_4 4^5$

48. $\log_{10} 10^8$

*See Example 8*

49. $\log_3 [\log_4 64]$

50. $\log_2 [\log_3 81]$

51. $\log_5 [\log_6 6]$

52. $\log_3 [\log_{10} 10^9]$

*Sound application*

Sound intensity $I$ is measured in *decibels* using the formula

$$I = 10 \log_{10} R$$

where $R$ is the ratio of the pressure of the given sound to the pressure of the least audible sound. Find $I$ in the following cases.

53. Interior of country house, $R = 100$.

54. Average office, $R = 10{,}000$.

55. Ordinary conversation, $R = 1{,}000{,}000$.

56. Jackhammer (at 20 feet), $R = 100{,}000{,}000$.

57. Airplane engine (at 15 feet), $R = 100{,}000{,}000{,}000$.

58. Loud rock band, $R = 1{,}000{,}000{,}000{,}000$.

*Photographic application*

The *transmission* $T$ of a film negative is the fraction of light that passes through the negative. The *density* of the negative is given by

$$D = \log_{10}\left(\frac{1}{T}\right)$$

Find the density in the following cases.

59. $T = \frac{1}{10}$

60. $T = \frac{1}{100}$

61. $T = \frac{1}{1000}$

## 11.2 PROPERTIES OF LOGARITHMS

We have the following properties of logarithms which are based upon the properties of exponents.

> **PROPERTY 1** For positive real numbers $b$, $N$, $M$ ($b \neq 1$),
>
> $$\log_b NM = \log_b N + \log_b M$$

**PROPERTY 2** For positive real numbers $b$, $N$, $M$ ($b \neq 1$),

$$\log_b \frac{N}{M} = \log_b N - \log_b M$$

**PROPERTY 3** For positive real numbers $b$ and $N$ ($b \neq 1$) and real number $r$,

$$\log_b (N^r) = r \log_b N$$

We sketch a proof only of Property 1. (The others are similar.) Let $x = \log_b M$ and $y = \log_b N$. Then

| | |
|---|---|
| Put into exponent form | $M = b^x$ and $N = b^y$ |
| ↓ Add exponents | $MN = b^x b^y = b^{x+y}$ |
| ↓ Put into logarithm form | $\log_b MN = x + y$ |
| ↓ Substitute $x = \log_b M$ and $y = \log_b N$ | $\log_b MN = \log_b M + \log_b N$ |

**EXAMPLE 9** The following use Property 1. The logarithm of a product is the sum of logarithms.

**(a)** $\log_4 3pq = \log_4 3 + \log_4 p + \log_4 q$

**(b)** $\log_{10} 7x^2y^5 = \log_{10} 7 + \log_{10} x^2 + \log_{10} y^5$

**EXAMPLE 10** The following use Property 2. The logarithm of a quotient is the difference of logarithms.

**(a)** $\log_7 \frac{t}{8} = \log_7 t - \log_7 8$

**(b)** $\log_2 \frac{m^5}{k^3} = \log_2 m^5 - \log_2 k^3$

| YES | NO |
|---|---|
| $\log_{10} 7a = \log_{10} 7 + \log_{10} a$ | ~~$\log_{10} 7a = \log_{10} 7 \cdot \log_{10} a$~~ |
| $\log_2 3 + \log_2 5 = \log_2 15$ | ~~$\log_2 3 + \log_2 5 = \log_2 8$~~ |
| $\log_3 \frac{m}{n} = \log_3 m - \log_3 n$ | ~~$\log_3 \frac{m}{n} = \frac{\log_3 m}{\log_3 n}$~~ |
| $\log_4 21 - \log_4 7 = \log_4 3$ | ~~$\log_4 21 - \log_4 7 = \log_4 14$~~ |

**EXAMPLE 11** The following use Property 3. The logarithm of a power is the exponent times the logarithm of the base. Notice that we must write roots in fractional-exponent form.

**(a)** $\log_5 x^3 = 3 \log_5 x$

**(b)** $\log_{10} \sqrt{T} = \log_{10} T^{1/2} = \frac{1}{2} \log_{10} T$

**EXAMPLE 12** Write $\log_2 \frac{x^3 y^5}{\sqrt[4]{z}}$ in terms of simpler logarithm terms.

**SOLUTION** Here, we will use all three logarithm properties. We write $\sqrt[4]{z} = z^{1/4}$.

| | |
|---|---|
| Property 2 | $\log_2 \frac{x^3 y^5}{z^{1/4}} = \log_2 x^3 y^5 - \log_2 z^{1/4}$ |
| ↓ Property 1 | $= \log_2 x^3 + \log_2 y^5 - \log_2 z^{1/4}$ |
| ↓ Property 3 | $= 3 \log_2 x + 5 \log_2 y - \frac{1}{4} \log_2 z$ |

**EXAMPLE 13** Write $T = \frac{1}{3} \log_6 k - 7 \log_6 m - \frac{1}{5} \log_6 n$ as a single logarithm expression.

**SOLUTION** This problem is the reverse of Example 12. We still use Properties 1, 2, and 3.

| | |
|---|---|
| Given | $T = \frac{1}{3}\log_6 k - 7\log_6 m - \frac{1}{5}\log_6 n$ |
| ↓ Property 3 | $= \log_6 k^{1/3} - \log_6 m^7 - \log_6 n^{1/5}$ |
| ↓ Properties 1 and 2 | $= \log_6 \frac{k^{1/3}}{m^7 n^{1/5}}$ |

**EXAMPLE 14** Solve $\log_6 3 + \log_6 x = 2$.

**SOLUTION** We must first rewrite this as a single logarithm term; then we can solve it.

| | |
|---|---|
| Given | $\log_6 3 + \log_6 x = 2$ |
| ↓ Property 1 | $\log_6 3x = 2$ |
| ↓ Put into exponent form | $3x = 6^2 = 36$ |
| ↓ Divide by 3 | $x = 12$ |

Therefore, the solution set is $\{12\}$.

**EXAMPLE 15** Solve $\log_3 x + \log_3 (x + 2) = 1$.

**SOLUTION** We again begin by combining the left side into a single logarithm term.

| | |
|---|---|
| Given | $\log_3 x + \log_3 (x + 2) = 1$ |
| Property 1 | $\log_3 x(x + 2) = 1$ |
| Put into exponent form | $x(x + 2) = 3^1 = 3$ |
| Put into standard form for quadratic equations | $x^2 + 2x - 3 = 0$ |
| Factor | $(x - 1)(x + 3) = 0$ |
| Set each factor = 0 | $x - 1 = 0$ or $x + 3 = 0$ |
| Solve each equation | $x = 1$ or $x = -3$ |

Since $\log_3 (-3)$ is not allowed, the solution set is simply $\{1\}$.

## PROBLEM SET 11.2

Write the following in terms of simpler logarithm expressions. (Use Properties 1, 2, and 3.)

*See Example 9*

1. $\log_5 4a$
2. $\log_7 2x$
3. $\log_{10} 7xyz$
4. $\log_3 10mnpq$

*See Example 10*

5. $\log_2 \dfrac{6}{r}$
6. $\log_7 \dfrac{a}{b}$
7. $\log_8 \dfrac{5}{3}$
8. $\log_4 \dfrac{t}{10}$

*See Example 11*

9. $\log_2 x^5$
10. $\log_3 a^4$
11. $\log_{12} 2^r$
12. $\log_{10} (1.05)^n$
13. $\log_7 \sqrt{y}$
14. $\log_8 \sqrt[5]{u}$

*See Example 12*

15. $\log_2 x^2y^5$
16. $\log_5 a^3b^4c^7$
17. $\log_4 \dfrac{u^9}{v^2}$
18. $\log_7 \dfrac{p^2q^3}{r^7}$
19. $\log_{10} \dfrac{\sqrt{a}}{b^3}$
20. $\log_2 \dfrac{x\sqrt[3]{y}}{z^4}$
21. $\log_3 \sqrt{\dfrac{mn}{k}}$
22. $\log_{10} \sqrt[4]{\dfrac{ab}{c^3}}$
23. $\log_{10} \dfrac{\sqrt[3]{xy^2}}{\sqrt{uv^3}}$
24. $\log_2 \dfrac{\sqrt[4]{pq^3}}{\sqrt[5]{rs^2t^3}}$

Write the following as single logarithm expressions.

*See Example 13*

25. $\log_2 x + \log_2 y$
26. $\log_{10} 3 + \log_{10} 4 + \log_{10} a$
27. $\log_4 a - \log_4 b$
28. $\log_2 7 - \log_2 x$
29. $2 \log_{10} x$
30. $\dfrac{1}{3} \log_7 (a + b)$
31. $\log_{10} x + \log_{10} y - \log_{10} z$
32. $\log_2 a - \log_2 b - \log_2 c$
33. $2 \log_3 u + 3 \log_3 v - 4 \log_3 w$
34. $\log_4 m - 2 \log_4 n - 6 \log_4 k$
35. $\dfrac{2}{3}\log_{10} x + \dfrac{4}{3}\log_{10} y - \dfrac{7}{3}\log_{10} z$
36. $-\dfrac{1}{2}\log_2 a - \dfrac{3}{2}\log_2 b - \dfrac{5}{2}\log_2 c$

Solve the following equations for $x$. (Be sure to check for extraneous solutions—see Example 15.)

*See Example 14*

37. $\log_4 2 + \log_4 x = 3$
38. $\log_{10} 5 + \log_{10} x = 2$
39. $\log_3 2 + \log_3 5 + \log_3 x = 2$
40. $\log_5 2 + \log_5 10 + \log_5 x = 3$
41. $\log_3 x - \log_3 5 = 3$
42. $\log_7 x - \log_7 2 = 2$

*See Example 15*

43. $\log_3 x + \log_3 (x + 6) = 3$
44. $\log_2 x + \log_2 (x - 2) = 3$
45. $\log_6 x + \log_6 (x - 5) = 2$
46. $\log_2 x + \log_2 2x = 5$
47. $\log_3 (x + 16) - \log_3 x = 2$
48. $3 \log_2 x - \log_2 x = 4$

*Psychology applications*

49. Fechner's law states that

$$\gamma = k(\log_{10} \beta - \log_{10} b)$$

where $\gamma$ is magnitude of a sensation, $k$ is a constant, $\beta$ is the magnitude of the stimuli, and $b$ is the least perceptible stimuli. Write this expression with a single logarithm.

50. Another psychological theory is that reaction time $R$ in a situation is given by

$$R = K \log_{10} n$$

where $n$ is the possible alternatives and $K$ is a constant. Write this as a single logarithm without a coefficient.

*Business application*

51. Company researchers have determined that its product's demand $Q$ is related to consumer income $I$, its price $P_a$, and its competitor's price $P_b$ by the equation

$$\log_e Q = \log_e b + 5 \log_e I - 0.5 \log_e P_a + 0.9 \log_e P_b$$

Write the expression on the right side of the equation with a single logarithm.

## 11.3 COMMON LOGARITHMS

Although we have discussed many possible logarithm bases, there are really only two that are used with great frequency. The first type uses a base of $e = 2.718 \ldots$. These are called *natural* logarithms and occur in the study and use of calculus. We will discuss them in more detail in Section 11.5.

The other type of logarithm uses a base of 10. These are called **common logarithms** and fit in nicely with our base-10 system of numeration. For simplicity, we write $\log x$ for $\log_{10} x$. (The base of 10 is understood.)

**EXAMPLE 16** The powers of 10 have simple common logarithms.

**(a)** $\log 100 = \log 10^2 = 2$

**(b)** $\log 1000 = \log 10^3 = 3$

**(c)** $\log 100{,}000 = \log 10^5 = 5$

**(d)** $\log 0.1 = \log 10^{-1} = -1$

**(e)** $\log 0.0001 = \log 10^{-4} = -4$

Notice that the logarithm is always the exponent of 10.

But what about all the other numbers? For numbers between 1 and 10, we use the **table of common logarithms** printed on the back endpapers. Suppose that we have a number such as 3.78. To find log 3.78 in the table, we find the 3.7 row and the 8 column. These meet at .5775; thus, log 3.78 = 0.5775. See the cutaway table below.

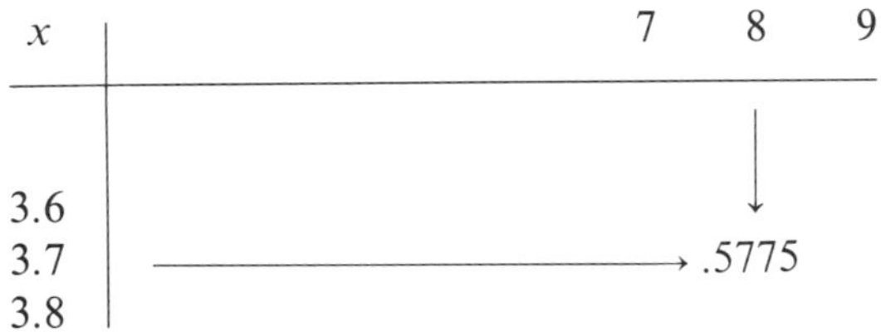


**EXAMPLE 17** These are further examples showing how the table of common logarithms is used.

**(a)** $\log 1.62 = .2095$

**(b)** $\log 8.13 = .9101$

**(c)** $\log 2.8 = \log 2.80 = .4472$

**(d)** $\log 6 = \log 6.00 = .7782$

Recall that with scientific notation (review Section 6.3), we can write any number $u$ as a product of a power of 10 and a number between 1 and 10.

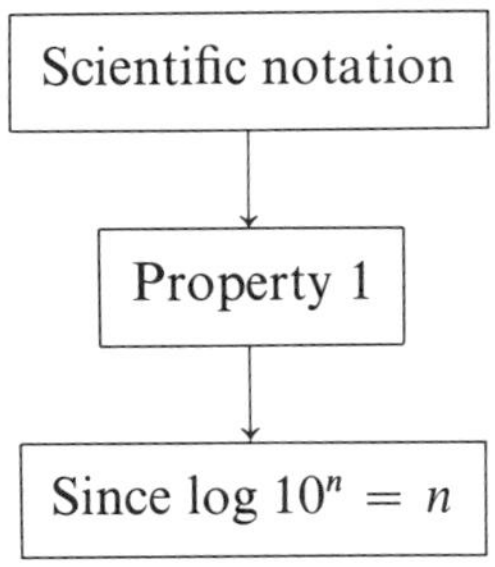


$$u = 10^n \times d$$

$$\log u = \log 10^n + \log d$$

$$\log u = n + \log d$$

Therefore, for any number $u$, $\log u$ has two parts: the integer part $n$, called the **characteristic,** and the decimal part $\log d$, called the **mantissa.**

> To find $\log u$, for any number $u$:
>
> 1. Write $u$ in scientific notation: $u = 10^n \times d$, where $d$ is between 1 and 10.
> 2. Look up $\log d$ in the table.
> 3. Write $\log u = n + \log d$.
> 4. If $n$ is negative, rewrite it as the difference of two positives; for instance, $-3 = 7 - 10$.

**EXAMPLE 18** Find log 23,400.

**SOLUTION** We first put 23,400 into scientific notation.

| | |
|---|---|
| Scientific notation | $23{,}400 = 10^4 \times 2.34$ |
| ↓ Use table | $\log 23{,}400 = 4 + .3692$ |
| ↓ Simplify | $\log 23{,}400 = 4.3692$ |

**EXAMPLE 19** Find log 7000.

**SOLUTION** We write this first in scientific notation.

| | |
|---|---|
| Scientific notation | $7000 = 10^3 \times 7.00$ |
| ↓ Use table | $\log 7000 = 3 + .8451$ |
| ↓ Simplify | $\log 7000 = 3.8451$ |

**EXAMPLE 20** Find log 0.000138.

**SOLUTION** We will get $-4$ as the characteristic. We rewrite this as $6 - 10$.

| | |
|---|---|
| Scientific notation | $0.000138 = 10^{-4} \times 1.38$ |
| ↓ Use table | $\log 0.000138 = -4 + .1399$ |
| ↓ Write $-4 = 6 - 10$ | $\log 0.000138 = 6.1399 - 10$ |

| YES | NO |
|---|---|
| $\log 0.000138 = 6.1399 - 10$ | ~~$\log 0.000138 = -4.1399$~~ |

Only the characteristic 4 is negative. By writing $-4.1399$, we would also (incorrectly) be making .1399 into $-.1399$. By writing $-4 = 6 - 10$, we are keeping .1399 positive.

Although $-4.1399$ is wrong, it is correct to add $-4 + .1399 = -3.8601$. Both answers are correct: $6.1399 - 10$ or $-3.8601$.

**EXAMPLE 21** Find log 0.0000009.

**SOLUTION** The characteristic here is $-7$, which we rewrite as $3 - 10$.

| | |
|---|---|
| Scientific notation | $0.0000009 = 10^{-7} \times 9.00$ |
| ↓ Use table | $\log 0.0000009 = -7 + .9542$ |
| ↓ Write $-7 = 3 - 10$ | $\log 0.0000009 = 3.9542 - 10$ |

As with Example 20, we can also write $3.9542 - 10$ as $-6.0458$ by adding.

We can redo Examples 18 to 21 using a hand calculator with a [log] key.

| PUNCH | DISPLAY | MEANING |
|---|---|---|
| [C] | 0. | Clear |
| [2] [3] [4] [0] [0] | 23400. | Example 18 |
| [log] | 4.3692159 | log 23400 |
| [7] [0] [0] [0] | 7000. | Example 19 |
| [log] | 3.845098 | log 7000 |
| [·] [0] [0] [0] [1] [3] [8] | 0.000138 | Example 20 |
| [log] | −3.8601209 | log 0.000138 |
| [·] [0] [0] [0] [0] [0] [0] [9] | 0.0000009 | Example 21 |
| [log] | −6.0457575 | log 0.0000009 |

Notice the two different ways in which the logarithms of the decimals were given.

| NUMBER | LOG (BY TABLE) | LOG (BY CALCULATOR) |
|---|---|---|
| 0.000138 | 6.1399 − 10 | −3.8601209 |
| 0.0000009 | 3.9542 − 10 | −6.0457575 |

Of course, the answers are the same, since $6.1399 - 10 = -3.8601$ and $3.9542 - 10 = -6.0458$.

We can reverse this process. Suppose that we are given $\log x = 3.6075$ and want to find $x$. We are looking for an **antilogarithm,** written antilog 3.6075.

**EXAMPLE 22** Find antilog 3.6075.

**SOLUTION** We reverse the logarithm process. Here, 3 is the characteristic and .6075 is the mantissa.

| | |
|---|---|
| Rewrite | $3.6075 = 3 + .6075$ |
| Since log 4.05 = .6075 | antilog $3.6075 = 10^3 \times 4.05$ |
| Put into usual form | antilog $3.6075 = 4050$ |

Here we find .6075 in the body of the table and see that it is in the 4.0 row and 5 column. Thus, $\log 4.05 = .6075$.

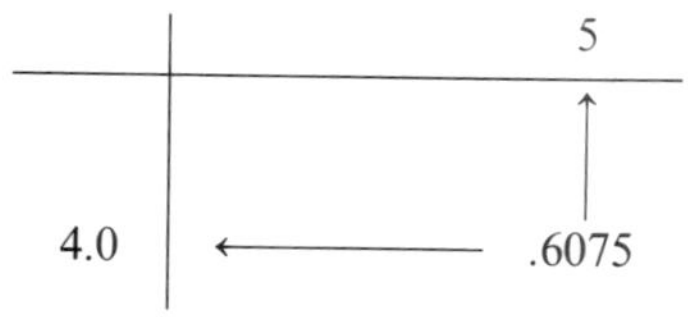


**EXAMPLE 23** Find antilog $4.8762 - 10$.

**SOLUTION** We again reverse the log process.

| | |
|---|---|
| Rewrite $4 - 10 = -6$ | $4.8762 - 10 = -6 + .8762$ |
| Use table | $\text{antilog}\,(4.8762 - 10) = 10^{-6} \times 7.52$ |
| Put into usual form | $\text{antilog}\,(4.8762 - 10) = 0.00000752$ |

Again, we find .8762 in the body of the table and see that it is in the 7.5 row and 2 column.

Suppose that we want to find $y = 10^x$, for a given $x$.

| | |
|---|---|
| Given | $y = 10^x$ |
| Put into logarithm form | $\log y = x$ |
| Definition of antilog | $y = \text{antilog}\ x$ |

Since $y = 10^x$ and $y = \text{antilog}\ x$, we have the following property.

> **PROPERTY 4** For any real number $x$,
>
> $$10^x = \text{antilog}\ x$$

**EXAMPLE 24** Find $10^{1.4}$.

**SOLUTION** We use Property 4 and find antilog 1.4.

| | |
|---|---|
| Property 4 | $10^{1.4} = \text{antilog}\ 1.4000$ |
| Find closest antilog in table | $= 10^1 \times 2.51$ |
| Put into usual form | $= 25.1$ |

In the table, .4000 does *not* appear exactly. So, we find the closest number to it: .3997, which has an antilog of 2.51.

We can rework Examples 22 to 24 with a hand calculator with a [log] key and an [INV] key. The [INV] key stands for inverse. Antilogarithms are the inverses of logarithms (review Section 9.5).

| PUNCH | DISPLAY | MEANING |
|---|---|---|
| [C] | 0. | Clear |
| [3] [·] [6] [0] [7] [5] | 3.6075 | Example 22 |
| [INV] [log] | 4050.4194 | antilog 3.6075 |
| [4] [·] [8] [7] [6] [2] [−] [1] [0] [=] | −5.1238 | Example 23 |
| [INV] [log] | .00000752 | antilog (4.8762 − 10) |
| [1] [·] [4] | 1.4 | Example 24 |
| [INV] [log] | 25.118864 | antilog 1.4 = $10^{1.4}$ |

## PROBLEM SET 11.3

Find the following common logarithms, using the table of common logarithms when necessary.

*See Example 16*

1. log 10
2. log 10,000
3. log 1,000,000
4. log 0.01
5. log 0.00001
6. log 1

*See Example 17*

7. log 6.52
8. log 3.02
9. log 5.3
10. log 8.8
11. log 4
12. log 8

*See Examples 18 and 19*

13. log 46,100
14. log 729,000
15. log 1,710,000
16. log 12
17. log 8900
18. log 63,000
19. log 200
20. log 40,000
21. log 7,000,000

*See Examples 20 and 21*

22. log 0.0623
23. log 0.521
24. log 0.000699
25. log 0.0073
26. log 0.29
27. log 0.000082
28. log 0.5
29. log 0.06
30. log 0.0003

Find the following antilogarithms. (Approximate if necessary.)

*See Example 22*

31. antilog 3.5977
32. antilog 5.7803
33. antilog 2.4742
34. antilog 1.9269
35. antilog 2.8892
36. antilog 3.4427

*See Example 23*

37. antilog 7.6656 − 10
38. antilog 5.7657 − 10
39. antilog 4.0294 − 10
40. antilog 2.6767 − 10
41. antilog 9.5 − 10
42. antilog 8.7 − 10

*See Example 24*

43. $10^{1.4099}$
44. $10^{2.8768}$
45. $10^{0.7007}$
46. $10^{2.9}$
47. $10^{3.1}$
48. $10^{4.6}$

*Electronics applications*

49. The power gain PG in an amplifier is measured by

$$PG = 10 \log_{10}\left(\frac{P_{\text{out}}}{P_{\text{in}}}\right)$$

where $P_{\text{out}}$ is power out and $P_{\text{in}}$ is the power in. Find the power gain PG if $P_{\text{in}} = 5$ and $P_{\text{out}} = 600$.

50. The impedance $Z_0$ in a coaxial TV cable line is given by

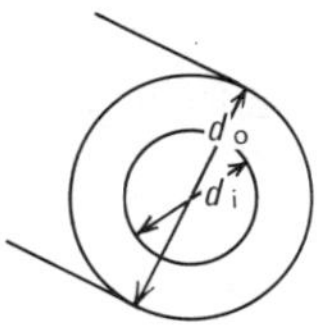


$$Z_0 = 138 \log_{10}\left(\frac{d_o}{d_i}\right)$$

where $d_i$ and $d_o$ are the inner and outer diameters of the cable. Find $Z_0$ if $d_i = 0.04$ meter and $d_o = 0.12$ meter.

*Photographic application*

The density of a film negative is given by

$$D = \log_{10}\left(\frac{1}{T}\right)$$

(see page 373). Find the density in the following cases.

51. $T = \frac{1}{2}$
52. $T = \frac{1}{16}$
53. $T = \frac{1}{50}$

*Chemical application*

The pH of a solution is given by

$$\text{pH} = -\log_{10}[\text{H}^+]$$

where $[H^+]$ is the molar concentration of hydrogen ions. Find the pH in the following cases. (*Hint:* First write logarithms as single numbers, rounded to the nearest tenth.)

54. Eyedrop solution: $[H^+] = 6.3 \times 10^{-8}$
55. Nonalkaline shampoo: $[H^+] = 1.6 \times 10^{-6}$
56. Detergent: $[H^+] = 3.2 \times 10^{-10}$
57. Fruit juice: $[H^+] = 8.0 \times 10^{-5}$

## 11.4 COMPUTATIONS WITH COMMON LOGARITHMS

Although many calculations are now done on electronic calculators, logarithms can still be used in many scientific calculations and applications. The rules for calculations with logarithms are based upon the three properties of logarithms, which we repeat here.

For real numbers $r$, $M > 0$, and $N > 0$,

1. $\log MN = \log M + \log N$.
2. $\log \frac{M}{N} = \log M - \log N$.
3. $\log N^r = r \log N$.

To multiply numbers using logarithms:

1. Find the logarithms of each number.
2. Add the logarithms.
3. Find the antilog of the sum.

**EXAMPLE 25** Compute $x = (24700)(7900)(0.036)$.

**SOLUTION** We use Property 1.

| | |
|---|---|
| Property 1 | $\log x = \log 24{,}700 + \log 7900 + \log 0.036$ |
| ↓ Use table | $\log x = 4.3927 + 3.8976 + (8.5563 - 10)$ |
| ↓ Simplify | $\log x = 16.8466 - 10 = 6.8466$ |
| ↓ Find antilog | $x = 7.02 \times 10^6 = 7{,}020{,}000$ |

> To divide numbers using logarithms:
>
> 1. Find the logarithms.
> 2. Subtract the logarithms.
> 3. Find the antilog of the difference.

**EXAMPLE 26** Compute $x = \dfrac{(460)(25{,}100)}{509}$.

**SOLUTION** We use Properties 1 and 2.

| | |
|---|---|
| Properties 1 and 2 | $\log x = \log 460 + \log 25{,}100 - \log 509$ |
| ↓ Use table | $\log x = 2.6628 + 4.3997 - 2.7067$ |
| ↓ Simplify | $\log x = 4.3558$ |
| ↓ Find antilog | $x = 2.27 \times 10^4 = 22{,}700$ |

**EXAMPLE 27** Compute $x = \dfrac{0.00172}{0.0504}$.

**SOLUTION** Here, the numerator will have a smaller characteristic than the denominator, so we will rewrite it. Instead of $-3 = 7 - 10$, we will use $-3 = 17 - 20$.

| | |
|---|---|
| Property 2 | $\log x = \log 0.00172 - \log 0.0504$ |
| Use table | $\log x = (7.2355 - 10) - (8.7024 - 10)$ |
| Rewrite $-3 = 17 - 20$ | $\log x = (17.2355 - 20) - (8.7024 - 10)$ |
| Simplify | $\log x = 8.5331 - 10$ |
| Find antilog | $x = 3.41 \times 10^{-2} = 0.0341$ |

> To compute exponents with logarithms:
>
> 1. Find the logarithm of the base.
> 2. Multiply this by the exponent.
> 3. Find the antilog of this product.

**EXAMPLE 28** Compute $x = (1.8)^{7.9}$.

**SOLUTION** Here, we use Property 3.

| | |
|---|---|
| Property 3 | $\log x = \log (1.8)^{7.9} = 7.9 (\log 1.8)$ |
| ↓ Use table | $\log x = 7.9(0.2553)$ |
| ↓ Simplify | $\log x = 2.01687$ |
| ↓ Find antilog | $x = 1.04 \times 10^2 = 104$ |

**EXAMPLE 29** Compute $x = \sqrt[3]{0.0175}$.

**SOLUTION** We first rewrite this as $x = (0.0175)^{1/3}$. The characteristic here is $-2$, which we usually write as $8 - 10$. Since 10 is not divisible by 3 but 30 is, we write this as $-2 = 28 - 30$.

| | |
|---|---|
| Property 3 | $\log x = \frac{1}{3}(\log 0.0175)$ |
| ↓ Use table; write $-2 = 28 - 30$ | $\log x = \frac{1}{3}(28.2430 - 30)$ |
| ↓ Distributive law | $\log x = 9.4143 - 10$ |
| ↓ Find antilog | $x = 2.60 \times 10^{-1} = 0.26$ |

**EXAMPLE 30** Compute $x = \sqrt{\dfrac{2550}{(142)(0.73)}}$.

**SOLUTION** We first write $x = \left(\dfrac{2550}{(142)(0.73)}\right)^{1/2}$. Now we use logarithm properties.

| | |
|---|---|
| Property 3 | $\log x = \frac{1}{2}\log\left(\frac{2550}{(142)(0.73)}\right)$ |
| ↓ | |
| Properties 1 and 2 | $\log x = \frac{1}{2}[\log 2550 - \log 142 - \log 0.73]$ |
| ↓ | |
| Use table | $\log x = \frac{1}{2}[3.4065 - 2.1523 - (9.8633 - 10)]$ |
| ↓ | |
| Simplify | $\log x = \frac{1}{2}[1.3909] = 0.69545$ |
| ↓ | |
| Find antilog | $x = 4.96$ |

## PROBLEM SET 11.4

Use logarithms to calculate the following.

*See Example 25*

1. $(7.5)(8.8)$
2. $(490)(62)$
3. $(6.9)(0.21)$
4. $(719)(0.02)$
5. $(0.003)(0.0172)$
6. $(0.0642)(0.00073)$
7. $(1700)(440)(0.052)$
8. $(49{,}000)(1200)(0.03)$

*See Examples 26 and 27*

9. $\dfrac{1720}{63}$
10. $\dfrac{405{,}000}{7300}$
11. $\dfrac{374}{11{,}800}$
12. $\dfrac{2030}{350{,}000}$
13. $\dfrac{(411)(2.33)}{105}$
14. $\dfrac{(2070)(0.05)}{0.00023}$
15. $\dfrac{7130}{(5000)(0.0195)}$
16. $\dfrac{0.063}{(41{,}000)(7360)}$

*See Example 28*

17. $1.2^{10}$
18. $2.1^{7}$
19. $(1.06)^{20}$
20. $(1.07)^{15}$
21. $2.7^{4.3}$
22. $3.1^{5.2}$

*See Example 29*

23. $\sqrt{49{,}500}$ 24. $\sqrt[3]{223{,}000}$

25. $\sqrt[3]{0.0628}$ 26. $\sqrt[4]{0.00531}$

27. $\sqrt{0.975}$ 28. $\sqrt[6]{0.834}$

*See Example 30*

29. $\sqrt{\dfrac{(251)(43)}{172}}$ 30. $\sqrt[3]{\dfrac{(443)(50{,}100)}{0.073}}$

31. $\left[\dfrac{(6100)(730{,}000)}{5440}\right]^{5/3}$ 32. $\left[\dfrac{(18{,}000)(2370)}{0.0667}\right]^{2/7}$

*Business application*

33. If a company invests \$50,000 at 7% interest, in 20 years its value $V$ will be

$$V = 50{,}000(1.07)^{20}$$

Use logarithms to compute this.

*Consumer application*

34. A family buys a \$6500 car. It depreciates 22% per year. After 6 years, its value $V$ is given by

$$V = 6500(0.78)^6$$

Use logarithms to compute this.

*Health application*

35. The chance that a patient will have no side effect from a drug is 99.9% (= 0.999). If 50 patients are given this drug, the chance that none of them will experience a side effect is $(0.999)^{50}$. Compute this with logarithms.

*Psychology application*

36. The time $t$ for a person to memorize $N$ symbols is predicted to be about

$$t = \frac{1}{120} \cdot N^{3/2}$$

Compute this for $N = 20$ with logarithms.

*Population application*

37. The world's population of approximately 4 billion is growing at 2% per year. In 20 years the population $P$ will be

$$P = 4{,}000{,}000{,}000(1.02)^{20}$$

Compute this with logarithms.

## 11.5 EXPONENTIAL EQUATIONS AND CHANGE OF BASE

An **exponential equation** is an equation that has an exponent or base as an unknown. As examples,

$$6^x = 16, \qquad 3^{2x+1} = 11, \qquad \text{and} \qquad x^{10} = 8$$

are exponent equations. To solve these equations, we use the following theorem.

> ***THEOREM 11.2*** Let $M$, $N$, and $b$ be positive real numbers ($b \neq 1$). If $M = N$, then $\log_b M = \log_b N$.

In words, *we can take the logarithm of both sides of an equation, and the results will also be equal.*

> To solve an exponential equation:
>
> 1. Take the common logarithm of both sides.
> 2. Simplify with Property 3 ($\log N^r = r \log N$).
> 3. Solve for the unknown.

**EXAMPLE 31** Solve $6^x = 16$.

**SOLUTION** We start by taking the logarithm of both sides.

| | |
|---|---|
| Given | $6^x = 16$ |
| ↓ Log of both sides | $\log 6^x = \log 16$ |
| ↓ Property 3 | $x \log 6 = \log 16$ |
| ↓ Divide by log 6 | $x = \dfrac{\log 16}{\log 6}$ |

If we find these logarithms in the table, we can obtain the approximate solution.

| | |
|---|---|
| Use table | $x = \dfrac{\log 16}{\log 6} \approx \dfrac{1.2041}{.7782} \approx 1.547$ |

Many hand calculators have a [log] key. The answer to this problem might be worked as follows on such a calculator.

| PUNCH | DISPLAY | MEANING |
|---|---|---|
| [C] | 0. | Clear |
| [1] [6] [log] | 1.20412 | log 16 |
| [÷] | 1.20412 | Divided by |
| [6] [log] | .77815125 | log 6 |
| [=] | 1.5474112 | Answer |

**EXAMPLE 32** Solve $3^{2x+1} = 11^x$.

**SOLUTION** We again take the logarithm of both sides. Then we isolate $x$.

| | |
|---|---|
| Given | $3^{2x+1} = 11^x$ |
| ↓ Log of both sides | $\log 3^{2x+1} = \log 11^x$ |
| ↓ Property 3 | $(2x + 1) \log 3 = x \log 11$ |
| ↓ Distributive law | $2x \log 3 + \log 3 = x \log 11$ |
| ↓ Subtract $2x \log 3$; distributive law | $\log 3 = x[\log 11 - 2 \log 3]$ |
| ↓ Solve for $x$ | $\dfrac{\log 3}{\log 11 - 2 \log 3} = x$ |

By using the table, we find the solution to be approximately 5.4747.

**EXAMPLE 33** Solve $x^{20} = 8$.

**SOLUTION** This problem is slightly different—the base is the unknown. But we still start by taking the logarithm of both sides and then isolate $x$.

| | |
|---|---|
| Given | $x^{20} = 8$ |
| Log of both sides | $\log x^{20} = \log 8$ |
| Property 3 | $20 \log x = \log 8$ |
| Divide by 20 | $\log x = \dfrac{\log 8}{20}$ |
| Use table; simplify | $\log x = \dfrac{.9031}{20} = 0.0452$ |
| Take antilog | $x = \text{antilog } 0.0452 = 1.11$ |

Therefore, the solution set is $\{1.11\}$.

Often, we wish to find the logarithm of a number to a base other than 10. To do this, we still use base-10 logarithms to help us. Suppose that we want $\log_b N$. Let us call this $x = \log_b N$ and solve for $x$.

| | |
|---|---|
| Given | $x = \log_b N$ |
| Put into exponent form | $b^x = N$ |
| Common log of both sides | $\log b^x = \log N$ |
| Property 3 | $x \log b = \log N$ |
| Divide by $\log b$ | $x = \dfrac{\log N}{\log b}$ |

This becomes our change-of-base property.

> **PROPERTY 5** For positive real numbers $b$ and $N$ ($b \neq 1$),
>
> $$\log_b N = \frac{\log N}{\log b}$$

Notice that this property is written in terms of base-10 logarithms.

**EXAMPLE 34** Find $\log_5 9$.

**SOLUTION** We use Property 5.

| | |
|---|---|
| $\log_b N = \dfrac{\log N}{\log b}$ | $\log_5 9 = \dfrac{\log 9}{\log 5}$ |
| ↓ Use table | $= \dfrac{0.9542}{0.6990}$ |
| ↓ Simplify | $= 1.365$ |

| YES | NO |
|---|---|
| $\log_5 9 = \dfrac{\log 9}{\log 5}$ | ~~$\log_5 9 = \log \dfrac{9}{5}$~~ |

We divide the common logarithms.

**EXAMPLE 35** Find $\log_e 7$.

**SOLUTION** As we mentioned in Section 11.3, logarithms to the base $e = 2.718\ldots$ are called **natural logarithms.** Although there are special tables for these logarithms, we will use Property 5 to find $\log_e 7$. (We use $\log e = 0.4343$.)

| | |
|---|---|
| $\log_b N = \dfrac{\log N}{\log b}$ | $\log_e 7 = \dfrac{\log 7}{\log e}$ |
| ↓ Use table | $= \dfrac{.8451}{.4343}$ |
| ↓ Simplify | $= 1.946$ |

## PROBLEM SET 11.5

Solve the following exponential equations.

See Example 31

1. $5^x = 12$
2. $4^x = 25$
3. $(1.07)^x = 2$
4. $(1.06)^x = 2.5$
5. $20^x = 13$
6. $48^x = 15$
7. $7^x = 0.5$
8. $5^x = 0.02$

See Example 32

9. $2^{x+1} = 9$
10. $5^{x-1} = 8$
11. $7^{3x-1} = 3$
12. $11^{2x+1} = 12$
13. $2^{x+1} = 3^{2x}$
14. $5^{x+1} = 7^x$
15. $6^{2x+1} = 4^{5x}$
16. $8^{3x-1} = 5^{x+4}$

See Example 33

17. $x^{10} = 3$
18. $x^{20} = 2.5$
19. $x^2 = 49$
20. $x^3 = 125$
21. $x^{1.7} = 3.2$
22. $x^{2.3} = 100$
23. $x^{15} = 3.7$
24. $x^7 = 2.2$

Find the following non-base-10 logarithms. (Use $\log e = 0.4343$.)

See Example 34

25. $\log_4 7$
26. $\log_8 13$
27. $\log_6 26.4$
28. $\log_{12} 21.2$
29. $\log_2 1.5$
30. $\log_5 3.2$
31. $\log_{6.5} 9.2$
32. $\log_{4.6} 3.1$

See Example 35

33. $\log_e 7$
34. $\log_e 25$
35. $\log_e 8.7$
36. $\log_e 1.8$

37. $\log_e 0.73$

38. $\log_e 0.0523$

*Business applications*

39. The time $t$ for money to double at 6% is the solution to the equation

$$(1.06)^t = 2$$

Solve this for $t$, the *doubling time.*

40. The rate of interest needed to double money in 10 years is the solution to

$$(1 + r)^{10} = 2$$

Solve for $r$. (*Hint:* First solve for $1 + r$; then subtract 1 to find $r$.)

*Science applications*

41. The work done by a certain expanding piston is given by $50 \log_e 9$. Compute this.

42. A radioactive substance starts out at 500 grams. The *half-life t* is the time it takes to decay to 250 grams. This is the solution to the equation

$$500e^{-0.4t} = 250$$

Solve this for half-life $t$. (Take the logarithm of both sides and use $\log e = 0.4343$.)

*Consumer applications*

43. A woman has a choice between two jobs: one starts at \$10,000 with 10% annual raises; the other starts at \$14,000 with 6% annual raises. Find the number of years $x$ it will take for the salaries to be equal by solving

$$10{,}000(1.10)^x = 14{,}000(1.06)^x$$

44. A family buys an \$800 TV set. Five years later they sell it for \$450. Find the rate of depreciation $r$ by solving

$$800(1 - r)^5 = 450$$

(*Hint:* First solve for $1 - r$; then find $r$.)

*Population application*

45. At 2% annual growth, the time $t$ that it will take the world's population of 4,000,000,000 to grow to 6,000,000,000 is given by

$$6{,}000{,}000{,}000 = 4{,}000{,}000{,}000(1.02)^t$$

Solve this for $t$.

## 11.6 EXPONENTIAL AND LOGARITHMIC FUNCTIONS

The **exponential function** has the form

$$y = f(x) = b^x$$

where $b > 0$ and $b \neq 1$. We graph exponential functions as we graph any function: make a table of $(x, y)$ pairs, plot these points, and draw a smooth curve through the points.

**EXAMPLE 36** Graph $y = 1.8^x$ for $-3 \leq x \leq 3$.

**SOLUTION** We use a hand calculator to help us approximate the powers of 1.8. Recall that

$$1.8^{-n} = \frac{1}{1.8^n}$$

| $x$ | $y = 1.8^x$ | $(x, y)$ |
|---|---|---|
| $-3$ | $\frac{1}{1.8^3} = \frac{1}{5.832} = 0.171$ | $(-3, 0.171)$ |
| $-2$ | $\frac{1}{1.8^2} = \frac{1}{3.24} = 0.309$ | $(-2, 0.309)$ |
| $-1$ | $\frac{1}{1.8^1} = \frac{1}{1.8} = 0.556$ | $(-1, 0.556)$ |
| 0 | $1.8^0 = 1$ | $(0, 1)$ |
| 1 | $1.8^1 = 1.8$ | $(1, 1.8)$ |
| 2 | $1.8^2 = 3.24$ | $(2, 3.24)$ |
| 3 | $1.8^3 = 5.832$ | $(3, 5.832)$ |

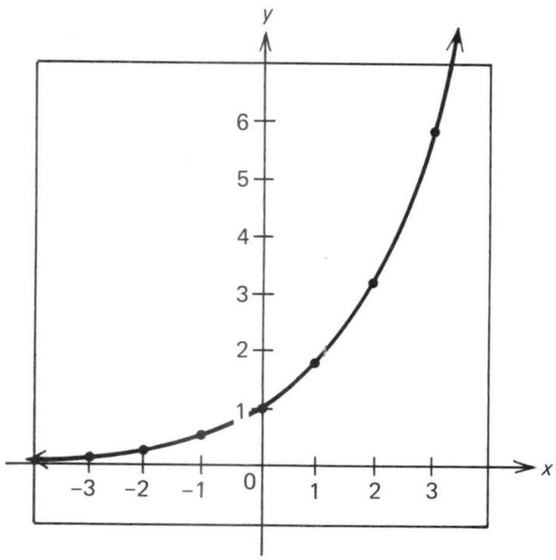

**EXAMPLE 37** Graph $y = 10 \cdot 2^{-x}$, for $-3 \le x \le 3$.

**SOLUTION** We again start with a table of $(x, y)$ pairs. We must be careful with the negative signs in the exponents.

Make table of $(x, y)$ pairs

| $x$ | $y = 10 \cdot 2^{-x}$ | $(x, y)$ |
|---|---|---|
| $-3$ | $10 \cdot 2^{-(-3)} = 10 \cdot 2^3 = 80$ | $(-3, 80)$ |
| $-2$ | $10 \cdot 2^{-(-2)} = 10 \cdot 2^2 = 40$ | $(-2, 40)$ |
| $-1$ | $10 \cdot 2^{-(-1)} = 10 \cdot 2^1 = 20$ | $(-1, 20)$ |
| $0$ | $10 \cdot 2^0 = 10 \cdot 1 = 10$ | $(0, 10)$ |
| $1$ | $10 \cdot 2^{-1} = 10 \cdot \frac{1}{2} = 5$ | $(1, 5)$ |
| $2$ | $10 \cdot 2^{-2} = 10 \cdot \frac{1}{4} = 2.5$ | $(2, 2.5)$ |
| $3$ | $10 \cdot 2^{-3} = 10 \cdot \frac{1}{8} = 1.25$ | $(3, 1.25)$ |

Plot points

Draw curve through points

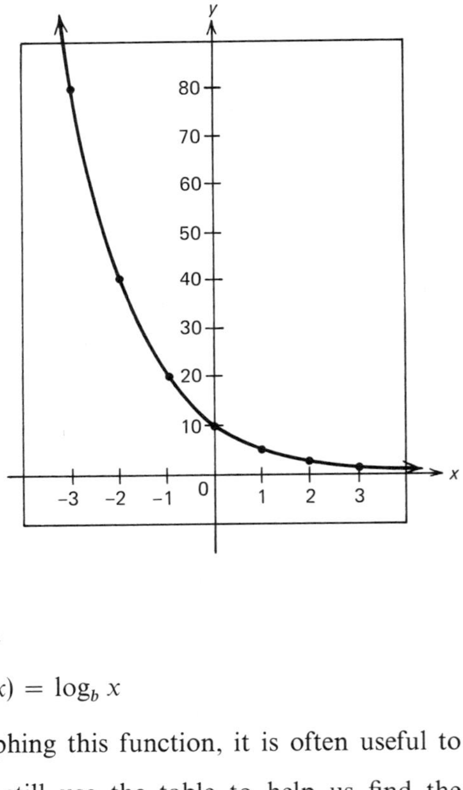


A **logarithmic function** has the form

$$y = f(x) = \log_b x$$

where $b > 0$ and $b \neq 1$. When graphing this function, it is often useful to use Property 5, $\log_b x = \dfrac{\log x}{\log b}$. We still use the table to help us find the logarithm values.

**EXAMPLE 38** Graph $y = \log_5 x$, for $0.1 \leq x \leq 10$.

**SOLUTION** We use the logarithm table to compute the $y$ values. We also use Property 5

$$\log_5 x = \frac{\log x}{\log 5} = \frac{\log x}{0.6990}$$

Use $\log_5 x = \dfrac{\log x}{\log 5} = \dfrac{\log x}{0.6990}$ to find $(x, y)$ pairs

| $x$ | $y = \log_5 x$ | $(x, y)$ |
|---|---|---|
| 0.1 | $\dfrac{-1}{0.6990} = -1.43$ | $(0.1, -1.43)$ |
| 0.5 | $\dfrac{9.6990 - 10}{.6990} = -0.43$ | $(0.5, -0.43)$ |
| 1.0 | 0 | $(1.0, 0.00)$ |
| 3.0 | $\dfrac{.4771}{.6990} = 0.68$ | $(3.0, 0.68)$ |
| 7.0 | $\dfrac{.8451}{.6990} = 1.21$ | $(7.0, 1.21)$ |
| 10.0 | $\dfrac{1.0000}{.6990} = 1.43$ | $(10.0, 1.43)$ |

Plot points

Draw curve through points

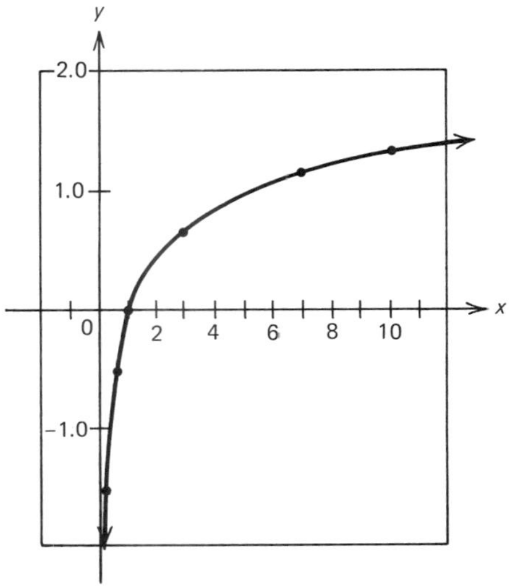

## PROBLEM SET 11.6

Graph the following exponential functions for $-3 \leq x \leq 3$. (Use a calculator or logarithm table, if necessary, to evaluate exponents.)

*See Example 36*

1. $y = 2^x$
2. $y = 3^x$
3. $y = 1.6^x$
4. $y = 2.1^x$
5. $y = 100(0.5)^x$
6. $y = 500(0.8)^x$

*See Example 37*

7. $y = 3^{-x}$
8. $y = 1.7^{-x}$
9. $y = 10(0.9)^{-x}$
10. $y = 200(0.7)^{-x}$

Graph the following logarithmic functions. (Use Property 5, $\log_b x = \dfrac{\log x}{\log b}$, and complete a table like the one below. Also, use $\log e = 0.4343$.)

| $x$ | 0.1 | 0.5 | 1.0 | 3.0 | 7.0 | 10.0 |
|---|---|---|---|---|---|---|
| $y = \log_b x$ | | | | | | |

*See Example 38*

11. $y = \log_{10} x$
12. $y = \log_9 x$
13. $y = \log_2 x$
14. $y = \log_7 x$
15. $y = \log_e x$
16. $y = \log_{1.7} x$

*Population application*

17. The population $y$ of a growing city is given by

$$y = 100{,}000(1.5)^{x/10}$$

where $x$ is number of years from now. Graph this equation using $x = 0$ 10, 20, 30 years.

*Health applications*

18. The alcohol content $y$ in a patient's body after $x$ hours is given by

$$y = 0.25(0.6)^x$$

Graph this equation for $x = 0, 1, 2, 3$.

19. The lead in the blood $y$ (in mg/100 ml) is related to the lead in the air (in $\text{mg/m}^3$) by the equation

$$y = 60 \log x - 20$$

Graph the equation for $x = 20, 40, 60, 80, 100$.

*Psychology application*

20. The number of tasks $y$ that a new worker can perform after $x$ days on the job is given by

$$y = 100(1 - 10^{-0.2x})$$

Graph this for $x = 0, 1, 2, 3, 4, 5$. (Use antilogarithms.) This is called a *learning curve.*

*Business applications*

21. A \$10,000 investment at 8% grows according to the formula

$$y = 10{,}000(1.08)^x$$

where $x$ is the number of years and $y$ the value of the investment. Graph this equation for $x = 0, 5, 10, 15, 20$.

22. A \$50,000 machine depreciates according to the formula

$$y = 50{,}000(0.9)^x$$

where $x$ is the number of years and $y$ the value of the investment. Graph this equation for $x = 0, 5, 10, 15, 20$.

## CHAPTER 11 SUMMARY

### Important Words and Phrases

antilogarithm (11.3)
characteristic (11.3).
common logarithm (11.3)
exponent form (11.1)
exponential equation (11.5)
exponential function (11.6)
logarithm (11.1)
logarithm form (11.1)
logarithmic function (11.6)
mantissa (11.3)
natural logarithm (11.5)
table of common logarithms (11.3)

### Important Theorems and Properties

For positive real numbers $b$ ($\neq 1$), $N$, $M$, and real numbers $x$ and $y$,

$\log_b N = x$ means $b^x = N$

If $b^x = b^y$, then $x = y$

If $M = N$, then $\log_b M = \log_b N$,

$\log_b MN = \log_b M + \log_b N$

$$\log_b \frac{M}{N} = \log_b M - \log_b N$$

$$\log_b (N^r) = r \log_b N$$

$$10^x = \text{antilog } x$$

$$\log_b N = \frac{\log N}{\log b}$$

## Important Procedures

*To find log u:*
1. Write $u$ as $10^n \times d$, where $1 \leq d < 10$.
2. Look up $\log d$ in the table.
3. Write $\log u = n + \log d$.
4. If $n$ is negative, then write $n$ as the difference of two positives (such as $-3 = 7 - 10$).

*To multiply (or divide) using logarithms:*
1. Find the logarithm of each number.
2. Add (or subtract) the logarithms.
3. Find the antilogarithm of the result.

*To compute exponents with logarithms:*
1. Find the logarithm of the base.
2. Multiply this by the exponent.
3. Find the antilogarithm of this product.

*To solve an exponential equation:*
1. Take the common logarithm of both sides.
2. Simplify by using $\log x^r = r \log x$.
3. Solve for the unknown.

# CHAPTER 11 REVIEW EXERCISES

1. Write $16^{3/4} = 8$ in logarithm form.
2. Write $\log_5 125 = 3$ in exponent form.
3. Solve $\log_4 x = 2$.
4. Solve $\log_x 1000 = 3$.
5. Simplify $\log_3 81$.
6. Simplify $\log_2 [\log_{10} 10{,}000]$.

Write the following in terms of simpler logarithm expressions.

7. $\log_8 6xyz$

8. $\log_7 \dfrac{T}{W}$

9. $\log_9 \sqrt{r}$

10. $\log_{10} \sqrt[4]{\dfrac{x}{yz^3}}$

11. Write $\frac{2}{5} \log_{10} x - \frac{4}{9} \log_{10} y$ as single logarithm expression.

12. Solve $\log_{10} 4 + \log_{10} x = 2$.

13. Solve $\log_2 x + \log_2 (x + 3) = 2$.

14. Find log 52,000.

15. Find log 0.0035.

16. Find antilog 1.7510.

17. Find antilog 7.5211 − 10.

18. Evaluate $10^{2.4}$.

Use logarithms to evaluate the following.

19. (4.51)(2590)(0.703)

20. $\dfrac{7350}{(443)(0.062)}$

21. $2.3^{3.5}$

22. $\sqrt[3]{0.529}$

23. $500(1.07)^{20}$

24. $\sqrt{\dfrac{421}{1727}}$

25. Solve $(1.06)^x = 2.5$.

26. Solve $x^{5.5} = 1000$.

27. Evaluate $\log_5 13$.

28. Evaluate $\log_e 20$.

29. Graph $y = 100(2^x)$ for $-3 \leq x \leq 3$.

30. Graph $y = \log_3 x$ for $x = 0.1, 0.5, 1.0, 3.0, 7.0, 10.0$.

# 12
# SEQUENCES AND SERIES

## 12.1 SEQUENCES

A **sequence** is a function, such as $s(n) = n^2$, in which the domain for $n$ is the natural numbers, 1, 2, 3, 4, and so on. For example, the sequence $s(n) = n^2$ appears as

$$1, 4, 9, 16, 25, \ldots$$

A sequence is like a string of numbers, usually defined by a rule or formula. The numbers in the sequence are called the **terms.** We usually write the terms of a sequence as

$$a_1, a_2, a_3, a_4, \ldots, a_n, \ldots$$

which we read "$a$ sub 1," "$a$ sub 2," and so on. The term $a_1$ is called the first term, $a_2$ is called the second, and so on. The term $a_n$ is called the $n$th term or the **general term.**

**EXAMPLE 1** Write out the first five terms of the sequence $a_n = 3n + 1$.

**SOLUTION** We substitute into the formula $a_n = 3n + 1$ with $n = 1, 2, 3, 4, 5$.

| Substitute $n = 1, 2, 3, 4, 5$, into formula for $a_n$ |
|---|

$$a_1 = 3(1) + 1 = 4$$
$$a_2 = 3(2) + 1 = 7$$
$$a_3 = 3(3) + 1 = 10$$
$$a_4 = 3(4) + 1 = 13$$
$$a_5 = 3(5) + 1 = 16$$

Thus, the sequence is 4, 7, 10, 13, 16, and so on.

**EXAMPLE 2** Write out the first five terms of

$$a_n = \frac{(-1)^n}{n + 1}$$

**SOLUTION** Again we substitute $n = 1, 2, 3, 4, 5$ into the formula. We must be careful with the term $(-1)^n$. Notice how this alternates between $-1$ (when $n$ is odd) and 1 (when $n$ is even).

| Substitute $n = 1, 2, 3, 4, 5$ into the formula for $a_n$ |
|---|

$$a_1 = \frac{(-1)^1}{1+1} = \frac{-1}{2}$$

$$a_2 = \frac{(-1)^2}{2+1} = \frac{1}{3}$$

$$a_3 = \frac{(-1)^3}{3+1} = \frac{-1}{4}$$

$$a_4 = \frac{(-1)^4}{4+1} = \frac{1}{5}$$

$$a_5 = \frac{(-1)^5}{5+1} = \frac{-1}{6}$$

Thus, the sequence is $\frac{-1}{2}, \frac{1}{3}, \frac{-1}{4}, \frac{1}{5}, \frac{-1}{6}$, and so on.

**EXAMPLE 3** Find the general term $a_n$ for the sequence

$$\frac{1}{1}, \frac{1}{4}, \frac{1}{9}, \frac{1}{16}, \frac{1}{25}, \ldots$$

**SOLUTION** This is the reverse problem. Here, we have the sequence, and we want the formula for $a_n$. We look for a pattern in the sequence. We see that each numerator is 1, while each denominator is a perfect square: $1^2, 2^2, 3^2, 4^2, 5^2, \ldots, n^2$, and so on. Thus, the general term is

| Write pattern in terms of $n$ |
|---|

$$a_n = \frac{1}{n^2}$$

**EXAMPLE 4** Find the general term $a_n$ for the sequence $5, 7, 9, 11, 13, 15, \ldots$.

**SOLUTION** Again, we have to find a pattern. Since the numbers are all two apart, we guess that this sequence is somehow related to the sequence of even numbers $2, 4, 6, 8, 10, \ldots, 2n, \ldots$. Each term of the given sequence is 3 greater than the terms of the sequence of even numbers. Therefore, we conclude:

| Write pattern in terms of $n$ |
|---|

$$a_n = 2n + 3$$

## PROBLEM SET 12.1

Write out the first five terms of the following sequences.

*See Examples 1 and 2*

1. $a_n = n + 2$
2. $a_n = 3n$
3. $a_n = n^2 + 1$
4. $a_n = n^3 - 1$
5. $a_n = 5n + 3$
6. $a_n = 6n - 4$
7. $a_n = (-1)^n$
8. $a_n = (-1)^n n^2$
9. $a_n = 2^n$
10. $a_n = \dfrac{1}{3^n}$
11. $a_n = \dfrac{1}{n}$
12. $a_n = \dfrac{n}{n+1}$
13. $a_n = \dfrac{n+1}{n^2}$
14. $a_n = \dfrac{n+3}{n^3}$
15. $a_n = 1 + \dfrac{1}{n}$
16. $a_n = 1 - \dfrac{1}{n}$

Find the general term $a_n$ for the following sequences.

*See Examples 3 and 4*

17. 1, 4, 9, 16, 25, . . .
18. 1, 8, 27, 64, 125, . . .
19. 3, 6, 9, 12, 15, 18, . . .
20. 5, 10, 15, 20, 25, . . .
21. 3, 9, 27, 81, 243, . . .
22. 10, 100, 1000, 10,000, . . .
23. $\frac{1}{2}, \frac{1}{4}, \frac{1}{6}, \frac{1}{8}, \frac{1}{10}, \ldots$
24. $\frac{1}{2}, \frac{1}{4}, \frac{1}{8}, \frac{1}{16}, \frac{1}{32}, \ldots$
25. $-1, 2, -3, 4, -5, \ldots$
26. $\frac{-1}{1}, \frac{1}{4}, \frac{-1}{9}, \frac{1}{16}, \frac{-1}{25}, \ldots$
27. $x, x^2, x^3, x^4, \ldots$
28. $x_1^2, x_2^2, x_3^2, x_4^2, \ldots$

*Population application*

29. One version of Pareto's law is that the population $a_n$ of the $n$th largest city is given by

$$a_n = \frac{P}{n}.$$

where $P = 7{,}500{,}000$, the population of New York. Write out the first six terms of this sequence.

*Physical application*

In atomic physics, the frequency $f$ of the light emitted by an excited hydrogen atom is given by the following four sequences. Write out the first four terms of each sequence. (Note where $n$ starts for each sequence; also, $Rc = 3.3 \times 10^{15}$.)

30. Lyman series: $f_n = Rc\left(1 - \frac{1}{n^2}\right), \quad n = 2, 3, 4, 5, \ldots$

31. Balmer series: $f_n = Rc\left(\frac{1}{2^2} - \frac{1}{n^2}\right), \quad n = 3, 4, 5, 6, \ldots$

32. Paschen series: $f_n = Rc\left(\frac{1}{3^2} - \frac{1}{n^2}\right), \quad n = 4, 5, 6, 7, \ldots$

33. Brackett series: $f_n = Rc\left(\frac{1}{4^2} - \frac{1}{n^2}\right), \quad n = 5, 6, 7, 8, \ldots$

## 12.2 SERIES

A **series** is a sum of the terms of a sequence. If there is a last term, it is a **finite series.** If there is no last term, it is an **infinite series.** As examples,

$$1 + 2 + 3 + 4 + 5 + 6 \qquad \text{is a finite series}$$

$$1 + \frac{1}{2} + \frac{1}{4} + \frac{1}{8} + \frac{1}{16} + \frac{1}{32} + \cdots \qquad \text{is an infinite series}$$

We deal only with finite series in this section.

We have a special compact notation for dealing with series. It is called **sigma notation** and uses the Greek letter $\sum$ (sigma). An example of this notation is

$$\sum_{n=1}^{6} a_n = a_1 + a_2 + a_3 + a_4 + a_5 + a_6$$

The $\sum$ stands for sum; the $n = 1$ below $\sum$ means start the sum at 1; the 6 above $\sum$ means add the sequence until $n = 6$; finally, the letter $n$ is called the **index of summation** (although any letter will do).

**EXAMPLE 5** Expand out $\sum_{n=1}^{8} n^2$.

**SOLUTION** We start with $n = 1$ and add the squares until $n = 8$.

| | |
|---|---|
| Substitute $n = 1$ to $n = 8$ for $n^2$ | $\sum_{n=1}^{8} n^2 = 1^2 + 2^2 + 3^2 + 4^2 + 5^2 + 6^2 + 7^2 + 8^2$ |
| ↓ | |
| $\sum$ means sum | $= 1 + 4 + 9 + 16 + 25 + 36 + 49 + 64$ |
| ↓ | |
| Simplify | $= 204$ |

**EXAMPLE 6** Expand out $\sum_{n=3}^{6} \frac{(-1)^n x^n}{n^3}$.

**SOLUTION** Here, the index of summation runs from 3 to 6.

| | |
|---|---|
| Substitute $n = 3$ to $n = 6$ in formula; $\sum$ means sum | $\sum_{n=3}^{6} \frac{(-1)^n x^n}{n^3} = \frac{(-1)^3 x^3}{3^3} + \frac{(-1)^4 x^4}{4^3} + \frac{(-1)^5 x^5}{5^3} + \frac{(-1)^6 x^6}{6^3}$ |
| ↓ | |
| Simplify | $= \frac{-x^3}{27} + \frac{x^4}{64} - \frac{x^5}{125} + \frac{x^6}{216}$ |

**EXAMPLE 7** Write the series $\frac{1}{10} + \frac{4}{100} + \frac{9}{1000} + \cdots + \frac{49}{10,000,000}$ in $\sum$ notation.

**SOLUTION** We must first discover a pattern for the general term. Notice that the numerators are perfect squares and the denominators are powers of 10. Let us rewrite the series as

$$\frac{1^2}{10^1} + \frac{2^2}{10^2} + \frac{3^2}{10^3} + \cdots + \frac{7^2}{10^7}$$

Now the pattern is clearer. The typical term is $n^2/10^n$, and $n$ runs from 1 to 7. Therefore, the series can be written

$$\sum_{n=1}^{7} \frac{n^2}{10^n}$$

**EXAMPLE 8** In statistics, the *mean m* and *variance* $\sigma^2$ of a set of data $x_1, x_2, x_3, \ldots, x_{100}$ are given by

$$m = \frac{1}{100}[x_1 + x_2 + \cdots + x_{100}]$$

$$\sigma^2 = \frac{1}{100}[(x_1 - m)^2 + (x_2 - m)^2 + \cdots + (x_{100} - m)^2]$$

Write these in $\sum$ notation.

**SOLUTION** For the mean $m$, the typical term is $x_n$; for the variance $\sigma^2$, the typical term is $(x_n - m)^2$. In both cases, the index $n$ runs from 1 to 100. Thus, we get

$$m = \frac{1}{100}\sum_{n=1}^{100} x_n$$

$$\sigma^2 = \frac{1}{100}\sum_{n=1}^{100} (x_n - m)^2$$

## PROBLEM SET 12.2

Expand out the following series. (In Problems 1 to 12, simplify to a single number.)

*See Example 5*

1. $\sum_{n=1}^{6} n$
2. $\sum_{n=1}^{8} (n + 2)$
3. $\sum_{n=1}^{5} 2n$
4. $\sum_{n=3}^{8} n^2$
5. $\sum_{n=4}^{7} n^3$
6. $\sum_{n=2}^{6} n^2 - 3$
7. $\sum_{n=1}^{5} \frac{1}{n}$
8. $\sum_{n=1}^{4} \frac{1}{n^2}$
9. $\sum_{n=1}^{6} \frac{1}{2^n}$
10. $\sum_{n=2}^{5} \frac{1}{3^n}$
11. $\sum_{n=1}^{4} \frac{n + 1}{n + 2}$
12. $\sum_{n=3}^{8} \frac{n - 1}{n + 1}$

*See Example 6*

13. $\sum_{n=1}^{4} x^n$
14. $\sum_{n=1}^{5} \frac{x^n}{n}$
15. $\sum_{n=1}^{5} n^x$
16. $\sum_{n=1}^{5} (x - n)$
17. $\sum_{n=1}^{6} \frac{(-1)^n}{x^n}$
18. $\sum_{n=1}^{5} \frac{(x - 10)^n}{3^n}$

Write the following series in $\sum$ notation.

*See Example 7*

19. $5 + 10 + 15 + 20 + 25$
20. $3 + 6 + 9 + 12 + 15 + 18 + 21$
21. $4 + 9 + 16 + 25 + 36$
22. $27 + 64 + 125$

23. $\frac{1}{3} + \frac{1}{4} + \frac{1}{5} + \frac{1}{6} + \frac{1}{7} + \frac{1}{8}$

24. $\frac{1}{5} + \frac{1}{25} + \frac{1}{125} + \frac{1}{625}$

25. $-1 + 2 - 3 + 4 - 5 + 6$

26. $\frac{-1}{2} + \frac{1}{4} - \frac{1}{8} + \frac{1}{16} - \frac{1}{32}$

*See Example 8*

27. $x + x^2 + x^3 + x^4 + x^5$

28. $\frac{1}{x^4} + \frac{1}{x^5} + \frac{1}{x^6} + \frac{1}{x^7}$

29. $\frac{x}{1} + \frac{x^2}{4} + \frac{x^3}{9} + \frac{x^4}{16}$

30. $x^2 + x^4 + x^6 + x^8$

31. $(x_1 - 5)^3 + (x_2 - 5)^3 + (x_3 - 5)^3$

32. $x_1 + x_2 + x_3 + x_4 + x_5$

*Population application*

33. According to Pareto's law, the total population of the top six cities is given by

$$\sum_{n=1}^{6} \frac{7{,}500{,}000}{n}$$

Compute this.

*Calculus application*

34. An approximate series for the natural logarithm (see Section 11.5) is given by

$$\log_e(1 + x) = x - \frac{x^2}{2} + \frac{x^3}{3} - \frac{x^4}{4} + \cdots - \frac{x^{20}}{20}$$

for $-1 < x < 1$. Write this in $\sum$ notation.

## 12.3 ARITHMETIC PROGRESSIONS

An **arithmetic progression** is a sequence for which the *difference* between any term (after the first) and the one before it is a constant. This difference is called the **common difference, *d*.**

**EXAMPLE 9** The following sequence is an arithmetic progression.

3, 9, 15, 21, 27, 33, . . .

We can subtract any two neighboring terms to find the common difference. For instance,

| $d$ = any term − previous term |
|---|

$d = 15 - 9 = 6$
$d = 27 - 21 = 6$

Notice that each term is 6 *more than* the one before it.

With only the first term $a_1$ and the common difference $d$, we can find the general term $a_n$ for the entire sequence. It works as follows.

| Each term is $d$ more than the previous term |
|---|

$$
\begin{aligned}
1 &\longrightarrow a_1 \\
2 &\longrightarrow a_2 = a_1 + d \\
3 &\longrightarrow a_3 = a_1 + d + d = a_1 + 2d \\
4 &\longrightarrow a_4 = a_1 + 2d + d = a_1 + 3d \\
5 &\longrightarrow a_5 = a_1 + 3d + d = a_1 + 4d \\
&\vdots \qquad \vdots \\
n &\longrightarrow a_n = a_1 + (n-1)d
\end{aligned}
$$

This is our pattern for the general term of an arithmetic sequence.

***THEOREM 12.1*** Let $a_1$ be the first term, $a_n$ the $n$th term and $d$ the common difference of an arithmetic sequence. Then

$$a_n = a_1 + (n-1)d$$

**EXAMPLE 10** Find the general term of the arithmetic progression with $a_1 = 16$ and $d = -2$.

**SOLUTION** Notice that the common difference is negative. This means that this sequence is decreasing:

$$16, 14, 12, 10, 8, \ldots$$

To find the general term, we substitute into Theorem 12.1.

| | |
|---|---|
| Theorem 12.1 | $a_n = a_1 + (n-1)d$ |
| ↓ Substitute $a_1 = 16$ and $d = -2$ | $a_n = 16 + (n-1)(-2)$ |
| ↓ Simplify | $a_n = 16 - 2n + 2 = 18 - 2n$ |

Thus, the general term is $a_n = 18 - 2n$.

> To find any term in any arithmetic progression:
>
> 1. Find the first term $a_1$ and common difference $d$.
> 2. Use Theorem 12.1 to find the needed term.

**EXAMPLE 11** Find $a_{25}$ for the arithmetic progression 2, 6, 10, 14, . . . .

**SOLUTION** The first term $a_1$ is obviously 2. The common difference $d$ is 4. We want the term for $n = 25$.

| | |
|---|---|
| Determine $a_1$, $d$ and $n$ | $a_1 = 2;\quad d = 4;\quad n = 25$ |
| ↓ $a_n = a_1 + (n - 1)d$ | $a_{25} = 2 + (25 - 1)(4)$ |
| ↓ Simplify | $a_{25} = 2 + (24)(4) = 98$ |

Now we wish to find the **sum of an arithmetic progression.**

Consider the task of adding the numbers from 1 to 100. Folklore has it that the mathematician Karl Gauss used the following procedure when he was seven years old. (Some say he was ten, others say the numbers were 1 to 1000, but this is not important; the procedure is.)

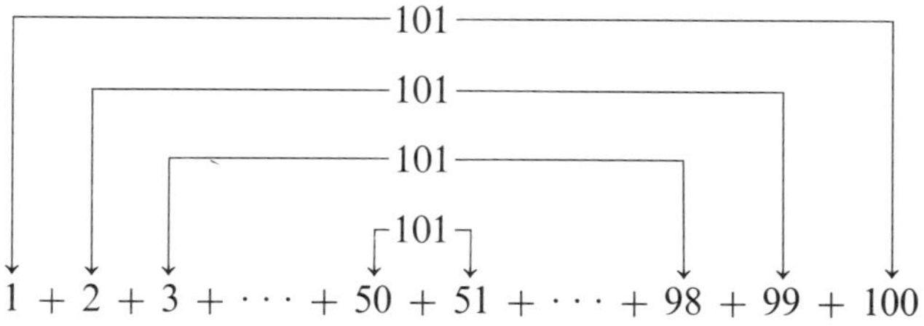


Instead of adding 1 + 2 + 3 and so on, he added 1 + 100 = 101, 2 + 99 = 101, . . . , 50 + 51 = 101. He had 50 such pairs each adding to 101; this makes (50)(101) = 5050.

This suggests the rule of *adding the first and last terms and then multiplying by half the number of terms.* As a theorem, this is stated:

> ***THEOREM 12.2*** Let $S_n$ be the sum of the first $n$ terms of an arithmetic progression, and $a_1$ and $a_n$ the first and $n$th terms. Then
>
> $$S_n = \frac{n}{2}(a_1 + a_n)$$

**EXAMPLE 12** Find the sum of the first 20 terms of 3, 8, 13, 18, 23, . . . .

**SOLUTION** The first term $a_1 = 3$. The common difference $d = 5$. We now need $a_{20}$, the last term.

| | |
|---|---|
| $a_n = a_1 + (n - 1)\,d$ | $a_{20} = 3 + (20 - 1)5$ |
| ↓ Simplify | $= 3 + (19)(5) = 98$ |

Now we use Theorem 12.2.

| | |
|---|---|
| Theorem 12.2 | $S_n = \frac{n}{2}(a_1 + a_{20})$ |
| ↓ Substitute $n = 20, a_1 = 3, a_{20} = 98$ | $S_{20} = \frac{20}{2}(3 + 98)$ |
| ↓ Simplify | $= 10(101) = 1010$ |

The sum of the first 20 terms is 1010.

## PROBLEM SET 12.3

Find the common difference for each of the following arithmetic progressions.

*See Example 9*

1. 2, 3, 4, 5, 6, . . .
2. 5, 7, 9, 11, 13, . . .
3. 3, 8, 13, 18, 23, . . .
4. 20, 17, 14, 11, 8, . . .

5. 9, 4, −1, −6, −11, . . .
6. −8, −4, 0, 4, 8, . . .
7. 4, $4\frac{1}{2}$, 5, $5\frac{1}{2}$, 6, . . .
8. 10, $9\frac{3}{4}$, $9\frac{1}{2}$, $9\frac{1}{4}$, 9, . . .

For the following arithmetic progressions, write out the first five terms and find the general term $a_n$.

*See Example 10*

9. $a_1 = 4;\quad d = 5$
10. $a_1 = 6;\quad d = 2$
11. $a_1 = 8;\quad d = -3$
12. $a_1 = 10;\quad d = -6$
13. $a_1 = -2;\quad d = 4$
14. $a_1 = -12;\quad d = 5$
15. $a_1 = 4;\quad d = \frac{1}{3}$
16. $a_1 = 5;\quad d = \frac{-2}{3}$

Find the indicated terms for the arithmetic progressions in Problems 1 to 8.

*See Example 11*

17. Find $a_{12}$ in Problem 1.
18. Find $a_{23}$ in Problem 2.
19. Find $a_{50}$ in Problem 3.
20. Find $a_{12}$ in Problem 4.
21. Find $a_{15}$ in Problem 5.
22. Find $a_{14}$ in Problem 6.
23. Find $a_{20}$ in Problem 7.
24. Find $a_{10}$ in Problem 8.

Find the indicated sums for the arithmetic progressions in Problems 1 to 8. (Use Problems 17 to 24.)

*See Example 12*

25. Find $S_{12}$ for Problem 1.
26. Find $S_{23}$ for Problem 2.
27. Find $S_{50}$ for Problem 3.
28. Find $S_{12}$ for Problem 4.
29. Find $S_{15}$ for Problem 5.
30. Find $S_{14}$ for Problem 6.
31. Find $S_{20}$ for Problem 7.
32. Find $S_{10}$ for Problem 8.

*Business application*

33. A machine is purchased for $60,000. It depreciates $5000 per year. Write a formula for $a_n$, the value after $n$ years. Also, write out the first four terms.

*Consumer applications*

34. A family buys $2000 worth of living-room furniture. They agree to make 40 monthly payments as follows: $70.00, $69.50, $69.00, $68.50, . . . . Find the last payment and sum of the 40 payments.

Tom Ascher has two job offers. Job 1 pays $12,000 with $800-per-year raises. Job 2 pays $14,000 with $500-per-year raises.

35. Write out the first four years' pay for job 1.
36. How much will job 1 pay in year 10?

37. How much *total* earning will Tom have made at job 1 after 10 years? (Add the first 10 years.)

38. Write out the first four years' pay for job 2.

39. How much will job 2 pay in year 10?

40. How much *total* earning will Tom have made at job 2 after 10 years? (Add the first 10 years.)

## 12.4 GEOMETRIC PROGRESSIONS

A **geometric progression** is a sequence for which the *ratio* of any term (after the first) to the one before it is a constant. This ratio is called the **common ratio, *r*.**

**EXAMPLE 13** The following is a geometric progression.

$$5, 10, 20, 40, 80, 160, \ldots$$

We can divide any two neighboring terms to find the common ratio. For instance,

> $r$ = any term ÷ previous term

$$r = \frac{20}{10} = 2$$

$$r = \frac{80}{40} = 2$$

Notice that each term is 2 *times* the term before it.

Just as with the arithmetic progressions, we can discover the general term for a geometric progression with only the first term $a_1$ and the common ratio $r$.

> Each term is $r$ times the previous term

$$\begin{aligned}
1 &\longrightarrow a_1 \\
2 &\longrightarrow a_2 = a_1 r \\
3 &\longrightarrow a_3 = (a_1 r)r = a_1 r^2 \\
4 &\longrightarrow a_4 = (a_1 r^2)r = a_1 r^3 \\
5 &\longrightarrow a_5 = (a_1 r^3)r = a_1 r^4 \\
n &\longrightarrow a_n = a_1 r^{n-1}
\end{aligned}$$

Therefore, the general term of a geometric progression can be given by

> ***THEOREM 12.3*** Let $a_1$ be the first term, $a_n$ the $n$th term, and $r$ the common ratio of a geometric progression. Then
>
> $$a_n = a_1 r^{n-1}$$

**EXAMPLE 14** Find the general term for the geometric progression with $a_1 = 6$ and $r = \frac{1}{3}$.

**SOLUTION** The common ratio is a fraction. This means that each term is $\frac{1}{3}$ of the previous term:

$$6, 2, \frac{2}{3}, \frac{2}{9}, \frac{2}{27}, \ldots$$

To find the general term, we substitute $a_1 = 6$ and $r = \frac{1}{3}$ into Theorem 12.3.

| $a_n = a_1 r^{n-1}$ | $a_n = 6(\frac{1}{3})^{n-1}$ |
|---|---|

Thus, the general term is $a^n = 6(\frac{1}{3})^{n-1}$.

> To find any term in a geometric progression:
>
> 1. Find the first term $a_1$ and the common ratio $r$.
> 2. Use Theorem 12.3 to find the needed term.

**EXAMPLE 15** Find $a_{11}$ for the geometric progression 24, 12, 6, 3, . . . .

**SOLUTION** The first term $a_1$ is 24. The common ratio is $\frac{12}{24} = \frac{1}{2}$. We want term for $n = 11$.

| | |
|---|---|
| Find $a_1$, $r$, and $n$ | $a_1 = 24;\quad r = \frac{1}{2};\quad n = 11$ |
| ↓ $a_n = a_1 r^{n-1}$ | $a_{11} = 24(\frac{1}{2})^{11-1} = 24(\frac{1}{2})^{10}$ |
| ↓ Simplify | $= 24 \cdot \frac{1}{1024} = \frac{3}{128}$ |

| YES | NO |
|---|---|
| 24, 12, 6, 3, . . . | ~~24, 12, 6, 3, . . .~~ |
| $r = \frac{12}{24} = \frac{1}{2}$ | ~~$r = \frac{24}{12} = 2$~~ |

To find $r$, we divide any term by the term *before* it.

As with arithmetic progressions, we want a formula to find the **sum of a geometric progression.** Let $S_n$ be the sum of the first $n$ terms. Then

| | |
|---|---|
| Given | $S_n = a_1 + a_1r + a_1r^2 + \cdots + a_1r^{n-1}$ |
| ↓ Multiply by $r$ | $rS_n = a_1r + a_1r^2 + \cdots + a_1r^{n-1} + a_1r^n$ |
| ↓ Subtract; cancel all common terms | $rS_n - S_n = (a_1r^n + \cancel{a_1r^{n-1}} + \cdots + \cancel{a_1r}) - (\cancel{a_1r^{n-1}} + \cdots + \cancel{a_1r} + a_1) = a_1r^n - a_1$ |
| ↓ Factor | $S_n(r - 1) = a_1(r^n - 1)$ |
| ↓ Divide by $r - 1$ | $S_n = \dfrac{a_1(r^n - 1)}{r - 1}$ |

This becomes our theorem.

> ***THEOREM 12.4*** Let $S_n$ be the sum of the first $n$ terms, $a_1$ the first term, and $r$ the common ratio of a geometric series. Then
>
> $$S_n = \frac{a_1(r^n - 1)}{r - 1}$$

**EXAMPLE 16** If \$1000 is deposited in the bank every year for 20 years at 6% interest, the total accumulation after 20 years is given by the geometric series

$$1000 + 1000(1.06) + 1000(1.06)^2 + 1000(1.06)^3 + \cdots + 1000(1.06)^{19}$$

Find the total.

**SOLUTION** We use Theorem 12.4 with $a_1 = 1000$, $r = 1.06$, and $n = 20$.

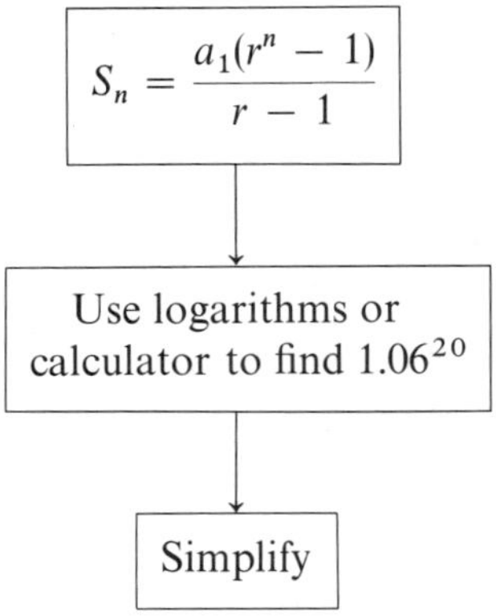


$$S_{20} = \frac{1000(1.06^{20} - 1)}{1.06 - 1}$$

$$= \frac{1000(3.207 - 1)}{1.06 - 1}$$

$$= \frac{1000(2.207)}{0.06} = 36{,}783.33$$

Thus, saving \$1000 per year for 20 years at 6% interest accumulates to about \$36,783.

## PROBLEM SET 12.4

Find the common ratio for each of the following geometric series.

*See Example 13*

1. 1, 3, 9, 27, . . .
2. 3, 6, 12, 24, . . .
3. 4, 40, 400, 4000, . . .
4. 2, 10, 50, 250, . . .
5. 60, 30, 15, $\frac{15}{2}$, . . .
6. 128, 32, 8, 2, . . .
7. 5, $-15$, 45, $-135$, . . .
8. 75, $-15$, 3, $\frac{-3}{5}$, . . .

For the following geometric progressions, write out the first five terms.

*See Example 14*

9. $a_1 = \frac{1}{2}$; $r = 4$
10. $a_1 = 4$; $r = 3$
11. $a_1 = 12$; $r = \frac{1}{3}$
12. $a_1 = 6000$; $r = \frac{1}{10}$
13. $a_1 = 10$; $r = -5$
14. $a_1 = \frac{1}{10}$; $r = -2$
15. $a_1 = \frac{1}{3}$; $r = \frac{1}{2}$
16. $a_1 = 18$; $r = -\frac{1}{3}$

Find the indicated terms for the geometric progressions in Problems 1 to 8.

*See Example 15*

17. Find $a_7$ in Problem 1.
18. Find $a_6$ in Problem 2.
19. Find $a_9$ in Problem 3.
20. Find $a_8$ in Problem 4.

21. Find $a_6$ in Problem 5.
22. Find $a_{10}$ in Problem 6.
23. Find $a_7$ in Problem 7.
24. Find $a_6$ in Problem 8.

Find the indicated sums for the geometric progressions in Problems 1 to 8.

*See Example 16*

25. Find $S_7$ in Problem 1.
26. Find $S_6$ in Problem 2.
27. Find $S_9$ in Problem 3.
28. Find $S_8$ in Problem 4.
29. Find $S_6$ in Problem 5.
30. Find $S_{10}$ in Problem 6.
31. Find $S_7$ in Problem 7.
32. Find $S_6$ in Problem 8.

*Musical application*

The frets on the neck of a guitar form a geometric progression. If the length from the bridge to the neck is $L$, the frets are at lengths $Lr$, $Lr^2$, $Lr^3$, and so on, from the bridge.

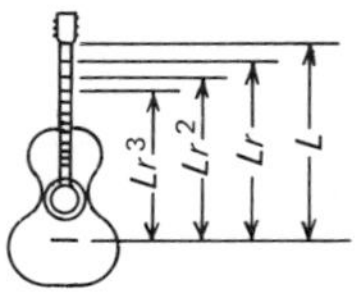


33. Find a formula for $a_{12}$, the length at the twelfth fret.
34. If the length at fret 12 is half the original length ($a_{12} = L/2$), solve for $r$. (Use Problem 33.)

*Life science application*

The eighteenth-century scientist Malthus claimed that populations grow geometrically (in a geometric progression). The monthly populations of a certain insect is given by 10,000, 15,000, 22,500, 33,750, . . . .

35. Find the common ratio $r$, which is here the growth rate.
36. Find $a_{10}$. (Use logarithms or a calculator.)

*Health applications*

37. While fasting, a person each day loses 1% of his or her body weight at the beginning of the day. In other words, the weight is 99% of what it was the day before. Write out the first four terms for a fasting 50-kilogram woman. What is the general term?

A hospital patient is being tested with a drug. He is given dosages in a geometric progression: 10, 20, 40, . . . .

38. What is the common ratio?
39. What is the tenth dosage?
40. What is the *total* of the first 10 dosages?

*Business applications*

41. A company buys a \$200,000 piece of equipment. It depreciates 20% per year. (It is worth 80% of last year's value.) Write out the first five terms and the general term.

If $P$ dollars are invested every year for $n$ years at $i$ rate of interest, after the $n$ years it will accumulate to

$$P + P(1 + i) + P(1 + i)^2 + \cdots + P(1 + i)^{n-1}$$

Use Theorem 12.4 to find the total accumulation in the following cases (see Example 16). (Use logarithms or a calculator for help.)

42. $P = \$2000$; $n = 10$ years; $i = 6\%\ (= 0.06)$
43. $P = \$500$; $n = 30$ years; $i = 5\%\ (= 0.05)$
44. $P = \$10{,}000$; $n = 15$ years; $i = 7\%\ (= 0.07)$
45. $P = \$3000$; $n = 20$ years; $i = 8\%\ (= 0.08)$

## 12.5 INFINITE GEOMETRIC PROGRESSIONS

What happens to a geometric progression with an infinite number of terms? That depends on the common ratio, $r$.

*Case 1.* If $|r| < 1$, the terms approach zero. For example, for $r = \frac{1}{3}$ and $a_1 = 1$,

$$1, \frac{1}{3}, \frac{1}{9}, \frac{1}{27}, \frac{1}{81}, \frac{1}{243}, \cdots \to 0$$

*Case 2.* If $|r| > 1$, the terms become larger and larger without bound. For example, for $r = 10$ and $a_1 = 20$,

$$20, 200, 2000, 20{,}000, 200{,}000, \ldots$$

*Case 3.* If $r = \pm 1$, the series is either constant or oscillating. As examples,

$$r = 1 \text{ and } a_1 = 2 \text{ gives } 2, 2, 2, 2, \ldots$$
$$r = -1 \text{ and } a_1 = 6 \text{ gives } 6, -6, 6, -6, \ldots$$

Case 1 is the only one for which we can add all the terms of the infinite series.

Recall that the formula for the sum of the first $n$ terms of a geometric series is

$$S_n = \frac{a_1(r^n - 1)}{r - 1}$$

In Case 1 where $|r| < 1$, the term $r^n$ gets smaller and smaller as $n$ gets larger. Eventually, it becomes so small that we can ignore it altogether. We call $S_\infty$

the **sum of an infinite geometric progression,** and we get

| | |
|---|---|
| $r^n$ approaches 0 | $S_\infty = \dfrac{a_1(0 - 1)}{r - 1}$ |
| Simplify | $= \dfrac{-a_1}{r - 1}$ |
| Multiply by $1 = \dfrac{-1}{-1}$ | $S_\infty = \dfrac{a_1}{1 - r}$ |

This is our theorem.

> ***THEOREM 12.5*** Let $a_1$ be the first term, $r$ the common ratio, and $S_\infty$ the sum of an infinite geometric progression. Then if $|r| < 1$,
>
> $$S_\infty = \frac{a_1}{1 - r}$$

**EXAMPLE 17** Find the sum of the infinite series

$$6 - 2 + \tfrac{2}{3} - \tfrac{2}{9} + \tfrac{2}{27} - \cdots$$

**SOLUTION** We can use Theorem 12.5 since the common ratio $r = \dfrac{-1}{3}$. Also, $a_1 = 6$.

| | |
|---|---|
| Theorem 12.5 | $S_\infty = \dfrac{a_1}{1 - r}$ |
| Substitute $a_1 = 6$ and $r = \dfrac{-1}{3}$ | $S_\infty = \dfrac{6}{1 - (-\frac{1}{3})}$ |
| Multiply by $1 = \dfrac{3}{3}$ | $= \dfrac{6}{(1 + \frac{1}{3})} \cdot \dfrac{3}{3}$ |
| Simplify | $= \dfrac{18}{3 + 1} = \dfrac{18}{4} = \dfrac{9}{2}$ |

**EXAMPLE 18** Find a fraction that is equal to the repeating decimal 0.37373737 . . . .

**SOLUTION** We write this as an infinite series:

| | |
|---|---|
| Separate terms | $0.37 + 0.0037 + 0.000037 + \cdots$ |
| Write as fractions | $= \frac{37}{100} + \frac{37}{10000} + \frac{37}{1,000,000} + \cdots$ |
| Identify terms | $a_1 = \frac{37}{100}; \quad r = \frac{1}{100}$ |
| $S_\infty = \frac{a_1}{1 - r}$ | $S_\infty = \frac{\frac{37}{100}}{1 - \frac{1}{100}}$ |
| Multiply by $1 = \frac{100}{100}$ | $= \frac{\frac{37}{100}}{(1 - \frac{1}{100})} \cdot \frac{100}{100}$ |
| Simplify | $= \frac{37}{100 - 1} = \frac{37}{99}$ |

Therefore, $0.373737\ldots = \frac{37}{99}$.

**EXAMPLE 19** Find the sum for the infinite series 2, 10, 50, 250, . . . .

**SOLUTION** Since the common ratio $r = 5$, which is greater than 1, we cannot use Theorem 12.5. There is *no* sum.

## PROBLEM SET 12.5

Find the sums of the following infinite geometric progressions. (*Be careful:* Some may not have sums—see Example 19.)

*See Example 17*

1. $1 + \frac{1}{2} + \frac{1}{4} + \frac{1}{8} + \cdots$
2. $1 + \frac{1}{3} + \frac{1}{9} + \frac{1}{27} + \cdots$
3. $1 - \frac{1}{2} + \frac{1}{4} - \frac{1}{8} + \cdots$
4. $1 - \frac{1}{3} + \frac{1}{9} - \frac{1}{27} + \cdots$
5. $3 + \frac{1}{2} + \frac{1}{12} + \frac{1}{72} + \cdots$
6. $\frac{1}{2}, \frac{3}{2}, \frac{9}{2}, \frac{27}{2}, \ldots$
7. $9 + 6 + 4 + \frac{8}{3} + \cdots$
8. $100 + 40 + 16 + 6.4 + \cdots$
9. $\frac{3}{5}, \frac{3}{4}, \frac{15}{16}, \frac{75}{64}, \ldots$
10. $\frac{3}{4}, -\frac{9}{20}, \frac{27}{100}, -\frac{81}{500}, \ldots$

Write each of the following repeating decimals as a fraction.

*See Example 18*

11. 0.44444 . . .
12. 0.77777 . . .
13. 0.010101 . . .
14. 0.232323 . . .
15. 0.174174174 . . .
16. 0.523523523 . . .

*Physical applications*

A turtle is walking across a 12-foot room. It first walks 6 feet, then 3 feet, then $1\frac{1}{2}$ feet, and so on.

17. What is the common ratio of this infinite geometric progression?
18. What is the sum of this infinite geometric progression?

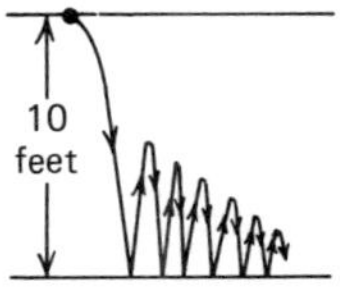


A ball is dropped from a height of 10 feet. Each time it rebounds to 0.6 of its previous height (see the sketch).

19. Write the downward trips as the infinite geometric progression 10.0, 6.0, 3.6, 2.16, . . . , and find the sum for total downward motion.
20. Write the upward trips as the infinite geometric progression 6.0, 3.6, 2.16, . . . , and find the sum for total upward motion.
21. Add the downward and upward total to find the overall total distance traveled.

*Economics application*

22. The government is considering a \$1000-per-family tax rebate. The government feels that each family will spend 70% of this (= \$700). The people receiving this \$700 will in turn spend 70% of it (= \$490), and so on. The resulting increase to the economy becomes

$$1000 + 700 + 490 + 343 + \cdots$$

This is called the *multiplier effect* and is an infinite geometric progression. Find the sum, which is the total effect of the original \$1000 on the economy.

## 12.6 THE BINOMIAL THEOREM

Recall from Chapter 3 that a **binomial** is a sum (or difference) of two terms, such as $x + y$ or $a - 3$. We are interested in the powers of binomials such as $x + y$. Notice the pattern that evolves.

$$
\begin{aligned}
(x + y)^0 &= 1 \\
(x + y)^1 &= x + y \\
(x + y)^2 &= x^2 + 2xy + y^2 \\
(x + y)^3 &= x^3 + 3x^2y + 3xy^2 + y^3 \\
(x + y)^4 &= x^4 + 4x^3y + 6x^2y^2 + 4xy^3 + y^4 \\
(x + y)^5 &= x^5 + 5x^4y + 10x^3y^2 + 10x^2y^3 + 5xy^4 + y^5
\end{aligned}
$$

There is a pattern for the coefficients. It appears as follows.

Row 0 ⟶ 1

Row 1 ⟶ 1 1

Row 2 ⟶ 1 2 1

Row 3 ⟶ 1 3 3 1

Row 4 ⟶ 1 4 6 4 1

Row 5 ⟶ 1 5 10 10 5 1

This triangle is called **Pascal's triangle.** Each entry is the sum of the two entries above it. For example, the 10 in row 5 is the sum of the 4 and 6 above it. This table continues as follows.

Row 6 ⟶ 1 6 15 20 15 6 1

Row 7 ⟶ 1 7 21 35 35 21 7 1

Row 8 ⟶ 1 8 28 56 70 56 28 8 1

and so on. Another pattern in expansions of $(x + y)^n$ is that the sum of the exponents in each term is $n$: $x^n$, $x^{n-1}y$, $x^{n-2}y^2$, $x^{n-3}y^3$, and so on. *Row $n$ of Pascal's triangle gives the coefficients in the expansion of* $(x + y)^n$.

**EXAMPLE 20** Use Pascal's triangle to expand out $(x + y)^7$.

**SOLUTION** We use row 7 of Pascal's triangle and arrange the terms so that the sums of the exponents are always 7.

$$(x + y)^7 = x^7 + 7x^6y + 21x^5y^2 + 35x^4y^3 + 35x^3y^4 + 21x^2y^5 + 7xy^6 + y^7$$

**EXAMPLE 21** Expand out $(a - 2b)^4$.

**SOLUTION** Here, we use row 4 of Pascal's triangle to find the coefficients: 1 4 6 4 1. We will let $x = a$ and $y = -2b$.

Use Pascal's triangle

$$(a - 2b)^4 = a^4 + 4a^3(-2b) + 6a^2(-2b)^2 + 4a(-2b)^3 + (-2b)^4$$

↓ Simplify

$$= a^4 + 4a^3(-2b) + 6a^2(4b^2) + 4a(-8b^3) + 16b^4$$

↓ Simplify

$$= a^4 - 8a^3b + 24a^2b^2 - 32ab^3 + 16b^4$$

In addition to Pascal's triangle, we have the following theorem for expanding $(x + y)^n$. It is known as the **Binomial Theorem.**

***THEOREM 12.6*** For real numbers $x$ and $y$ and positive integer $n$,

$$(x + y)^n = x^n + nx^{n-1}y + \frac{n(n-1)}{1 \cdot 2}x^{n-2}y^2 + \frac{n(n-1)(n-2)}{1 \cdot 2 \cdot 3}x^{n-3}y^3 + \cdots + y^n$$

We can use either Pascal's triangle or Theorem 12.6 to find the coefficients in $(x + y)^n$. Both will give the same answer.

**EXAMPLE 22** Expand out $(x + y)^7$ using the Binomial Theorem 12.6.

**SOLUTION** We use the Binomial Theorem with $n = 7$. (Compare with Example 20.)

Binomial Theorem

$$(x + y)^7 = x^7 + 7x^6y + \frac{7 \cdot 6}{1 \cdot 2}x^5y^2 + \frac{7 \cdot 6 \cdot 5}{1 \cdot 2 \cdot 3}x^4y^3 + \frac{7 \cdot 6 \cdot 5 \cdot 4}{1 \cdot 2 \cdot 3 \cdot 4}x^3y^4 + \frac{7 \cdot 6 \cdot 5 \cdot 4 \cdot 3}{1 \cdot 2 \cdot 3 \cdot 4 \cdot 5}x^2y^5 + \frac{7 \cdot 6 \cdot 5 \cdot 4 \cdot 3 \cdot 2}{1 \cdot 2 \cdot 3 \cdot 4 \cdot 5 \cdot 6}xy^6 + \frac{7 \cdot 6 \cdot 5 \cdot 4 \cdot 3 \cdot 2 \cdot 1}{1 \cdot 2 \cdot 3 \cdot 4 \cdot 5 \cdot 6 \cdot 7}y^7$$

↓

Simplify

$$(x + y)^7 = x^7 + 7x^6y + 21x^5y^2 + 35x^4y^3 + 35x^3y^4 + 21x^2y^5 + 7xy^6 + y^7$$

| YES | NO |
|---|---|
| $(x + y)^7 = x^7 + 7x^6y + 21x^5y^2 + \cdots$ | ~~$(x + y)^7 = x^7 + y^7$~~ |

We must use the Binomial Theorem or Pascal's triangle to expand $(x + y)^7$.

**EXAMPLE 23** Write out the first four terms of $(a + 3b)^{10}$.

**SOLUTION** We use the Binomial Theorem with $x = a$, $y = 3b$, and $n = 10$.

Binomial Theorem

$$(a + 3b)^{10} = a^{10} + 10a^9(3b)^1 + \frac{10 \cdot 9}{1 \cdot 2}a^8(3b)^2 + \frac{10 \cdot 9 \cdot 8}{1 \cdot 2 \cdot 3}a^7(3b)^3 + \cdots$$

↓

Simplify

$$= a^{10} + 10a^9(3b) + 45a^8(9b^2) + 120a^7(27b^3) + \cdots$$

↓

Simplify

$$= a^{10} + 30a^9b + 405a^8b^2 + 3240a^7b^3 + \cdots$$

**EXAMPLE 24** Write out the first four terms of $(m - n)^{20}$.

**SOLUTION** We work this as we did the others, using special caution with the negative signs.

Binomial Theorem

$$(m - n)^{20} = m^{20} + 20m^{19}(-n) + \frac{20 \cdot 19}{1 \cdot 2} m^{18}(-n)^2 + \frac{20 \cdot 19 \cdot 18}{1 \cdot 2 \cdot 3} m^{17}(-n)^3 + \cdots$$

Simplify

$$= m^{20} - 20m^{19}n + 190m^{18}n^2 - 1140m^{17}n^3 + \cdots$$

Notice that we treated $(m - n)$ as $[m + (-n)]$ and used $y = -n$ in the Binomial Theorem.

## PROBLEM SET 12.6

Use Pascal's triangle to expand out the following expressions completely.

*See Examples 20 and 21*

1. $(x + y)^4$
2. $(a + b)^5$
3. $(m + n)^6$
4. $(p + q)^7$
5. $(u - v)^6$
6. $(h - k)^7$

Write out the first four terms of the following.

*See Examples 22 and 23*

7. $(x + 3)^{12}$
8. $(a + 2)^9$
9. $(u + 5v)^8$
10. $(m + 4k)^6$
11. $(2a + 3b)^5$
12. $(2x + 5y)^7$

*See Example 24*

13. $(x - y)^9$
14. $(u - v)^8$
15. $(x - 6)^5$
16. $(y - 3)^{10}$
17. $(2x - 3y)^4$
18. $(2a - b)^8$

*Genetics application*

In a certain family there is a $\frac{1}{4}$ chance of a child having blue eyes. If the family has 3 children, the likelihoods for blue-eyed children are given by

$$(\tfrac{1}{4} + \tfrac{3}{4})^3 = (\tfrac{1}{4})^3 + 3(\tfrac{1}{4})^2(\tfrac{3}{4}) + 3(\tfrac{1}{4})(\tfrac{3}{4})^2 + (\tfrac{3}{4})^3$$

$$= \tfrac{1}{64} + \tfrac{9}{64} + \tfrac{27}{64} + \tfrac{27}{64}$$

| ↑ | ↑ | ↑ | ↑ |
|---|---|---|---|
| 3 blue-eyed children | 2 blue-eyed children | 1 blue-eyed child | 0 blue-eyed children |

Follow the same procedure (with the Binomial Theorem) to find the likelihoods of blue-eyed children in the following circumstances.

19. Four children: $(\frac{1}{4} + \frac{3}{4})^4$
20. Five children: $(\frac{1}{4} + \frac{3}{4})^5$
21. Six children: $(\frac{1}{4} + \frac{3}{4})^6$

# CHAPTER 12 SUMMARY

## Important Words and Phrases

arithmetic progression (12.3)
binomial (12.6)
Binomial Theorem (12.6)
common difference (12.3)
common ratio (12.4)
finite series (12.2)
general term (12.1)
geometric progression (12.4)
index of summation (12.2)
infinite geometric progression (12.5)
infinite series (12.2)
Pascal's triangle (12.6)
sequence (12.1)
series (12.2)
$\sum$ (12.2)
sigma notation (12.2)
sum of an arithmetic progression (12.3)
sum of a geometric progression (12.4)
sum of an infinite geometric progression (12.5)
terms of a sequence (12.1)

## Important Theorems and Definitions

*For an arithmetic series:*

common difference: $d = a_{n+1} - a_n$
general term: $a_n = a_1 + (n - 1)d$

sum of first $n$ terms: $S_n = \frac{n}{2}(a_1 + a_n)$

*For a geometric progression:*

common ratio: $r = \frac{a_{n+1}}{a_n}$

general term: $a_n = a_1 r^{n-1}$

sum of first $n$ terms: $S_n = \frac{a_1(r^n - 1)}{r - 1}$

sum of infinite series: $S_\infty = \frac{a_1}{1 - r}$ (only if $|r| < 1$)

*Binomial Theorem:*

$$(x + y)^n = x^n + nx^{n-1}y + \frac{n(n-1)}{1 \cdot 2}x^{n-2}y^2 + \frac{n(n-1)(n-2)}{1 \cdot 2 \cdot 3}x^{n-3}y^3 + \cdots + y^n$$

## Important Procedures

*To use $\sum$ notation:*

1. Discover a pattern for the typical term $a_n$.
2. Put the first and last indices below and above the $\sum$.

## CHAPTER 12 REVIEW EXERCISES

Write out the first five terms for the following sequences.

1. $a_n = n^2 + 3$
2. $a_n = \dfrac{(-1)^n}{n}$

Find the general term $a_n$ for the following sequences.

3. $\frac{1}{3}, \frac{1}{9}, \frac{1}{27}, \frac{1}{81}, \ldots$
4. $\dfrac{1}{x}, \dfrac{1}{x^2}, \dfrac{1}{x^3}, \dfrac{1}{x^4}, \ldots$

Expand out the following series, and simplify.

5. $\displaystyle\sum_{n=1}^{5} \frac{1}{n}$
6. $\displaystyle\sum_{n=1}^{6} \frac{(-1)^n}{x^n}$

Write the following series in $\sum$ notation.

7. $1 + 8 + 27 + 64 + 125$
8. $\dfrac{1}{x} + \dfrac{4}{x^2} + \dfrac{9}{x^3} + \dfrac{16}{x^4} + \cdots + \dfrac{100}{x^{10}}$

For the arithmetic progression 20, 17, 14, 11, . . . :

9. Find the common difference, $d$.
10. Find $a_{10}$.
11. Find $S_{10}$.
12. For an arithmetic series with $a_1 = 6$ and $d = 3$, find $a_n$.

For the geometric progression 4, 8, 16, 32, . . . :

13. Find the common ratio, $r$.
14. Find $a_{10}$.
15. Find $S_{10}$.
16. For a geometric progression with $a_1 = 6$ and $r = \frac{1}{2}$, find $a_n$.
17. Find the sum of the infinite geometric progression $10, 5, \frac{5}{2}, \frac{5}{4}, \frac{5}{8}, \ldots$.
18. Write $0.171717\ldots$ as a fraction.

Write out the first four terms for the following.

19. $(a + 2)^{10}$
20. $(2x - y)^7$

# ANSWERS TO SELECTED EXERCISES

## PROBLEM SET 1.1

**1.** {2, 4, 6, 8, 10, 12} **3.** {Alaska, Alabama, Arizona, Arkansas} **7.** ∅
**9.** {odd numbers between 1 and 13, inclusive} **11.** {multiples of 10, from 10 to 60}
**13.** {first five U.S. presidents} **15.** {work days of the week} **17.** True **19.** True **21.** False
**23.** {1, 2, 3, 4, 5, 6, 8, 10} **25.** {1, 2, 3, 4, 5, 6, 7, 8} **27.** {1, 3, 4, 5, 7, 10, 13} **29.** {2, 4, 6}
**31.** {7} **33.** ∅ **35.** Alcohol addicts: 247 **37.** Drug addicts: 178
**39.** Drug-and-tobacco addicts: 51 **41.** Alcohol-or-tobacco addicts: 603
**43.** Alcohol-or-drug-or-tobacco addicts: 680 **45.** Television purchasers: 262
**47.** Automobile-and-bedroom-furniture purchasers: 22
**49.** Automobile-and-bedroom-furniture-and-television purchasers: 3
**51.** Automobile-or-bedroom-furniture purchasers: 460

## PROBLEM SET 1.2

**1.** True **3.** False **5.** True **7.** True **9.** True **11.** False **13.** 3 **15.** 0
**17.** 5.2
**19–23.**

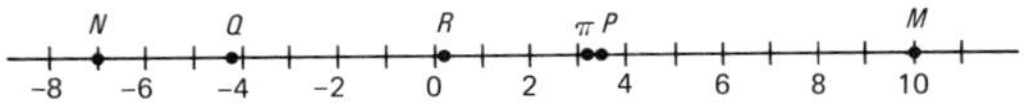


**25.** Transitive **27.** Trichotomy

**29.** Transitive **31.** True **33.** False **35.** False **37.** True **39.** True **41.** True

**43.**

**45.**

**47.**

**49.**

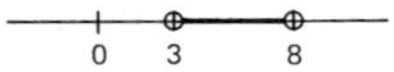


**51.**

**53.**

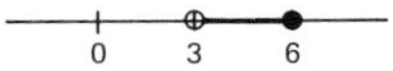


**55.** $e < E_w$

**57.** Yes, for protein; yes, for calories

## PROBLEM SET 1.3

**1.** Commutative **3.** Multiplication property of equality **5.** Associative **7.** Associative
**9.** Zero multiplier **11.** Closure **13.** Identity **15.** Zero product **17.** Commutative
**19.** Zero multiplier **21.** $x$ **23.** 10 **25.** 8 **27.** 3 **29.** 0 **31.** Real number
**33.** 1 **35.** $x + 3 = 0$ **37.** 8 **39.** $-8; \frac{1}{8}$ **41.** $\frac{-2}{5}; \frac{5}{2}$ **43.** $-3; \frac{-1}{3}$ **45.** $3; -3$

## PROBLEM SET 1.4

**1.** 10 **3.** 2.5 **5.** 6.9 **7.** $-7$ **9.** 6 **11.** 0 **13.** 7 **15.** 18.8 **17.** $-20.6$
**19.** 1 **21.** 4 **23.** 5 **25.** $-2$ **27.** $-9$ **29.** $-7$ **31.** 14 **33.** 38 **35.** $-4$
**37.** 25 **39.** 6 **41.** $-12$ **43.** $-23$ **45.** 17 **47.** $-18$ **49.** 40,000 **51.** 0
**53.** $-2000$ **55.** $-3000$

## PROBLEM SET 1.5

**1.** $-20$ **3.** 21 **5.** 90 **7.** 10.5 **9.** 16.5 **11.** $\frac{1}{15}$ **13.** 2 **15.** 6 **17.** $-8$
**19.** $-5$ **21.** $-20$ **23.** $-4000$ **25.** 24 **27.** $-110$ **29.** $-5$ **31.** $-4$ **33.** 4
**35.** 25 **37.** 16 **39.** 1000 **41.** $-64$ **43.** 10,000 **45.** 0.36 **47.** $\frac{4}{25}$ **49.** $x^6$
**51.** $(p + q)^2$ **53.** $\left(\frac{1}{a}\right)^3$ **55.** 1.4025517 **57.** 5.9604645 **59.** 112,642.4 **61.** 89,359.145
**63.** 0 **65.** 0 **67.** 0 **69.** 0.349

## PROBLEM SET 1.6

**1.** 4 **3.** 11 **5.** 3 **7.** 19 **9.** 14 **11.** 250 **13.** 219
**15.** 67 **17.** 40 **19.** 162 **21.** 56 **23.** 62 **25.** 4 **27.** 2 **29.** $\frac{20}{19}$ **31.** $\frac{1}{2}$
**33.** $\frac{13}{12}$ **35.** $-1$ **37.** 160 **39.** $-247$ **41.** 99 **43.** $-459$
**45.** TC = 464,000; TR = 172,000 **47.** 0.0859 **49.** 796.92 **51.** 0.04 **53.** 25

## CHAPTER 1 REVIEW EXERCISES

**1.** $\{-4, -3, -2, -1, 0, 1, 2, 3, 4, 5\}$ **2.** {Multiples of 100, from 100 to 500} **3.** Yes **4.** No
**5.** $\{a, b, c, d, e\}$ **6.** $\{b, c\}$ **7.** True **8.** True **9.** $-1$ **10.** 3.5

**11–12.** 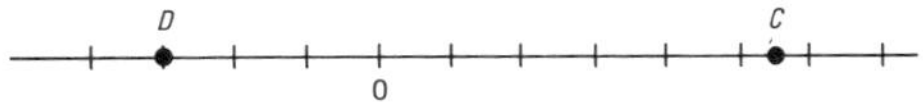

**13.** False **14.** True

**15.** 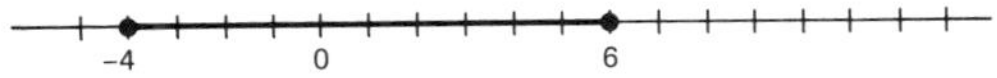

**16.** Transitive **17.** Substitution

**18.** Associative **19.** Commutative **20.** Distributive **21.** Zero product **22.** $\frac{9}{5}$ **23.** 0
**24.** 8.2 **25.** $-3$ **26.** $-13$ **27.** $-13$ **28.** $-20$ **29.** $-4$ **30.** 9 **31.** $-30$
**32.** 80 **33.** $-5$ **34.** 10 **35.** $-5$ **36.** $-125$ **37.** 30 **38.** $-28$ **39.** 25 **40.** 3
**41.** 124 **42.** $-4$ **43.** 8 **44.** $\frac{1}{5}$ **45.** $\frac{3}{2}$

## PROBLEM SET 2.1

**1.** 18 **3.** $-9$ **5.** 8 **7.** 72 **9.** 5 **11.** $-1$ **13.** $-4$ **15.** $\frac{19}{9}$ **17.** $\frac{-13}{8}$

**19.** 4 **21.** $-2$ **23.** $-7$ **25.** No solution **27.** $\frac{16}{11}$ **29.** 3 **31.** $-1$ **33.** 2

**35.** 0 **37.** $-3$ **39.** 4 **41.** $-3$ **43.** 4 **45.** No solution **47.** $\frac{-21}{5}$ **49.** 24,000

**51.** 77 **53.** 14,776.12 **55.** 5

## PROBLEM SET 2.2

**1.** $\frac{7}{a}$ **3.** $\frac{15}{g}$ **5.** $\frac{q+8}{p}$ **7.** $\frac{j-h}{k}$ **9.** $\frac{5}{b-a}$ **11.** $\frac{8}{p-q+r}$ **13.** $4b - 3a$

**15.** $\frac{3n+9m+2}{15}$ **17.** $\frac{-58-5u}{-5v-6w-7y}$ **19.** $\frac{C}{2\pi}$ **21.** $\frac{S-2\pi r^2}{2\pi r}$ **23.** $\frac{2A-hB}{h}$ **25.** $\frac{I}{pt}$

**27.** $\frac{C-10{,}000}{400}$ **29.** $\frac{V-50{,}000}{-4000}$ **31.** $\frac{E}{c^2}$ **33.** $\frac{V}{R}$ **35.** $\frac{uv}{u+v}$ **37.** $\frac{F_0B-aV}{V+B}$

**39.** $\frac{1-0.08y}{0.05}$

## PROBLEM SET 2.3

**1.** $x > 3$
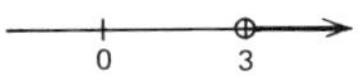


**3.** $k \leq 10$

**5.** $w > 6$

**7.** $3 < p$

**9.** $x \le 21$

**11.** $60 \le r$

**13.** $x > 4$

**15.** $m \ge 5$

**17.** $y > 1$

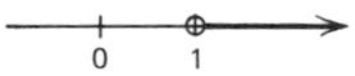

**19.** $\frac{11}{2} \le f$

**21.** $\frac{-7}{10} < p$

**23.** $x < 4$

**25.** $\frac{-2}{3} < g$

**27.** $1 \ge t$

**29.** $1 > r$

**31.** $\frac{1}{6} < k$

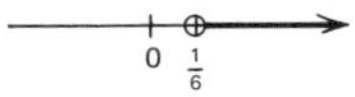

**33.** $x < 8$

**35.** $y > \frac{9}{8}$

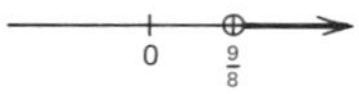

**37.** $0 \le w$

**39.** $m > 5$

**41.** $\frac{5}{2} < k$

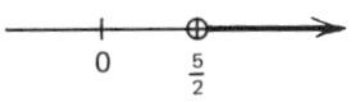

**43.** $3 < x < 7$

**45.** $-10 \le a \le -2$

**47.** $5 \le p \le 9$

**49.** $\frac{1}{4} \le t \le \frac{13}{4}$

**51.** $4 > u > -1$

**53.** $-25 \le x \le 35$

**55.** $-80 < a < -20$

**57.** $4 \le x \le 10$

**59.** $8 \le x \le 9$

**61.** $1 < r < 3$

**63.** $\frac{7}{3} < x < 4$

**65.** $y \le 60$

**67.** $y \le 1.5$

**69.** $72.06 \le x \le 77.94$

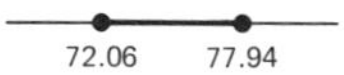

**71.** $6.906 \le x \le 7.494$

## PROBLEM SET 2.4

**1.** $-6, 6$ **3.** $-\frac{1}{2}, \frac{1}{2}$ **5.** $\varnothing$ (no solution) **7.** $-12, 12$ **9.** $-11, 13$ **11.** $-17, -3$

**13.** $\varnothing$ (no solution) **15.** $-4, 2$ **17.** $4$ **19.** $\frac{-1}{3}, 1$ **21.** $-7, -3$ **23.** $\varnothing$ (no solution)

**25.** $\frac{1}{7}$ **27.** $\frac{4}{5}, \frac{-8}{5}$

## PROBLEM SET 2.5

**1.** $-4 < x < 4$

**3.** $-\frac{1}{2} \le m \le \frac{1}{2}$

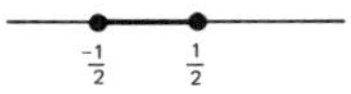


**5.** $6 < u < 12$

**7.** $-16 \le t \le -6$

**9.** $-6 < b < 10$

**11.** $-1 < y < 2$

**13.** $\frac{-9}{7} \le z \le \frac{1}{7}$

**15.** $\frac{2}{3} < p < 6$

**17.** $\frac{-7}{9} \le q \le \frac{5}{9}$

**19.** $x < -10$ or $x > 10$

**21.** $a \le \frac{-2}{3}$ or $a \ge \frac{2}{3}$

**23.** $r < -9$ or $r > -5$

**25.** $s \le -2$ or $s \ge 18$

**27.** $v < -6$ or $v > 4$

**29.** $r \le \frac{-5}{3}$ or $r \ge 3$

**31.** $k < \frac{-16}{5}$ or $k > \frac{4}{5}$

**33.** $x \le \frac{6}{11}$ or $x \ge \frac{8}{11}$

**35.** $y < \frac{-3}{2}$ or $y > \frac{1}{6}$

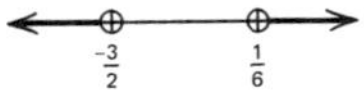


**37.** $2.499 < w < 2.501$

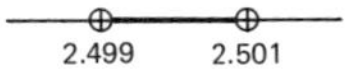


## PROBLEM SET 2.6

**1.** 17 and 24 **3.** 12 **5.** 360 **7.** 15 and 16 **9.** 22 and 24 **11.** 17, 19, and 21
**13.** \$6000 at 7%; \$9000 at 6% **15.** \$3000 at 8%; \$5500 at 6% **17.** 8% and 9%
**19.** \$6666.67 at 8%; \$3333.33 at 5% **21.** 8 oz at 25%; 12 oz at 15% **23.** 25 lb at 90%; 75 lb at 70%
**25.** 7.5 **27.** $t \ge 79$ **29.** $10° \le C \le 25°$

## CHAPTER 2 REVIEW EXERCISES

**1.** 16 **2.** 4 **3.** $-5$ **4.** $-8$ **5.** $\frac{-10}{3}$ **6.** $\frac{c+b}{a}$ **7.** $\frac{5b+3a}{10}$ **8.** $\frac{pV}{nR}$

**9.** $k \le -7$

**10.** $q < 5$

**11.** $t \le -3$

**12.** $y < -5$

**13.** $\frac{-31}{2} < u$

**14.** $\frac{5}{4} < m < \frac{5}{2}$

**15.** $-5 \leq n \leq 11$

**16.** $2 < x < 6$

**17.** $-10, 10$ **18.** $-5, 11$ **19.** $\frac{5}{2}, -1$ **20.** $1 < f < 5$

**21.** $-1 \leq w \leq \frac{9}{5}$

**22.** $r < -8$ or $r > 2$

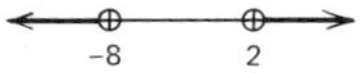


**23.** $v \leq \frac{-6}{7}$ or $v \geq \frac{-4}{7}$

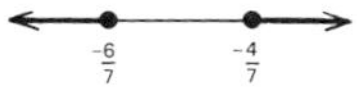


**24.** 9 **25.** 25, 27, 29

**26.** \$2000 at 9%; \$11,000 at 6% **27.** 50 ml of 40%; 150 ml of 20% **28.** $t \geq 46$

## PROBLEM SET 3.1

**1.** 7 **3.** $-8$ **5.** 1 **7.** $-1$ **9.** 5; binomial **11.** 2; trinomial **13.** 7; trinomial **15.** $2n$; trinomial **17.** 13, 4, 8 **19.** $-59$, 5, 11 **21.** 21, 1, 33 **23.** 57, 1, 3 **25.** 77,774,907 **27.** 70,895,853 **29.** 20,000 **31.** 6500 **33.** 132,000 **35.** 414,000 **37.** 3.8 **39.** 3.8

## PROBLEM SET 3.2

**1.** $6x$ **3.** $-16pq^2$ **5.** $8x^2 - 2x - 7$ **7.** $11p^3 - 4p^2 - 8p - 6$ **9.** $13m^2 - 4m - 4$ **11.** $5x^3 - 2x^2 + 7x - 3$ **13.** $24z^2 - 2z + 2$ **15.** $11x^4 + 5x^3 + 3x^2 + x + 5$ **17.** $2x^2 - 9x + 1$ **19.** $-y^3 - 13y^2 + 4y + 1$ **21.** $m^4 - 3m^3 + m^2 + 11m + 20$ **23.** $2t^2 - 11t - 3$ **25.** $w^4 + 4w^3 - 14w^2 + 7w + 12$ **27.** $-a^5 + 8a^4 + 7a^3 - 6a^2 - 4a + 1$ **29.** $9a - 7$ **31.** $4x + 11$ **33.** $4a - 7b$ **35.** $5t + 3$ **37.** $-0.001Q^3 - 4820Q - 131{,}000$ **39.** $0.02Q^2 + 60Q + 1100$

## PROBLEM SET 3.3

**1.** $x^{11}$ **3.** $r^{17}$ **5.** $10x^5$ **7.** $-30a^3b^6$ **9.** $-9a^4b^5c^7$ **11.** $15xy^4z^4t$ **13.** $-12a^8b^6c^5$ **15.** $24r^5s^6t^3u^7$ **17.** $x^7$ **19.** 1 **21.** $\frac{1}{h^6}$ **23.** $2x^5$ **25.** $\frac{-5y}{x^3}$ **27.** $-2m^6$ **29.** $\frac{-7z^3}{x^3}$ **31.** $-1$ **33.** $(2x + 1)^2$ **35.** 1 **37.** $x^{3n}$ **39.** $b^{2n+1}$

## PROBLEM SET 3.4

**1.** $32x^2 - 12xy$ **3.** $15p^2q^2 - 6p^2q^3 - 9pq^3$ **5.** $-14m^3n^4 + 35m^6n^3 - 56m^2n^6$ **7.** $x^2 + 3x - 10$ **9.** $6x^2 - 19x + 15$ **11.** $5y^2 + 32y - 21$ **13.** $a^2 + 4ab + 3b^2$ **15.** $15x^2 - 7xy - 2y^2$ **17.** $r^2 - 64$ **19.** $16x^2 - 9y^2$ **21.** $100x^2 - 9a^2$ **23.** $t^2 - 4t + 4$ **25.** $x^2 + 6x + 9$ **27.** $p^2 + 4pq + 4q^2$ **29.** $x^3 + 3x^2 - 7x + 15$ **31.** $6r^3 - 11r^2 + 11r - 12$ **33.** $m^5 + m^3 + 3m^2 - 20m + 15$ **35.** $x^4 - 3x^3 - x + 21$

**37.** $10a^6 - 4a^5 - 20a^4 + 15a^3 + 2a^2 - 24a - 6$ **39.** $x^3 + 6x^2 + 11x + 6$ **41.** $6r^3 - 5r^2 - 44r + 15$
**43.** $x^4 - x^3 + 2x^2 + x - 3$ **45.** $x^3 - 3x - 2$ **47.** $p^4 - 9p^2 - 4p + 12$
**49.** $\text{TR} = -25Q^2 + 1000Q + 190{,}000$ **51.** $W = 6t^4 - 3t^3 + 109t^2 - 50t + 150$

**53.** $y = \dfrac{-P}{96EI}(11x^3 - 24Lx^2 + 15L^2x - 2L^3)$

## PROBLEM SET 3.5

**1.** $4x^3 - 3x^2 + 5x$ **3.** $3a^8 - 4a^4 - 5$ **5.** $2m^3n - m^4 + 3n^3$ **7.** $4s^3 - s + \dfrac{2}{s}$

**9.** $2ab^3 - 3a^4b + \dfrac{4}{b}$ **11.** $x - 3$ **13.** $a^2 - 5a - 2$ **15.** $3r^2 - r - 5$

**17.** $5b^2 + 8b + 21 + \dfrac{65}{b - 3}$ **19.** $t^3 - 8t^2 + 12t - 29 + \dfrac{52}{t + 2}$ **21.** $a^4 + a^3 + a^2 + a + 1$

**23.** $r^6 + r^5 + r^4 + r^3 + r^2 + r + 1 + \dfrac{1}{r - 1}$ **25.** $h^2 - 2h - 3$ **27.** $m^2 - m + 6 + \dfrac{-m + 24}{m^2 - 2m - 3}$

**29.** $3k^2 + k + 2 + \dfrac{-2k - 1}{2k^2 - k + 1}$

## CHAPTER 3 REVIEW EXERCISES

**1.** $-1$ **2.** 3 **3.** Binomial; 4 **4.** Trinomial; 9 **5.** 9, 1, 6 **6.** $a^2b$ **7.** $6r - 14s$
**8.** $9t^3 - 2t^2 + 3t - 9$ **9.** $5a^4 - 3a^3 + a^2 - a - 7$ **10.** $16u^3 - u^2 + 10u + 1$

**11.** $-5n^3 - 7n^2 + 15n + 2$ **12.** $42a^{16}$ **13.** $36x^6y^{12}z^7$ **14.** $-7t^3$ **15.** $\dfrac{8pq^3}{r^2}$

**16.** $\dfrac{1}{(a + 3b)^6}$ **17.** $-8x^3y^3 + 12x^2y^8 + 16x^5y^5$ **18.** $12h^2 - 7hk - 10k^2$ **19.** $64u^2 - 9$

**20.** $4c^2 - 20c + 25$ **21.** $2m^4 + m^3 - 23m - 20$ **22.** $t^3 - 2t^2 - 13t - 10$ **23.** $5a^3 - 4a^2 + 6$

**24.** $2y^4 - x - \dfrac{5y^6}{x}$ **25.** $r^2 + r + 1$ **26.** $3s^2 - 8s + 26 + \dfrac{-81s + 81}{s^2 + 2s - 3}$

## PROBLEM SET 4.1

**1.** $8(a + b)$ **3.** $4(m - 2n)$ **5.** $6(2r - 3s + 4t)$ **7.** $a(2 + 3a)$ **9.** $4k(4k - 3)$
**11.** $x^2(2x^2 - 3x - 11)$ **13.** $2b^2(4b^6 - 3b^4 - 2b^2 + 1)$ **15.** $9b(9a - b)$ **17.** $6rs^2(2s + 3r)$
**19.** $5xy^2(5x^4 + 2xy - 6y^5)$ **21.** $7ab^2c(3c^2 - 2a^2)$ **23.** $2r^3p^4(8q^5 - 9r)$ **25.** $5x(6x^3y^2 - 5xz^2 + 8yz)$
**27.** $8rs^2t^3u^2(4u^2 - 3rs)$ **29.** $2y(2w^2x^2y + 3x^3z^2 + 4w^4y^3z^3)$ **31.** $(x + 5)(a + b)$ **33.** $(6 - m)(p - q)$
**35.** $(8 - r)(2x - 1)$ **37.** $[a(x + y) - b](x + y)$ **39.** $-a + 7$ **41.** $-x + x^2y + x^3$

**43.** $-st + s^2$ **45.** $p(1 - d)$ **47.** $y = \dfrac{L^3}{EI}\left(\dfrac{P}{3} + \dfrac{wL}{8}\right)$

## PROBLEM SET 4.2

**1.** $(x + 1)(x + 5)$ **3.** $(w - 3)(w - 4)$ **5.** Cannot be factored **7.** $(x - 7)(x - 8)$
**9.** $(p + 3)(p + 10)$ **11.** $(a + 8b)(a - 2b)$ **13.** $(u + 15v)(u - 2v)$ **15.** $(x - 5y)(x - 7y)$

**17.** $(u + 13v)(u - 12v)$ **19.** Cannot be factored **21.** $(3x + 1)(2x + 1)$ **23.** $(3t + 2)(4t - 1)$
**25.** $(3m + 1)(3m - 2)$ **27.** $(5y - 3)(2y + 3)$ **29.** $(4w - 1)(5w - 3)$ **31.** $(3x + 7)(5x - 2)$
**33.** $(3p + 7q)(2p - 9q)$ **35.** $(3r + 8s)(4r - s)$ **37.** $(5c - 7d)(c - 6d)$ **39.** Cannot be factored
**41.** $(a^2 + 5)(a^2 + 2)$ **43.** $(x^2 - 2)(x^2 + 7)$ **45.** $(t^2 + 9s^2)(t^2 + 2s^2)$ **47.** $(4r^2 - 1)(5r^2 + 3)$
**49.** $(6a^2 - b^2)(a^2 - 3b^2)$

## PROBLEM SET 4.3

**1.** $(x + 5)(x - 5)$ **3.** $(1 - p)(1 + p)$ **5.** $(4m - 7n)(4m + 7n)$ **7.** $(9 - ab)(9 + ab)$
**9.** Cannot be factored **11.** $(\frac{1}{3}uv - ab)(\frac{1}{3}uv + ab)$ **13.** $(a + b + 1)(a + b - 1)$
**15.** $(m + n + a + b)(m + n - a - b)$ **17.** $(a^2 + b^2)(a - b)(a + b)$ **19.** $(x^2 + 9y^2)(x - 3y)(x + 3y)$
**21.** $(x + 1)^2$ **23.** $(t - 5)^2$ **25.** $(r - 9)^2$ **27.** $(a - 3b)^2$ **29.** $(2x - \frac{1}{3})^2$ **31.** $(2r + 9s)^2$
**33.** $(5a - 6bc)^2$ **35.** $(a^2 + 9b^2)^2$ **37.** $(x + y)(x^2 - xy + y^2)$ **39.** $(a + 1)(a^2 - a + 1)$
**41.** $(m - 5n)(m^2 + 5mn + 25n^2)$ **43.** $(4w + 1)(16w^2 - 4w + 1)$ **45.** $\left(a - \frac{1}{3}\right)\left(a^2 + \frac{a}{3} + \frac{1}{9}\right)$
**47.** $(3ab + 10cd)(9a^2b^2 - 30abcd + 100c^2d^2)$ **49.** $(x + y)(x^2 - xy + y^2)(x - y)(x^2 + xy + y^2)$
**51.** $(u - v)(u^2 + uv + v^2)(u^6 + u^3v^3 + v^6)$ **53.** $(m^4 + n^4)(m^8 - m^4n^4 + n^8)$

## PROBLEM SET 4.4

**1.** $(a + 2)(b + 3)$ **3.** $(a + b)(c + d)$ **5.** $(x - y)(x - 8)$ **7.** $(u - 7v)(v + 8)$
**9.** $(m - 5)(2m + n)$ **11.** $(3k - 5h)(4k - 7)$ **13.** Cannot be factored **15.** $(2x - 5)(a - 2c)$

## PROBLEM SET 4.5

**1.** $x^3(x + 1)(x - 1)$ **3.** $5x^2(x - 4)(x + 4)$ **5.** $t^2(t^4 + 1)(t^2 + 1)(t + 1)(t - 1)$
**7.** $b(b - 1)(b^2 + b + 1)$ **9.** $3xy(2x + 3y)(4x^2 - 6xy + 9y^2)$
**11.** $a(1 - 2ab)(1 + 2ab + 4a^2b^2)(1 + 2ab)(1 - 2ab + 4a^2b^2)$ **13.** $4a(a - 2)(a + 3)$
**15.** $5x^3(x - 6)(x - 1)$ **17.** $3a^2(3a - 2)(2a - 1)$ **19.** $3s^2t(6s - 23t^2)$ **21.** $c(a + 7)(b - 2)$
**23.** $4pq(p + 2q)(p - 5)$ **25.** $0.0005(q - 1000)^2$ **27.** $-(x - 300)(x - 100)$ **29.** $3000(t + 2)^2$

## CHAPTER 4 REVIEW EXERCISES

**1.** $2x(1 - 5x^6)$ **2.** $5y^2(4y^6 - 3y^4 + 2y^2 - 1)$ **3.** $3ab^2c(c^2 - 2ac^2 + 3a^2)$ **4.** $(8r - 7s)(a - 2b)$
**5.** $(m + 8)(m - 5)$ **6.** $(p - 4q)(p + 3q)$ **7.** $(3t - 1)(8t + 5)$ **8.** $(5a - 2b)(3a - 4b)$
**9.** $(h^2 - 3)(h^2 + 9)$ **10.** $(3r^2 - 2)(2r^2 - 1)$ **11.** $(1 - 7n)(1 + 7n)$
**12.** $(5abc^2 - 6uv^2)(5abc^2 + 6uv^2)$ **13.** $(x + 2y + z)(x + 2y - z)$ **14.** $(2x - 7y)^2$
**15.** $(a + 5)(a^2 - 5a + 25)$ **16.** $(2u - 10v)(4u^2 + 20uv + 100v^2)$ **17.** $(a - 2b)(a + 3)$
**18.** $(r - s)(s - t)$ **19.** $x^2(x^2 + 1)(x + 1)(x - 1)$ **20.** $xy^2(x + y)(x^2 - xy + y^2)$
**21.** $2ab(a - 2b)(a^2 + 2ab + 4b^2)(a + b)(a^2 - 2ab + 4b^2)$ **22.** $4r^3(r - 5)(r + 4)$ **23.** $3t(4s - 1)(2s - 3)$
**24.** $8p(p - q)(p - 3)$

## PROBLEM SET 5.1

**1.** $\frac{5}{7}$ **3.** $\frac{4}{15}$ **5.** $\frac{3}{4x}$ **7.** $\frac{6a}{7c}$ **9.** $\frac{5y^5z^{10}}{3x^3}$ **11.** $\frac{-9a^2c^7}{8b^4d^{10}}$ **13.** $\frac{6}{x}$ **15.** $\frac{7 + x}{y + x^2}$

**17.** $-1$ **19.** $\frac{b + x}{b - x}$ **21.** $\frac{a^2 + ax + x^2}{b}$ **23.** $\frac{x - y}{y}$ **25.** $\frac{x - 1}{8}$ **27.** $\frac{x + 7}{x + 5}$ **29.** $\frac{x + 5}{x - 2}$

**31.** $\frac{x + 7}{x + 6}$ **33.** $\frac{a + b}{(a - b)^2}$ **35.** $\frac{3x + 1}{2x + 1}$ **37.** $\frac{2x + 3}{3x + 2}$ **39.** $\frac{x + y}{2x - y}$ **41.** $\frac{x + y}{x - 6}$

**43.** $\frac{a^2 + 2a + 4}{a + 2}$

## PROBLEM SET 5.2

**1.** $\frac{1}{9}$ **3.** $\frac{3}{143}$ **5.** $\frac{2b}{3c}$ **7.** $\frac{-7an^3}{8c^2m^6}$ **9.** $\frac{-40ad^2}{21c^2x^2y}$ **11.** $\frac{3ab^2}{20x^5}$ **13.** $\frac{1}{x(x - 5)}$

**15.** $\frac{(x - 2)^2(x - 3)}{x + 2}$ **17.** $x$ **19.** $\frac{x - 7}{x - 4}$ **21.** $\frac{a - 5}{a - 9}$ **23.** $\frac{3x + 1}{x + 5}$ **25.** $\frac{3x + 1}{x - 2}$

**27.** $\frac{x + 5}{x^2(x - 3)}$ **29.** $\frac{(x - 2)(x - 4)}{x^4(x + 5)}$

## PROBLEM SET 5.3

**1.** $\frac{3}{4}$ **3.** $\frac{22}{27}$ **5.** $\frac{3x^5y^3a}{2b^2}$ **7.** $\frac{-5cx^2y}{2a}$ **9.** $\frac{2b^3c}{3d^2e}$ **11.** $\frac{3a^3d^3x^2y^2z^5}{4cw}$ **13.** $\frac{x^4}{x - 4}$

**15.** $\frac{x - 5}{x + 3}$ **17.** $\frac{x(x + 3)}{x + 7}$ **19.** $\frac{p + 4}{p - 3}$ **21.** $\frac{(x + 3)(x + 7)}{(x + 5)(x - 2)}$ **23.** $\frac{x - 7}{x - 2}$ **25.** $\frac{4x - 3}{x + 7}$

**27.** $\frac{x}{x + 4}$ **29.** $\frac{x^2 + 2x + 4}{x + 9}$

## PROBLEM SET 5.4

**1.** $\frac{7}{11}$ **3.** $\frac{-1}{45}$ **5.** $\frac{24x + 5}{60}$ **7.** $\frac{4}{ax}$ **9.** $\frac{-2}{6a}$ **11.** $\frac{p^4 + p^2 + 1}{p^7}$ **13.** $\frac{3m - 5n}{m^2n^2}$

**15.** $\frac{4a^2bc^2 + 3ac^3 - 5b^2}{a^3b^2c^4}$ **17.** $\frac{x^3 + y^3}{x^2y^3}$ **19.** $\frac{8x + 25}{(x + 4)(x - 3)}$ **21.** $\frac{r^2 - 7r - 15}{(r + 3)(r - 2)}$

**23.** $\frac{5k^2 - 8k + 3}{(2k - 1)k}$ **25.** $\frac{-5h}{x(x + h)}$ **27.** $\frac{4x^2 + 5x - 39}{(x + 5)(x - 4)}$ **29.** $\frac{13x - 19}{(x + 1)(x - 1)(x - 2)}$

**31.** $\frac{-2x - 26}{(x + 6)(x + 6)(x - 1)}$ **33.** $\frac{2x^2 - x - 20}{x(x - 7)(x + 4)}$ **35.** $\frac{5x^2 + 10x + 16}{(x - 2)(x + 2)(x^2 + 2x + 4)}$

**37.** $\frac{4x + 20}{(x + 1)(x + 2)(x + 3)}$ **39.** $\frac{2x^2 + 10x + 9}{(x + 1)(x + 7)(2x + 1)}$ **41.** $\frac{6x + 1}{x}$ **43.** $\frac{x^2 + 2}{x}$

**45.** $\frac{4x^2 + 20x - 3}{x + 5}$ **47.** $\frac{2x^2 + 12x + 5}{2x + 1}$ **49.** $\frac{x^2 + 2x + 4}{x^2}$ **51.** $\frac{7x^2 + 83x + 244}{(x + 5)(x + 7)}$

**53.** $\frac{x^3 + 4x^2 - x - 24}{x(x + 4)}$ **55.** $\frac{10{,}000t^2 + 20{,}000t + 1{,}000{,}000}{100 + t^2}$ **57.** $\frac{(n - 1)(R_2 - R_1)}{R_1R_2}$ **59.** $\frac{Rc(n^2 - 4)}{4n^2}$

**61.** $\frac{Rc(n^2 - 16)}{16n^2}$ **63.** $\frac{C_2C_3 + C_1C_3 + C_1C_2}{C_1C_2C_3}$ **65.** $\frac{A + Akx + B}{1 + kx}$

## PROBLEM SET 5.5

**1.** $\frac{5}{6}$ **3.** $\frac{9y}{8x}$ **5.** $\frac{ad}{bc}$ **7.** $\frac{15c}{14a}$ **9.** $\frac{28}{45}$ **11.** $\frac{5x+1}{3x+1}$ **13.** $\frac{x^3+x}{7x^2+1}$ **15.** $\frac{b+a}{b-a}$

**17.** $\frac{18-4x}{360x+3x^2}$ **19.** $\frac{2a^2x^2+yx^3}{3ab-4a^3x}$ **21.** $\frac{abx-bcy+acz}{5abc^2-6ab^2c+7a^2bc}$ **23.** $\frac{2a^2+a-21}{4a^2-5a-26}$

**25.** $\frac{10p^2+8p}{6p^2-5p-5}$ **27.** $\frac{3x^2+2x-33}{x+7}$ **29.** $\frac{-P\Delta Q}{Q\Delta P}$ **31.** $\frac{Pz\sigma_m+MA\sigma_p}{Az\sigma_p\sigma_m} \leq 1$

## PROBLEM SET 5.6

**1.** 3 **3.** $\frac{3}{28}$ **5.** $\frac{14}{11}$ **7.** 2 **9.** 1 **11.** 8 **13.** $\frac{-4}{3}$ **15.** 2 **17.** $\varnothing$ **19.** 2

**21.** $-1$ **23.** 2 **25.** $\frac{-3}{7}$ **27.** 10 **29.** $\frac{-13}{4}$ **31.** 18 **33.** $-1$ **35.** $\varnothing$

**37.** $\frac{ab}{a+b}$ **39.** $\frac{-n}{An-1}$ **41.** $\frac{sl-a}{s-l}$ **43.** $\frac{s^2f_2}{f_2-s^2}$ **45.** $\frac{sY_2-Z}{Y_2-Y_1}$ **47.** 8 **49.** $\frac{s_2P_1}{P_1-H}$

**51.** $\frac{m_x}{1+\frac{1}{2}m}$ **53.** $\frac{fu}{f-u}$ **55.** $\frac{CC_2C_3}{C_2C_3-CC_3-CC_2}$

## PROBLEM SET 5.7

**1.** 1 **3.** 7 **5.** 5, 6 **7.** $\frac{6}{5}$ **9.** $\frac{60}{47}$ **11.** $\frac{50}{3}$ **13.** 30, 50
**15.** 500 at 50 mph; 2500 at 100 mph **17.** 2 at 400 mph; 2.5 at 500 mph **19.** 5 **21.** 5 **23.** 2727

## CHAPTER 5 REVIEW EXERCISES

**1.** $\frac{4x^2z^3}{3y^4}$ **2.** $\frac{7-x}{y+x^4}$ **3.** $\frac{m-6}{m-3}$ **4.** $\frac{p-3}{p^2-3p+9}$ **5.** $\frac{3a^2bc^8e^6}{2d^7f}$ **6.** $x+1$ **7.** $\frac{x+1}{x+5}$

**8.** $\frac{-3n^2t^2}{4m^2p^2r^2s^2}$ **9.** $a+1$ **10.** $\frac{x-3}{x+5}$ **11.** $\frac{x^2(x+7)}{x-2}$ **12.** $\frac{x^2+y^2}{x^3y^2}$ **13.** $\frac{6a^2+7a-8}{a^3}$

**14.** $\frac{14x-16}{(x-5)(x+1)(x+4)}$ **15.** $\frac{-6a^2+13a-30}{(a+2)(a^2-2a+4)(a-2)}$ **16.** $\frac{8r^2-24r+7}{r-3}$

**17.** $\frac{m^3+3m^2-6m+3}{m(m+3)}$ **18.** $\frac{12}{5a}$ **19.** $\frac{3r+18}{4r^2-27r}$ **20.** $\frac{2x^2+x-15}{10-2x}$ **21.** 45 **22.** $-1$

**23.** $-43$ **24.** $\frac{ms}{n-m}$ **25.** 7 **26.** $\frac{40}{9}$ **27.** 50 (Sue); 45 (Carol) **28.** 40

## PROBLEM SET 6.1

**1.** $x^{11}$ **3.** $t^{12}$ **5.** $-20x^2y^{10}z^4$ **7.** $r^4$ **9.** $\frac{1}{m^3}$ **11.** $\frac{8a}{bc^3}$ **13.** $x^{10}$ **15.** $6^{18}$

**17.** $10^{3n}$ **19.** $32a^5$ **21.** $25m^2$ **23.** $49x^2y^4$ **25.** $64m^{12}n^{18}$ **27.** $40{,}000a^8b^{2n}$

**29.** $a^{3n}b^{3n+3}c^{3n+6}$ **31.** $\frac{x^4}{81}$ **33.** $\frac{125}{t^3}$ **35.** $\frac{-27a^6}{64b^3}$ **37.** $\frac{64x^9y^6}{z^6w^{12}}$ **39.** $\frac{1000x^6}{-y^{3n}}$ **41.** $\frac{16a^{4k}}{b^{4k+4}}$

**43.** $a^{10}b^{11}$ **45.** $500m^{16}n^7$ **47.** $\frac{a^4}{b^2c^7}$ **49.** $\frac{2a^4}{b}$ **51.** $\frac{9n^7}{2m^8}$ **53.** $\frac{a^4b^{16}}{c^2}$ **55.** $16x^{32}y^4$

**57.** $\frac{1}{8p^{39}q^6r^6}$ **59.** $\frac{p^6}{q^{14}r^8}$ **61.** $\frac{-8a^3}{b^{18}c^6}$ **63.** $\frac{q^4}{p^8r^{14}}$ **65.** $(1.015)^{20}$ **67.** $2^{46}$ or $4^{23}$

**69.** $2^3 = 8$ **71.** $\frac{1}{2^3}$ **73.** $10^4$

## PROBLEM SET 6.2

**1.** 1 **3.** $\frac{1}{9}$ **5.** 1 **7.** $x^5$ **9.** 1 **11.** $\frac{b^4}{a^3}$ **13.** $\frac{p^4r^7}{q^3}$ **15.** $\frac{10c^4b^3}{a^2d^5}$ **17.** $\frac{80}{27}$

**19.** $\frac{1}{x^6}$ **21.** $m^2$ **23.** $\frac{1}{b^5}$ **25.** $\frac{1}{x^7}$ **27.** $r^3$ **29.** $f^8$ **31.** $\frac{y^2}{x^4}$ **33.** $\frac{64v^{12}}{q^9}$ **35.** 1

**37.** $\frac{b^{11}}{a^4}$ **39.** $\frac{u^{11}}{v^{14}w^{14}}$ **41.** $\frac{a^3c^{15}}{b^6}$ **43.** $\frac{y^8}{x^4q^4}$ **45.** 1 **47.** $\frac{65}{8}$ **49.** $\frac{k+1}{k^2}$ **51.** $\frac{a+1}{a^3}$

**53.** $\frac{u^2+y^2}{uv}$ **55.** $\frac{r^2s^2}{s^2-r^2}$ **57.** $\frac{a^2b^2+a^3}{b^2-a^3}$ **59.** 2,500,000 **61.** 1,500,000 **63.** 1,071,429

**65.** 833,333 **67.** $\frac{100}{(1.07)^{10}}$ **69.** $\frac{4t^5}{e^{2t}\cdot 15}$ **71.** $\frac{40}{3}$ **73.** $\frac{160}{15}$

## PROBLEM SET 6.3

**1.** $5.1 \times 10^4$ **3.** $4.05 \times 10^7$ **5.** $6.25 \times 10^{11}$ **7.** $4 \times 10^{-3}$ **9.** $7.05 \times 10^{-4}$
**11.** $4.2 \times 10^0$ **13.** 30,000 **15.** 42,100,000,000 **17.** 8.2 **19.** 0.04 **21.** 0.291
**23.** 0.00000333 **25.** $6 \times 10^8$ **27.** $9.66 \times 10^{-1}$ **29.** $1.736 \times 10^9$ **31.** $1.846 \times 10^2$
**33.** $4 \times 10^2$ **35.** $3 \times 10^6$ **37.** $9 \times 10^{-12}$ **39.** 1.4 **41.** $8 \times 10^{15}$ **43.** $6.4 \times 10^{-11}$
**45.** $1.6105 \times 10^{20}$ **47.** $9.3 \times 10^7$ **49.** $4.5 \times 10^9$ **51.** $1.06 \times 10^{11}$ **53.** $4.4 \times 10^{-8}$
**55.** $2 \times 10^{12}$ **57.** $6 \times 10^{-4}$ **59.** 600,000,000,000,000,000,000,000 **61.** 30,000,000,000
**63.** 0.0000000000000002 **65.** 10,000,000,000,000,000,000,000,000,000,000,000,000,000

## PROBLEM SET 6.4

**1.** $\sqrt{x}$ **3.** $\sqrt[5]{m+n}$ **5.** $\sqrt[3]{t^2}$ **7.** $\sqrt[9]{(x+2y)^8}$ **9.** $\frac{1}{\sqrt[3]{p}}$ **11.** $\frac{1}{\sqrt[5]{k^4}}$ **13.** $m^{1/3}$ **15.** $r^{1/10}$

**17.** $u^{5/2}$ **19.** $k^{4/3}$ **21.** $z^{-1/4}$ **23.** $f^{-1/5}$ **25.** 8 **27.** 2 **29.** No real number

**31.** $10x$ **33.** $\frac{1}{2r^3}$ **35.** $\frac{2r}{5s^2}$ **37.** 9 **39.** 2 **41.** $-2$ **43.** 6 **45.** $\frac{1}{2}$ **47.** $\frac{4}{5}$

**49.** 4 **51.** 27 **53.** $\frac{16}{81}$ **55.** $\frac{1}{256}$ **57.** $\frac{1}{216}$ **59.** $\frac{64}{27}$ **61.** $x + x^{5/6}$ **63.** $t^{8/15} - t^{9/20}$

**65.** $y^{7/12} + y^{1/12}$ **67.** $u^{-1/6} + u^{-3/4} + u^{-3/10}$ **69.** $x^{1/8}y^{1/12}$ **71.** $u^{2/9}v^{-1/6}$ **73.** $\frac{x^{1/12}}{y^{6/35}}$

**75.** $\frac{v^{15/16}}{u^{5/8}w^{3/4}}$ **77.** $S = 2W^{2/5}H^{7/10}$ **79.** 1.76 **81.** 1.56 **83.** 112 **85.** 28 **87.** 400

**89.** 43.92 **91.** 32.94

## PROBLEM SET 6.5

**1.** $5\sqrt{2}$ **3.** $2\sqrt{3}$ **5.** $-10\sqrt{10}$ **7.** $7\sqrt{2}$ **9.** $40\sqrt{15}$ **11.** $5\sqrt[3]{2}$ **13.** $3\sqrt[3]{5}$
**15.** $3\sqrt[4]{2}$ **17.** $2\sqrt[6]{10}$ **19.** $2x\sqrt{2x}$ **21.** $3r^3\sqrt{5r}$ **23.** $2pq^2r^3\sqrt{14pq}$ **25.** $3a^2b^3c^4\sqrt{3bc}$
**27.** $40hk^3m^4n^5\sqrt{5hkmn}$ **29.** $5r^2\sqrt{4r}$ **31.** $10uv^2\sqrt[3]{10u^2v}$ **33.** $3xyz^2\sqrt[4]{4xy^3z}$ **35.** $2mn^2k^3\sqrt[6]{5m^5nk}$
**37.** $\frac{\sqrt{19}}{2}$ **39.** $\frac{3\sqrt{7}}{10}$ **41.** $\frac{2\sqrt{6}}{5}$ **43.** $\frac{\sqrt[3]{11}}{2}$ **45.** $\frac{3\sqrt[3]{3}}{5}$ **47.** $\frac{\sqrt[5]{2}}{5}$ **49.** $\frac{a^2\sqrt{a}}{b^2}$ **51.** $\frac{\sqrt{u}}{3v^2}$
**53.** $\frac{rs\sqrt{r}}{10pq^4}$ **55.** $\frac{x^2\sqrt[3]{x}}{y^3}$ **57.** $\frac{2ab\sqrt[3]{2ab^2}}{3c^2d^3}$ **59.** $\frac{y\sqrt[5]{x^2y^3}}{3zt^2}$ **61.** $\frac{4\sqrt{11}}{11}$ **63.** $\sqrt{3}$ **65.** $\frac{5\sqrt{10a}}{a}$
**67.** $\frac{\sqrt{39}}{13}$ **69.** $\frac{\sqrt{15}}{5}$ **71.** $\frac{\sqrt{22a}}{2a^2}$ **73.** $\frac{7\sqrt[4]{4}}{2}$ **75.** $\frac{\sqrt[3]{5x^2}}{x}$ **77.** $\frac{2\sqrt[3]{3u^2v^2}}{3u^2v}$ **79.** $\frac{\sqrt[3]{12}}{3}$
**81.** $\frac{\sqrt[3]{6ab^2}}{2b}$ **83.** $\frac{\sqrt[5]{30ux^3y^2z^4}}{3xyz}$ **85.** $\frac{\sqrt{2krca(k-r)}}{a(k-r)}$ **87.** $\frac{\sqrt{LC}}{2\pi LC}$ **89.** $\frac{k\sqrt{D}}{D}$

## PROBLEM SET 6.6

**1.** $14\sqrt{2}$ **3.** $15\sqrt{3x}$ **5.** $(-2+3x)\sqrt[3]{x}$ **7.** $6\sqrt{2}$ **9.** $17\sqrt{5}$ **11.** $-\sqrt{2}$ **13.** $16\sqrt{6}$
**15.** $6\sqrt{x}$ **17.** $16\sqrt{2t}$ **19.** $7\sqrt[3]{2}$ **21.** $21\sqrt[3]{2}$ **23.** $24\sqrt[3]{3x}$ **25.** $(1+x+x^2)\sqrt[3]{x}$
**27.** $(1+4p+9p^3)\sqrt[3]{2p}$ **29.** $3\sqrt{7}+3\sqrt{5}$ **31.** $\sqrt{30}-2\sqrt{3}$ **33.** $-3$
**35.** $3+\sqrt{3}+2\sqrt{2}+\sqrt{6}$ **37.** $25+4\sqrt{7}$ **39.** $2-6\sqrt{5}$ **41.** $\sqrt{15}-\sqrt{10}+3\sqrt{2}-2\sqrt{3}$
**43.** $6+2\sqrt{5}$ **45.** $5+2\sqrt{6}$ **47.** $11-2\sqrt{30}$ **49.** $-1$ **51.** 4 **53.** 8 **55.** 40
**57.** $\frac{5\sqrt{6}+15}{-3}$ **59.** $\frac{40+8\sqrt{10}}{15}$ **61.** $\frac{5\sqrt{6}+6\sqrt{10}}{-21}$ **63.** $\frac{3-2\sqrt{2}}{-1}$
**65.** $\frac{10-5\sqrt{10}-2\sqrt{7}+\sqrt{70}}{-6}$ **67.** $\frac{2\sqrt{11}-2\sqrt{3}}{8}$ **69.** $\frac{\sqrt{30}+5}{1}$ **71.** $\frac{\sqrt{255}+\sqrt{17}+\sqrt{285}+\sqrt{19}}{-2}$
**73.** $\frac{\sqrt{10}-\sqrt{14}+\sqrt{15}-\sqrt{21}}{-2}$ **75.** $t=\frac{1+\sqrt{p}}{1-p}$ **77.** $\Delta T=\frac{10u\sqrt{u}-10u}{u-1}$

## PROBLEM SET 6.7

**1.** 25 **3.** No solution **5.** 36 **7.** 48 **9.** 4 **11.** 10 **13.** 1 **15.** 9 **17.** 6
**19.** No solution **21.** 125 **23.** 16 **25.** 63 **27.** 2 **29.** 13 **31.** 604.7
**33.** $T=\frac{f^2L^2m}{k^2}$

## CHAPTER 6 REVIEW EXERCISES

**1.** $-24a^7b^3$ **2.** $\frac{-2x^2}{y^2}$ **3.** $81u^8v^{12}$ **4.** $\frac{125m^6}{n^9k^3}$ **5.** $288x^8y^7$ **6.** $\frac{p^{65}}{q^{50}}$ **7.** 1 **8.** 9
**9.** $\frac{b^5}{a^4c}$ **10.** $\frac{1}{a^7}$ **11.** $k^4$ **12.** $\frac{x^{12}}{y^4}$ **13.** $\frac{xy}{y-x}$ **14.** $1.75\times 10^{10}$ **15.** $6.21\times 10^{-5}$
**16.** 98,000 **17.** 0.0000037 **18.** $1.054\times 10^6$ **19.** $4\times 10^4$ **20.** $3.2\times 10^{-14}$ **21.** $\sqrt[7]{a}$
**22.** $\sqrt[5]{b^4}$ **23.** $\frac{1}{\sqrt[9]{c^2}}$ **24.** $m^{1/3}$ **25.** $(a+b)^{2/7}$ **26.** $x^{-1/6}$ **27.** $5x^2$ **28.** $-7$

**29.** No real number **30.** 64 **31.** $\frac{64}{27}$ **32.** $x^{5/6} - x^{7/12}$ **33.** $\frac{a^{1/8}c^{3/16}}{b^{1/6}}$ **34.** $6\sqrt{3}$

**35.** $2\sqrt[3]{100}$ **36.** $4xy^3\sqrt{5xy}$ **37.** $3pqr^2\sqrt[3]{2pq^2}$ **38.** $\frac{\sqrt{10}}{5}$ **39.** $\frac{\sqrt[3]{6ax^2}}{2x}$ **40.** $-15\sqrt{2}$

**41.** $-7\sqrt{5}$ **42.** $(6 - 4p)\sqrt[3]{5p}$ **43.** $6 - 2\sqrt{7} + 3\sqrt{5} - \sqrt{35}$ **44.** $11 - 6\sqrt{2}$ **45.** 3

**46.** $\frac{\sqrt{10} + 2\sqrt{15}}{-5}$ **47.** $\frac{6\sqrt{3} - 6\sqrt{5} - \sqrt{6} + \sqrt{10}}{-2}$ **48.** 24 **49.** 2 **50.** 20

## PROBLEM SET 7.1

**1.** 3, 4 **3.** −7, 2 **5.** $\frac{7}{2}$, 8 **7.** 2, 3 **9.** −8, 5 **11.** $\frac{1}{2}$, 6 **13.** $\frac{4}{3}, \frac{-1}{2}$ **15.** −4, 3

**17.** 7, 9 **19.** $\frac{-5}{3}, \frac{3}{2}$ **21.** $\frac{1}{2}$, 6 **23.** −2, −5 **25.** 10, −4 **27.** 12, −1 **29.** −5, 1

**31.** −8, −2 **33.** 3, 11 **35.** $\frac{-4}{3}$, 8 **37.** $\frac{5}{6}$, 4 **39.** 6 **41.** 9 **43.** $\frac{5}{2}$, 6 **45.** 2, 8

**47.** 10 **49.** 4

## PROBLEM SET 7.2

**1.** 9, −9 **3.** $-\sqrt{15}, \sqrt{15}$ **5.** $\frac{-4}{5}, \frac{4}{5}$ **7.** 3, −7 **9.** $\frac{17}{5}, \frac{-13}{5}$ **11.** $\sqrt{7} + 5, -\sqrt{7} + 5$

**13.** $\frac{3}{2}, \frac{-5}{2}$ **15.** $\frac{\sqrt{3} + 2}{3}, \frac{-\sqrt{3} + 2}{3}$ **17.** $\frac{1 + \sqrt{2}}{30}, \frac{1 - \sqrt{2}}{30}$ **19.** 7, −11 **21.** $4 + \sqrt{37}, 4 - \sqrt{37}$

**23.** $\frac{1 + \sqrt{29}}{2}, \frac{1 - \sqrt{29}}{2}$ **25.** $\frac{3 + \sqrt{89}}{8}, \frac{3 - \sqrt{89}}{8}$ **27.** $\frac{8}{3}$, −1 **29.** $\frac{2 + \sqrt{14}}{2}, \frac{2 - \sqrt{14}}{2}$

**31.** $\frac{1 + \sqrt{21}}{5}, \frac{1 - \sqrt{21}}{5}$ **33.** $\frac{3 + \sqrt{89}}{8}, \frac{3 - \sqrt{89}}{8}$ **35.** 2000

## PROBLEM SET 7.3

**1.** $7i$ **3.** $5i$ **5.** $\frac{2}{3}i$ **7.** $2i\sqrt{6}$ **9.** $-4i$ **11.** $-2i\sqrt{3}$

**13.** Real part = 4; imaginary part = 6 **15.** Real part = 7; imaginary part = $\frac{-\sqrt{2}}{2}$

**17.** Real part = 12; imaginary part = 0 **19.** $15 - 2i$ **21.** $3 + 9i$ **23.** $6 + 6i$ **25.** $6 - 4i$
**27.** $4 + 2i$ **29.** $22 - 19i$ **31.** $-7i$ **33.** $7 - 14i$ **35.** $1 + 5i$ **37.** $-6 + 4i$
**39.** $7 + i$ **41.** $23 - 2i$ **43.** $-2 + 11i$ **45.** $13 - i$ **47.** $-5 + 12i$ **49.** $21 - 20i$

**51.** 25 **53.** 29 **55.** 101 **57.** $\frac{11}{10} + \frac{7}{10}i$ **59.** $\frac{9}{15} - \frac{17}{15}i$ **61.** $\frac{8}{5} - \frac{17}{10}i$ **63.** $\frac{-2}{5} - \frac{1}{5}i$

**65.** $\frac{-5}{4} - \frac{3}{2}i$ **67.** $-1 - 10i$ **69.** $\frac{14}{5} - \frac{7}{5}i$ **71.** $\frac{24}{37} - \frac{4}{37}i$ **73.** $-i$ **75.** $i$ **77.** $i$

**79.** −1 **81.** −1500 **83.** 1803

## PROBLEM SET 7.4

**1.** $3, \frac{-5}{2}$ **3.** $\frac{-5+\sqrt{73}}{8}, \frac{-5-\sqrt{73}}{8}$ **5.** $\frac{3+\sqrt{29}}{2}, \frac{3-\sqrt{29}}{2}$ **7.** $\frac{1+\sqrt{21}}{5}, \frac{1-\sqrt{21}}{5}$

**9.** $\frac{1+\sqrt{61}}{6}, \frac{1-\sqrt{61}}{6}$ **11.** $\frac{-3+\sqrt{13}}{4}, \frac{-3-\sqrt{13}}{4}$ **13.** $\frac{-3+\sqrt{41}}{8}, \frac{-3-\sqrt{41}}{8}$ **15.** $\frac{\sqrt{6}}{3}, \frac{-\sqrt{6}}{3}$

**17.** $\frac{3+\sqrt{229}}{2}, \frac{3-\sqrt{229}}{2}$ **19.** $\frac{-1+i\sqrt{14}}{3}, \frac{-1-i\sqrt{14}}{3}$ **21.** $5+\sqrt{13}, 5-\sqrt{13}$

**23.** $1+\sqrt{2}, 1-\sqrt{2}$ **25.** $6.275, -1.275$ **27.** $5.260, -0.760$ **29.** $-0.869, 1.535$

**31.** $7.162, 0.838$ **33.** $\frac{-5+\sqrt{37}}{2}$ **35.** $5+\sqrt{22}, 5-\sqrt{22}$ **37.** $\frac{7+\sqrt{145}}{8}$ **39.** $\frac{1}{2}$

**41.** $\frac{1+\sqrt{5}}{2}$ **43.** $\frac{15+\sqrt{305}}{4}$ **45.** $25+5\sqrt{21}, 25-5\sqrt{21}$ **47.** $4+2\sqrt{2}, 4-2\sqrt{2}$

## PROBLEM SET 7.5

**1.** 3 and 11 **3.** 2 and 8 **5.** 3 and $\frac{1}{3}$ **7.** 11 and 13 **9.** 4 or $\frac{1}{2}$ **11.** 6 and 11
**13.** 6 and 8 **15.** $\frac{1}{2}$ **17.** 2.5 **19.** 3 (returning), 4 (going) **21.** 40 mph (for 40 mi), 50 mph (for 100 mi)

## PROBLEM SET 7.6

**1.** $x < 1$ or $x > 2$

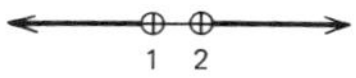

**3.** $\frac{-1}{2} \le r \le 3$

**5.** $\frac{-5}{4} < z < \frac{2}{3}$

**7.** $u \le 3$ or $u \ge 4$

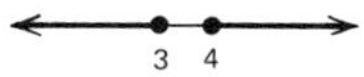

**9.** $k < -9$ or $k > 2$

**11.** $-5 \le x \le \frac{-1}{2}$

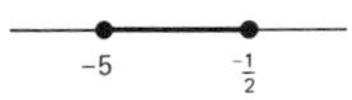

**13.** $\frac{-2}{3} < t < \frac{1}{2}$

**15.** $x \le -3$ or $x \ge 3$

**17.** $-2 < z$ or $z > 5$

**19.** $1 \le k \le 10$

**21.** $-3 < x < \frac{5}{2}$

**23.** $2 < t < 3$

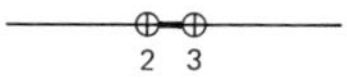

**25.** $t < 3$ or $t > 7$

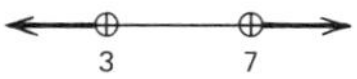

## CHAPTER 7 REVIEW EXERCISES

**1.** $5, \frac{-7}{2}$ **2.** $-5, 3$ **3.** $6, \frac{-1}{2}$ **4.** $-2, 3$ **5.** $-7, 1$ **6.** $2, -9$ **7.** $\frac{-2}{9}, \frac{2}{9}$

**8.** $\frac{4+\sqrt{3}}{12}, \frac{4-\sqrt{3}}{12}$ **9.** $-4+\sqrt{17}, -4-\sqrt{17}$ **10.** $2, \frac{-1}{3}$ **11.** $8i$ **12.** $-5i$

**13.** Real part $= -6$, imaginary part $= 10$ **14.** $10 + 3i$ **15.** 14 **16.** $34 + 13i$ **17.** 65

**18.** $2 + i$ **19.** $\frac{5 - 14i}{17}$ **20.** $-i$ **21.** 1 **22.** $3 + \sqrt{19}, 3 - \sqrt{19}$ **23.** $\frac{-7 + \sqrt{89}}{5}, \frac{-7 - \sqrt{89}}{5}$

**24.** $\frac{1 + \sqrt{61}}{6}, \frac{1 - \sqrt{61}}{6}$ **25.** $\frac{-3 + \sqrt{13}}{2}$ **26.** 1 and 7 **27.** 1 **28.** 40 and 30

**29.** $x < \frac{-5}{2}$ or $x > 7$ 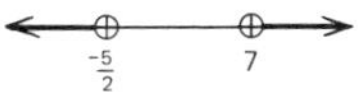


**30.** $\frac{-1}{2} \le a \le 5$ 

## PROBLEM SET 8.1

**1–12.**

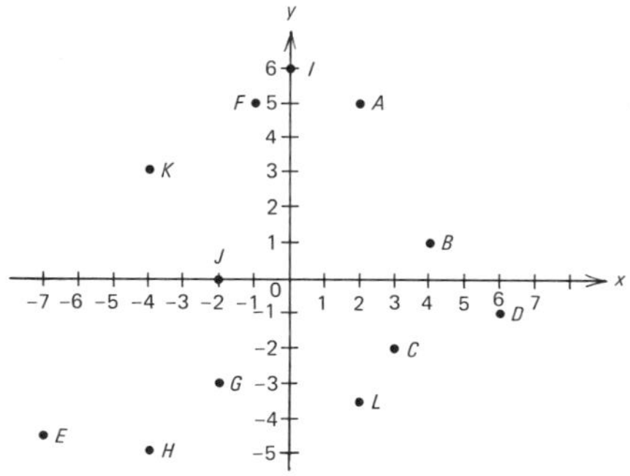


**13.** (1, 2) **15.** (3, −1) **17.** (0, −3)

**19.** (−4.5, −1) **21.** (−4.5, 2.5) **23.** (0, 4) **25.** (6, 0); (0, 6)

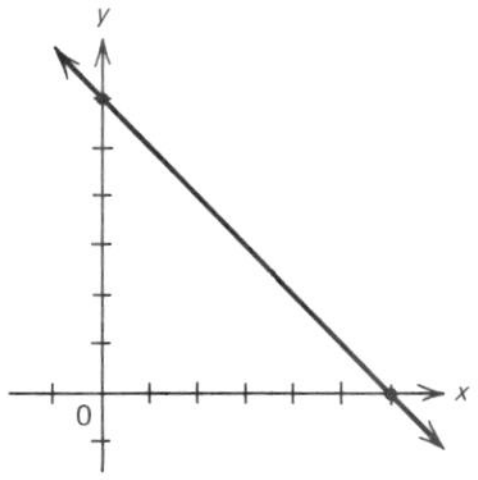

**27.** (7, 0); (0, −7) **29.** (5, 0); (0, 10) **31.** (3, 0); (0, −2) **33.** $(\frac{13}{5}, 0); (0, \frac{13}{2})$

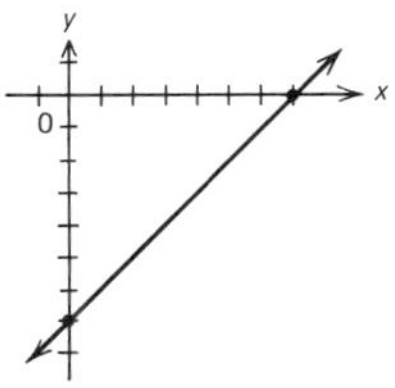

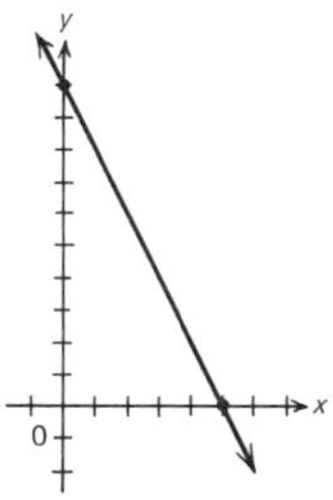

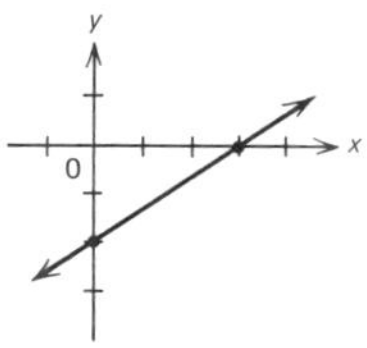

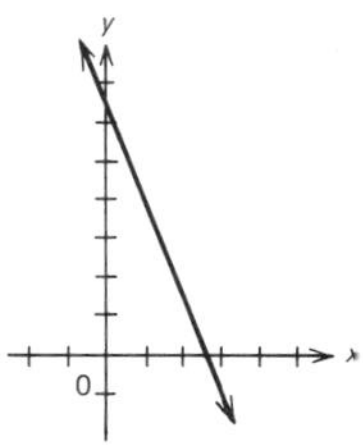

**35.** $(\frac{8}{3}, 0); (0, -4)$

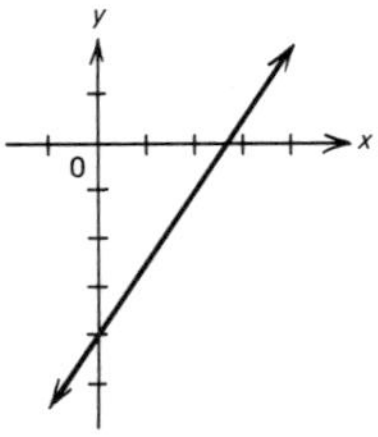


**37.** $(0, 0)$

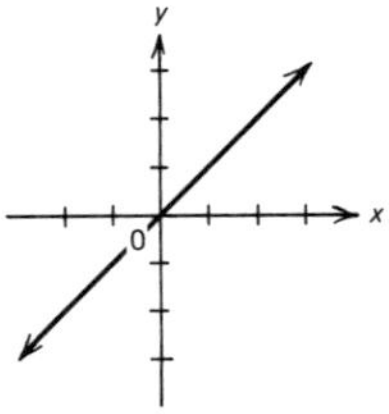


**39.** $(0, 0); (1, -2)$

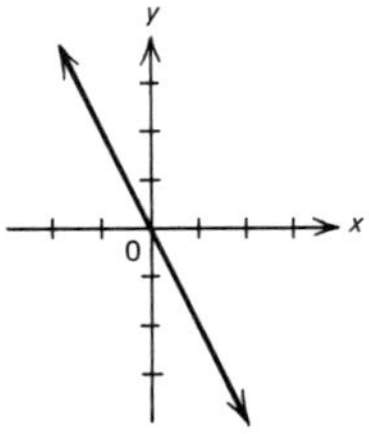


**41.**

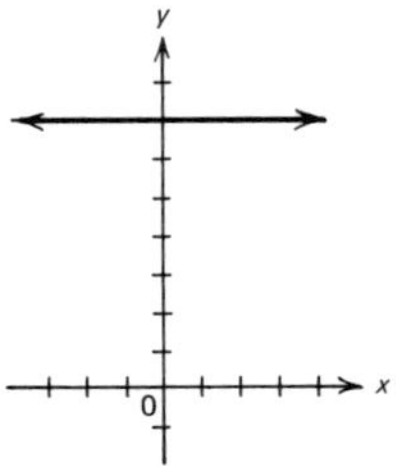


**43.**

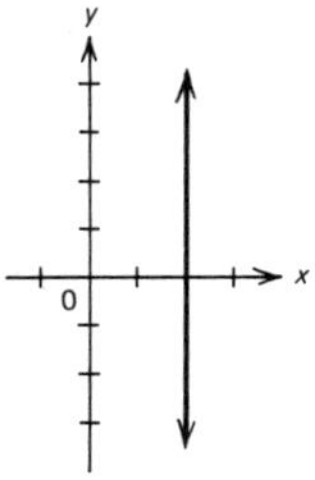


**45.**

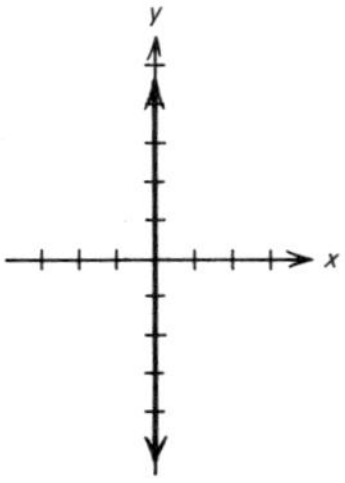


**47.**

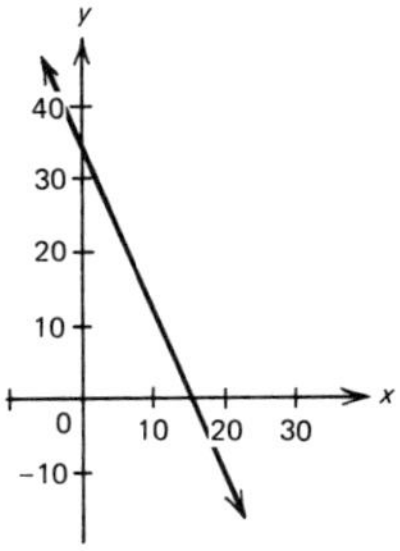


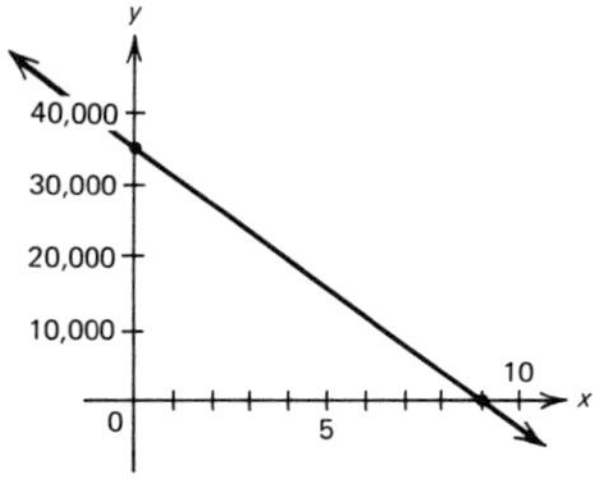


## PROBLEM SET 8.2

**1.** $\sqrt{20}$ **3.** $\sqrt{68}$ **5.** $\sqrt{221}$ **7.** 5 **9.** 17 **11.** 6 **13.** 2 **15.** $-4$ **17.** $\dfrac{-10}{11}$

**19.** 0 **21.** Undefined **23.** Undefined **25.** $\overleftrightarrow{AB}$ parallel to $\overleftrightarrow{CD}$ **27.** $\overleftrightarrow{AC}$ parallel to $\overleftrightarrow{BD}$

**29.** $\overleftrightarrow{AB}$ perpendicular to $\overleftrightarrow{CD}$ **31.** $65 \dfrac{\$}{\text{item}}$ **33.** $-4 \dfrac{\$}{\text{item}}$ **35.** $10 \dfrac{\text{m}}{\text{sec}}$ **37.** $43.3 \dfrac{\text{mi}}{\text{hr}}$

## PROBLEM SET 8.3

**1.** $y = 5x - 8$ **3.** $y = 10x - 13$ **5.** $y = 6x + 2$ **7.** $y = -x + 5$ **9.** $y = \dfrac{2}{5}x - \dfrac{17}{5}$

**11.** $y = 2x + 4$ **13.** $y = -4x + 19$ **15.** $y = \dfrac{-10}{11}x - \dfrac{67}{11}$ **17.** $y = 8$ **19.** $y = -4$

**21.** $m = 7$; $y$-int. $= -5$ **23.** $m = -6$; $y$-int. $= -2$ **25.** $m = \dfrac{2}{3}$; $y$-int. $= \dfrac{-7}{3}$

**27.** $m = 0$; $y$-int. $= 6$ **29.** $m = -2$; $y$-int. $= 5$ **31.** $m = \frac{-6}{5}$; $y$-int. $= \frac{2}{5}$ **33.** $m = 9$, $y$-int. $= -7$

**35.** $m = \frac{-2}{5}$; $y$-int. $= \frac{7}{5}$ **37.** $y = 7x - 40$ **39.** $y = \frac{-1}{2}x + \frac{15}{2}$ **41.** $y = 2x + 7$

**43.** $y = \frac{5}{4}x + \frac{15}{4}$ **45.** $m = \frac{3}{97}$; $y = \frac{3}{97}x + \frac{9119}{97}$ **47.** $y = \frac{-1}{1000}x + 17$ **49.** $y = 0.09x + 900$

## PROBLEM SET 8.4

**1.**

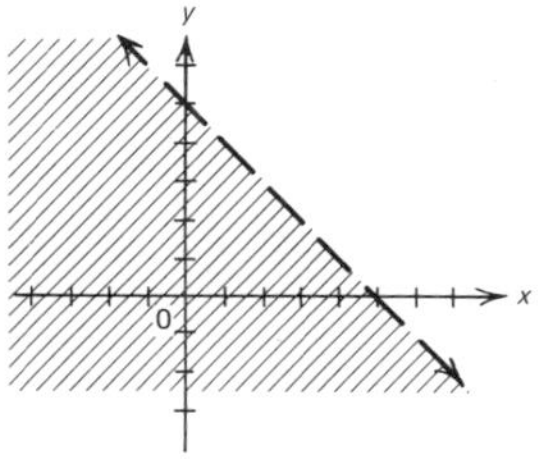


**3.**

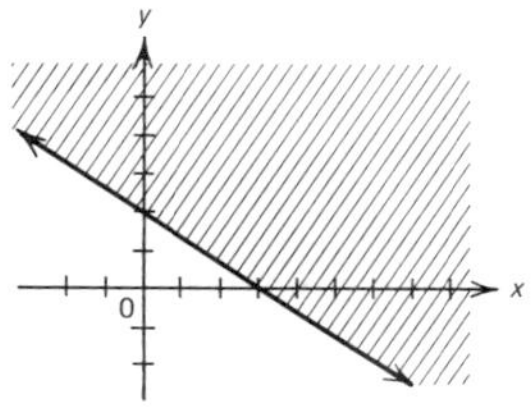


**5.**

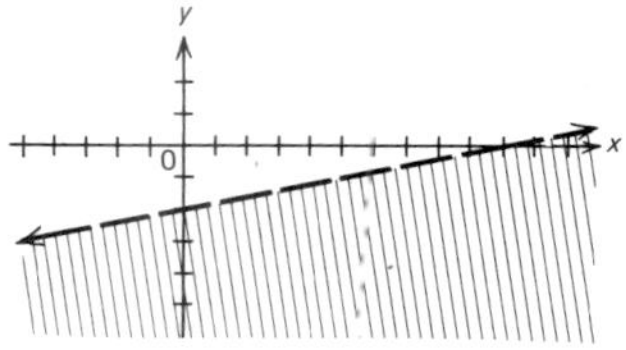


**7.**

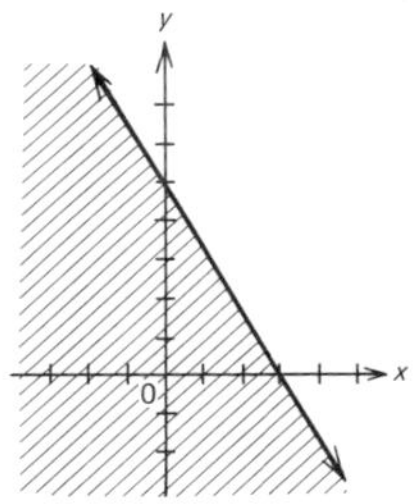


**9.**

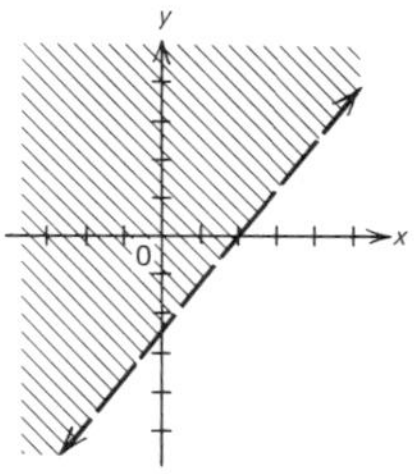


**11.**

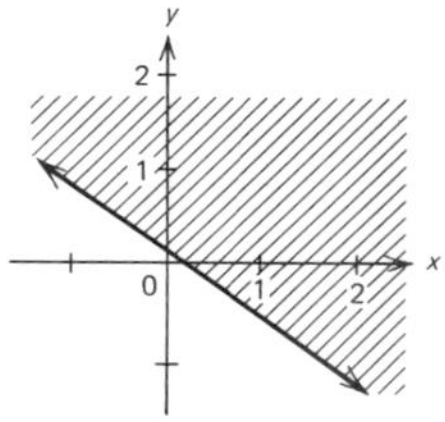


**13.**

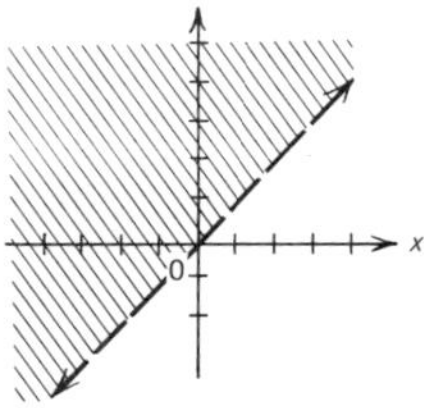


**15.**

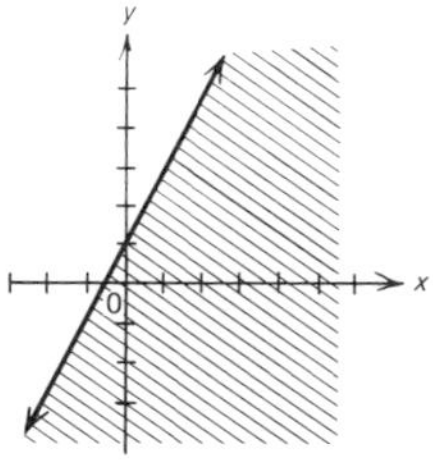


**17.**

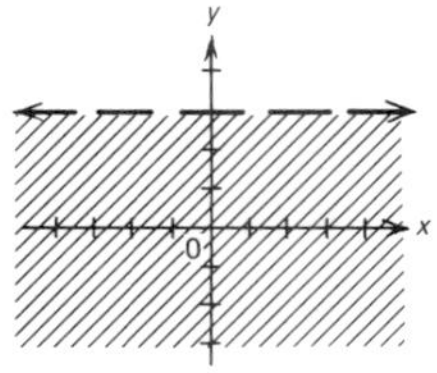


**19.**

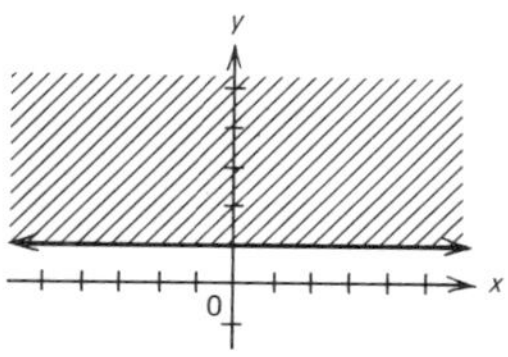


**21.**

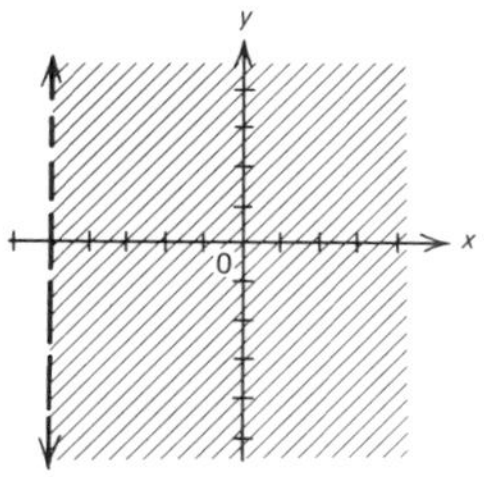


**23.**

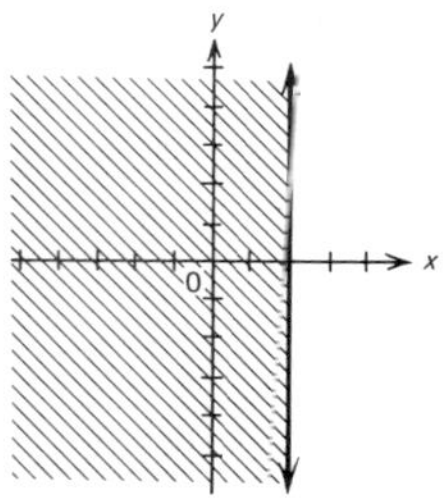

**25.**

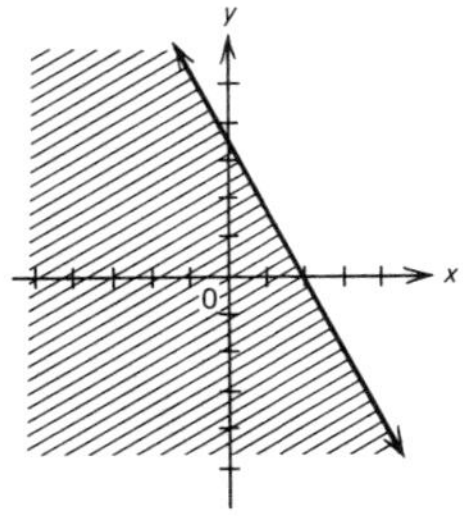


**27.**

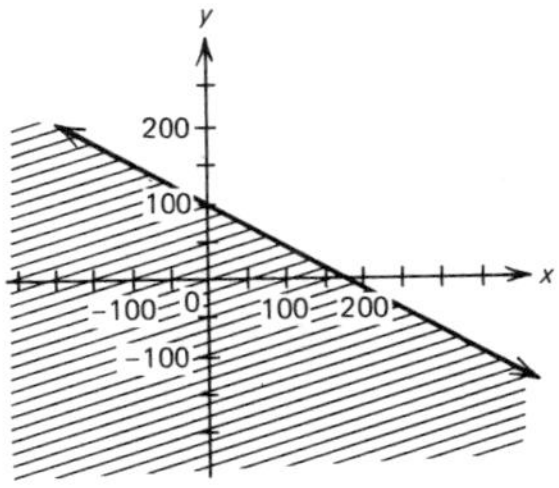


## PROBLEM SET 8.5

**1.** $x = ky$ **3.** $a = kb^4$ **5.** $z = \dfrac{k}{x}$ **7.** $L = \dfrac{k}{\sqrt{g}}$ **9.** $a = kxyz^3$ **11.** $x = \dfrac{k}{ab^4}$ **13.** 5

**15.** 5 **17.** 40 **19.** 18 **21.** 3 **23.** 1215 **25.** 30 **27.** 405 **29.** 8 **31.** $\dfrac{18}{5}$

**33.** 405 **35.** $\dfrac{1215}{64}$ **37.** 66,667 **39.** 1 **41.** 35 **43.** 64 **45.** 2700

## CHAPTER 8 REVIEW EXERCISES

**1–2.**

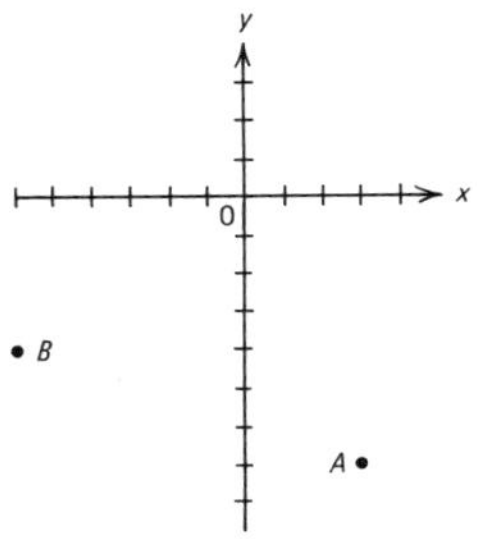


**3.** (1, 3) **4.** (5, −2) **5.** (7, 0) **6.** (0, −2)

**7.**

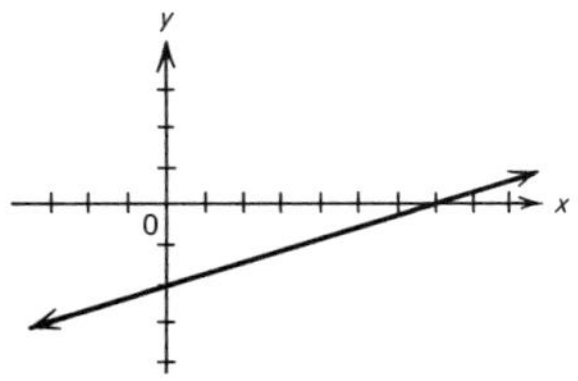


**8.**

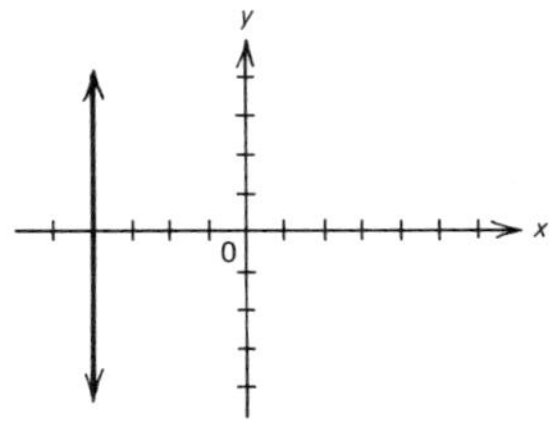


**9.** 5 **10.** $\dfrac{-4}{3}$

**11.** $AB$ perpendicular to $CD$ **12.** $y = -8x - 2$ **13.** $y = -5x + 9$ **14.** $m = \dfrac{2}{3}$

**15.** $y = \frac{-1}{2}x - 2$

**16.**

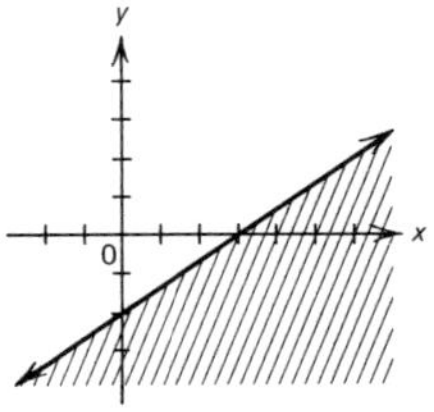

**17.**

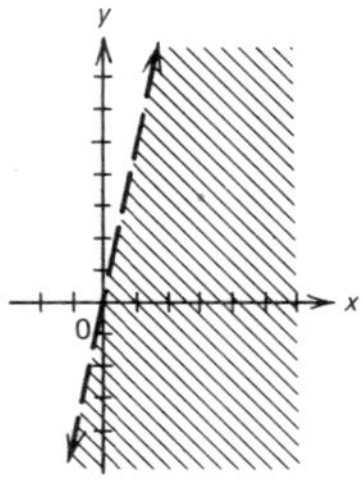

**18.**

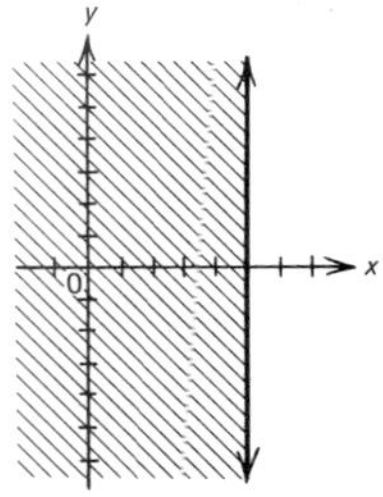

**19.** $x = \frac{ky\sqrt{z}}{a^2\sqrt[3]{b}}$ **20.** 7 **21.** $\frac{7}{2}$

## PROBLEM SET 9.1

**1.** Domain = {1, 2, 3, 4, 5, 6}; range = {5, 7, 10, 14, 19, 25}; a function

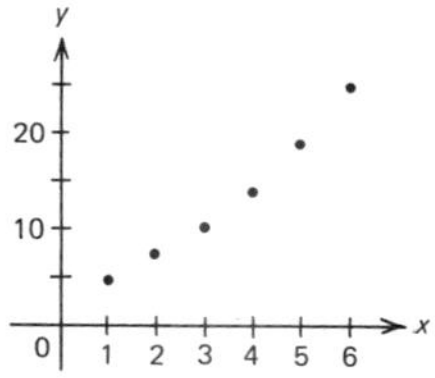

**3.** Domain = {1}; range = {5, 6, 7, 8, 9}; not a function

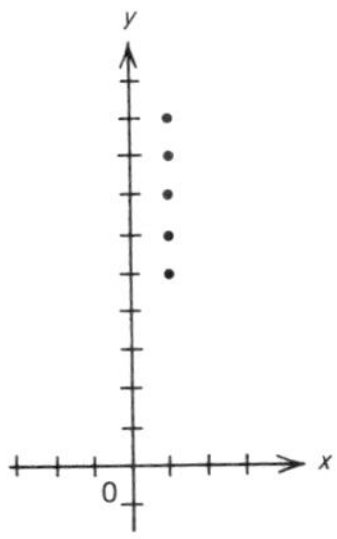

**5.** Domain = {1, 2, 3, 4, 7, 10}; range = {4, 6}; a function

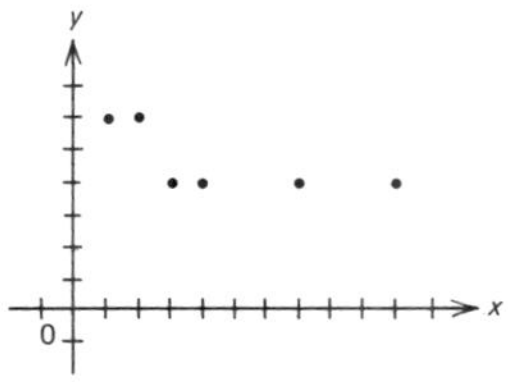

**7.**
$3 \to 27$
$4 \to 64$
$-2 \to -8$
$-5 \to -125$

**9.**
$3 \to \frac{1}{3}$
$4 \to \frac{1}{4}$
$-8 \to \frac{-1}{8}$
$-10 \to \frac{-1}{10}$

**11.** **(a)** 5 **(b)** 3 **(c)** 8 **(d)** 15 **(e)** 13 **(f)** $a + b + 5$

**13.** **(a)** 6 **(b)** 8 **(c)** 3 **(d)** $-4$ **(e)** 3 **(f)** $6 - a - b$ **15.** **(a)** 0 **(b)** 8 **(c)** 18 **(d)** 200 **(e)** 648 **(f)** $2a^2 + 4ab + 2b^2$ **17.** **(a)** 4 **(b)** 14 **(c)** 4 **(d)** 74 **(e)** 8 **(f)** $a^2 + 2ab + b^2 - 3a - 3b + 4$
**19.** **(a)** $2x + 2h + 5$ **(b)** $2h$ **(c)** 2 **21.** **(a)** $2x^2 + 4xh + 2h^2$ **(b)** $4xh + 2h^2$ **(c)** $4x + 2h$
**23.** **(a)** $x^2 + 2xh + h^2 - 3x - 3h + 4$ **(b)** $2xh + h^2 - 3h$ **(c)** $2x + h - 3$
**25.** A function

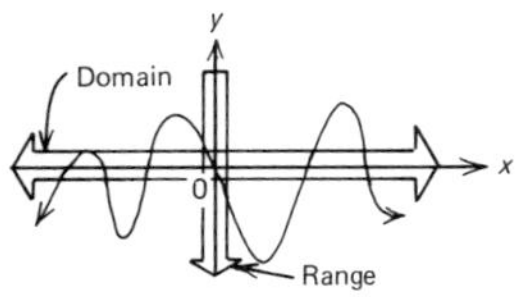


**27.** Not a function

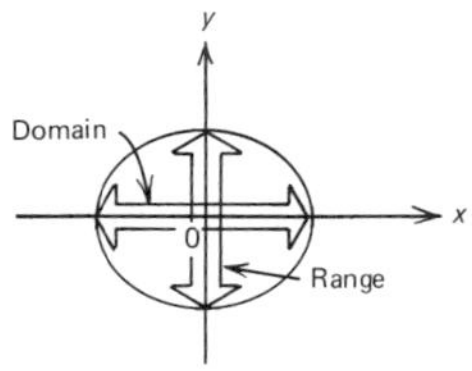


**29.** A function

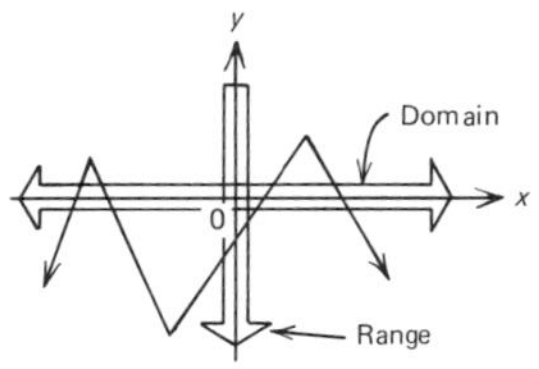


**31.** 3000 **33.** 300 **35.** 50 **37.** 64 **39.** $16x^2 + 32xh + 16h^2$ **41.** $32x + 16h$

## PROBLEM SET 9.2

**1.**

| x | y |
|---|---|
| −3 | 9 |
| −2 | 4 |
| −1 | 1 |
| 0 | 0 |
| 1 | 1 |
| 2 | 4 |
| 3 | 9 |

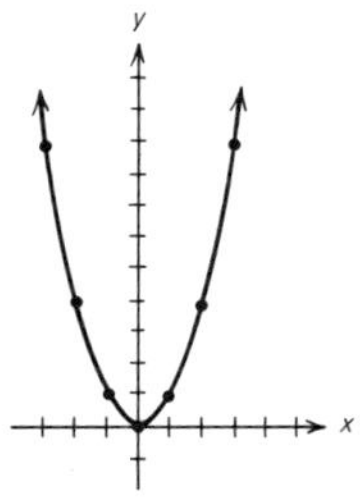


**3.**

| x | y |
|---|---|
| −1 | 2 |
| 0 | 3 |
| 1 | 2 |
| 2 | −1 |
| 3 | −6 |
| 4 | −13 |
| 5 | −22 |

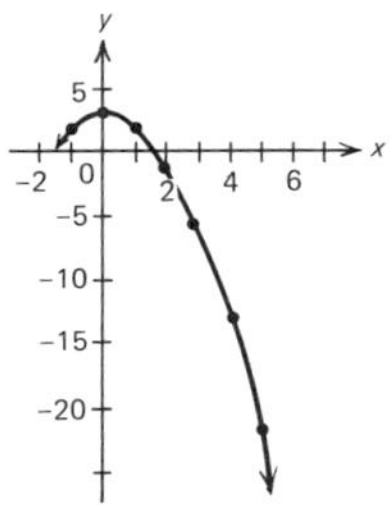


**5.**

| x | y |
|---|---|
| −2 | 3 |
| −1 | 1 |
| 0 | 1 |
| 1 | 3 |
| 2 | 7 |
| 3 | 13 |
| 4 | 21 |

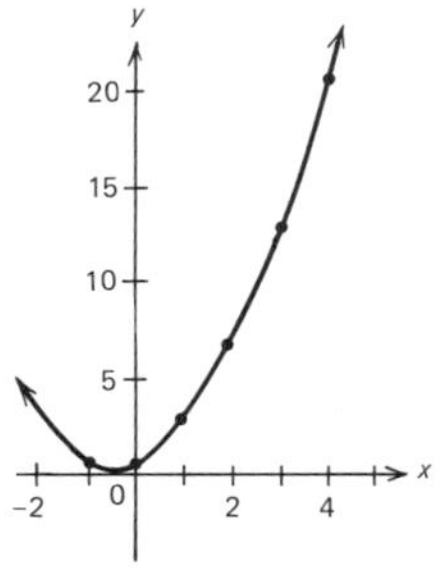


**7.**

| x | y |
|---|---|
| 0 | 2 |
| 1 | 3 |
| 2 | 8 |
| 3 | 17 |
| 4 | 30 |
| 5 | 47 |
| 6 | 68 |

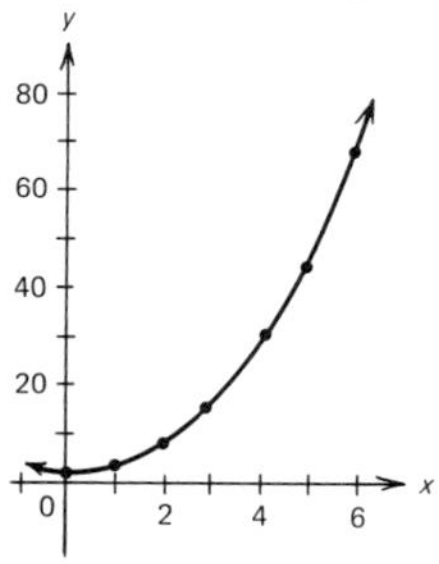


**9.**

| x | y |
|---|---|
| −1 | 12 |
| 0 | 8 |
| 1 | 2 |
| 2 | −6 |
| 3 | −16 |
| 4 | −28 |
| 5 | −42 |

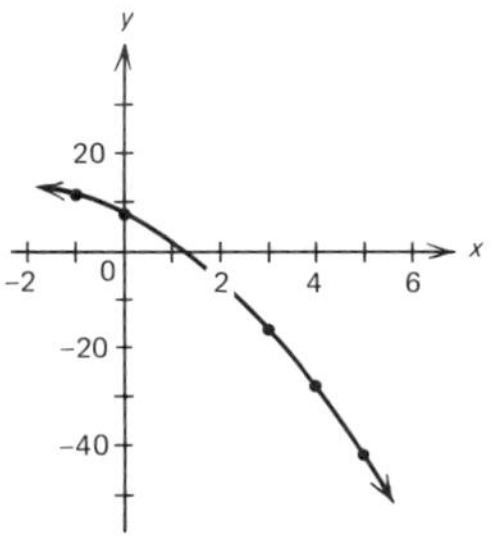


**11.**

| x | y |
|---|---|
| −3 | 8 |
| −2 | −2 |
| −1 | −6 |
| 0 | −4 |
| 1 | 4 |
| 2 | 18 |
| 3 | 38 |

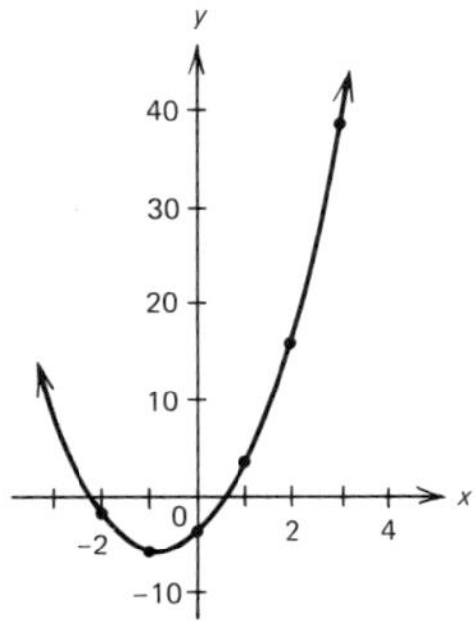


**13.** **(a)** (1, 0); (5, 0) **(b)** (0, 5) **(c)** (3, −4)

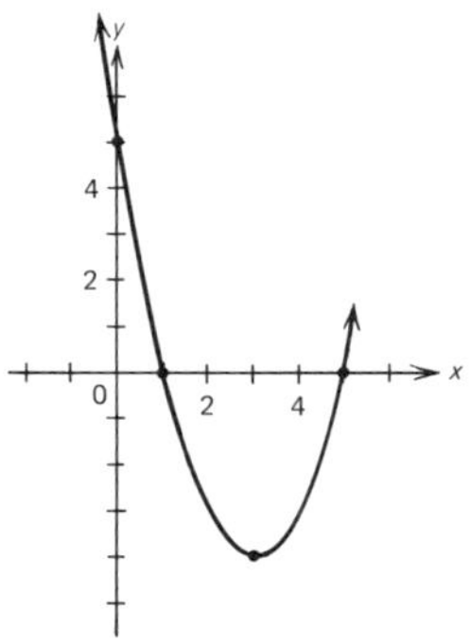

**15.** **(a)** $(2, 0)$; $(-2, 0)$ **(b)** $(0, -4)$ **(c)** $(0, -4)$

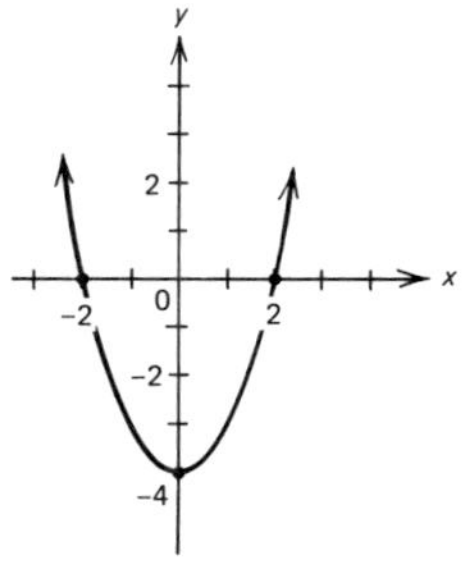


**17.** **(a)** $(4, 0)$; $(-2, 0)$ **(b)** $(0, 8)$ **(c)** $(1, 9)$

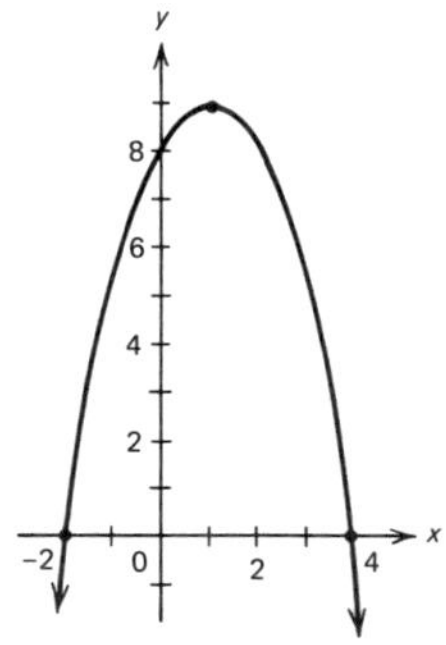


**19.** **(a)** $(2, 0)$; $(-1, 0)$ **(b)** $(0, -4)$ **(c)** $(\frac{1}{2}, -4\frac{1}{2})$

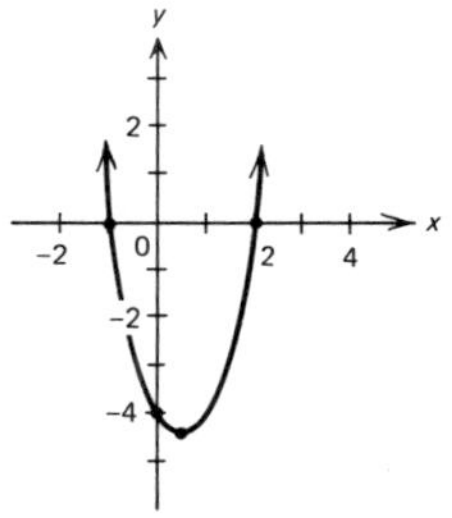


**21.** **(a)** $(0, 0)$; $(6, 0)$ **(b)** $(0, 0)$ **(c)** $(3, 9)$

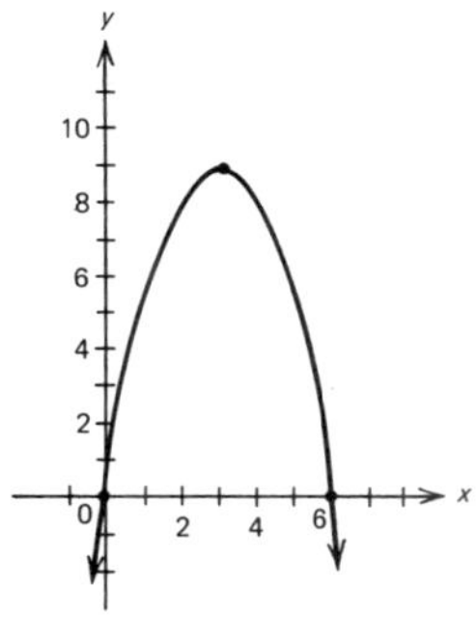


**23.** **(a)** $(2, 0)$; $(5, 0)$ **(b)** $(0, 30)$ **(c)** $(3\frac{1}{2}, -6\frac{3}{4})$

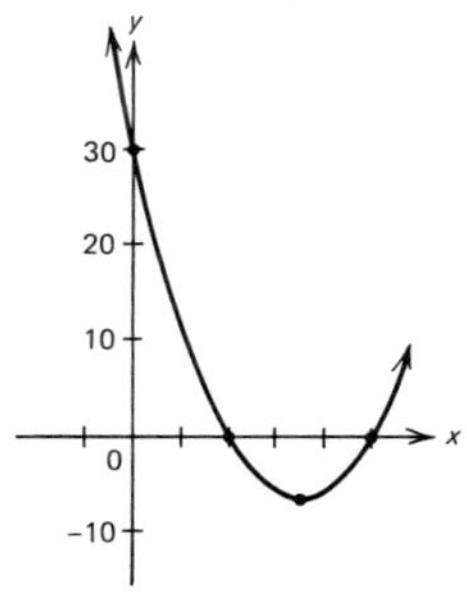


**25.**

| x | y |
|---|---|
| 0 | 0 |
| 1 | 64 |
| 2 | 96 |
| 3 | 96 |
| 4 | 64 |
| 5 | 0 |

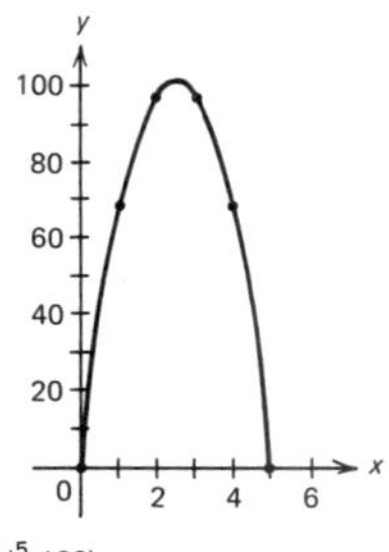


Vertex$=(\frac{5}{2}, 100)$

**27.**

| x | y |
|---|---|
| 0 | 10,000 |
| 1 | 10,875 |
| 2 | 11,500 |
| 3 | 11,875 |
| 4 | 12,000 |
| 5 | 11,875 |
| 6 | 11,500 |
| 7 | 10,875 |
| 8 | 10,000 |

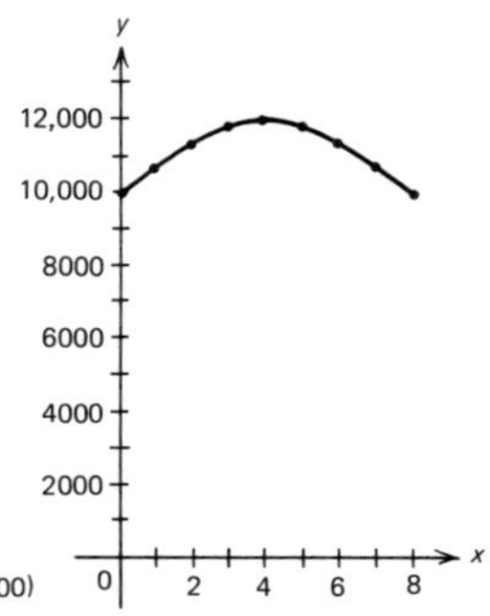


Vertex=(4,12000)

**29.**

| x | y |
|---|---|
| 0 | −100 |
| 10 | 1700 |
| 20 | 3100 |
| 30 | 4100 |
| 40 | 4700 |
| 50 | 4900 |
| 60 | 4700 |

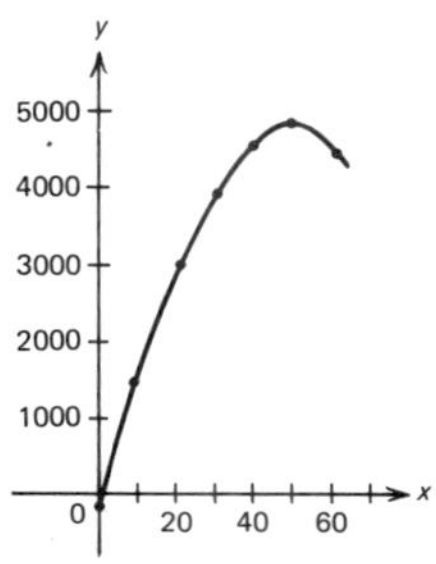


Vertex=(50,4900)

## PROBLEM SET 9.3

1.

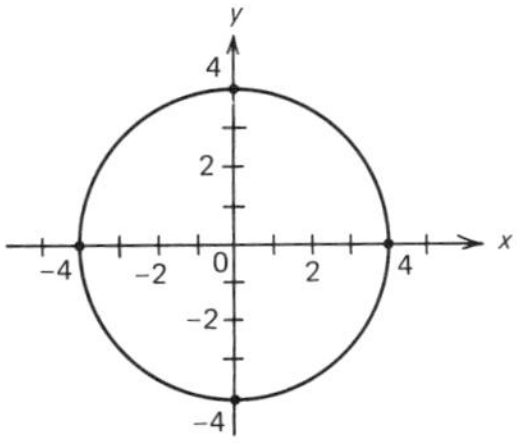


3.

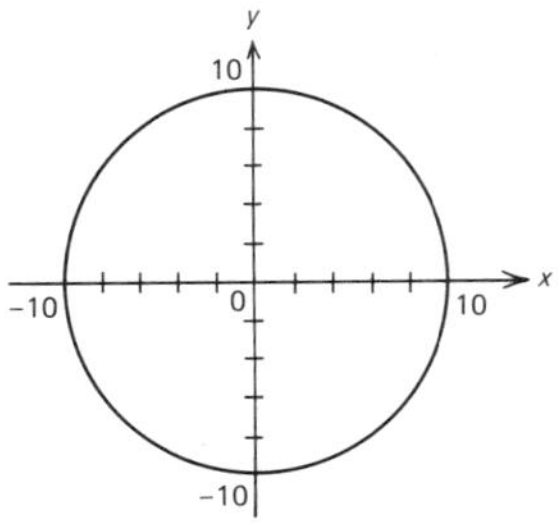


5.

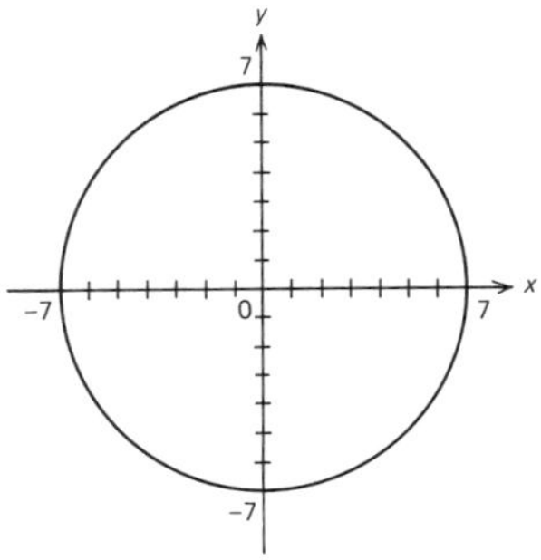


7.

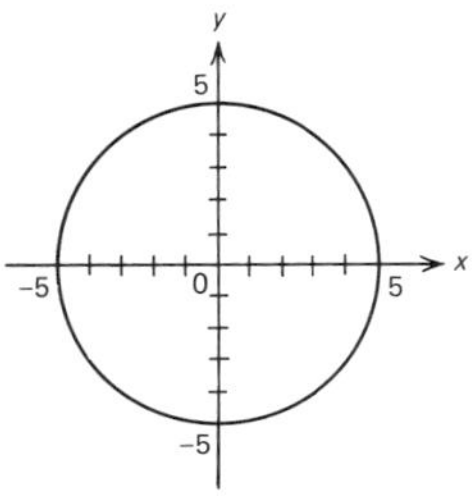


9.

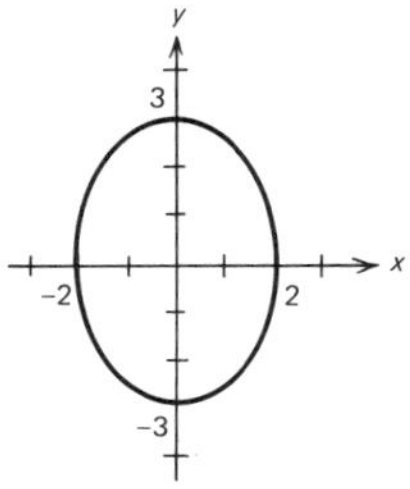


11.

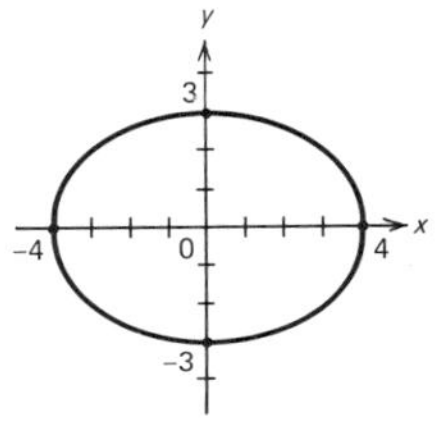


13.

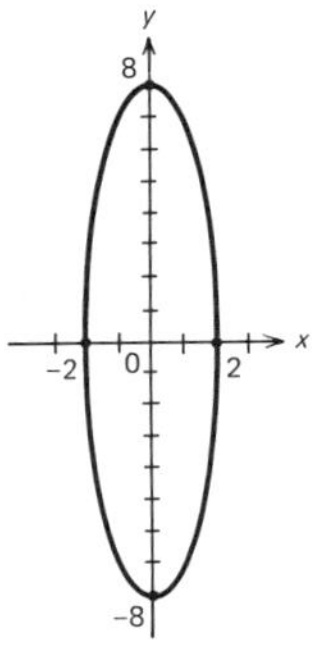


15.

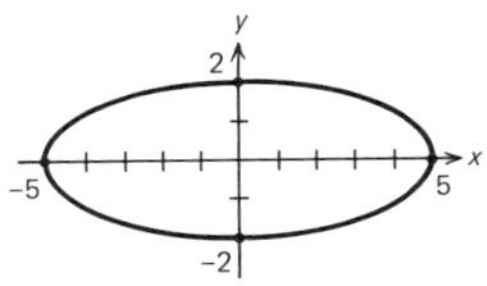


17.

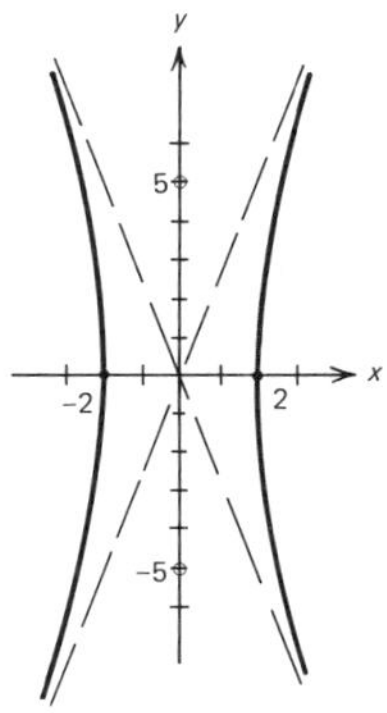


19.

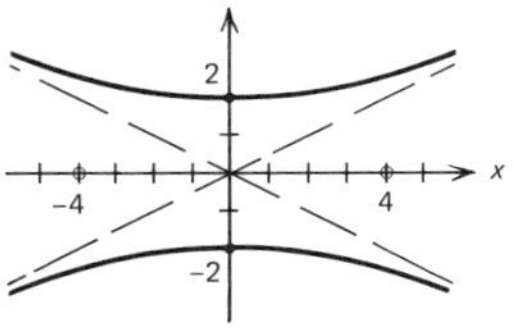


21.

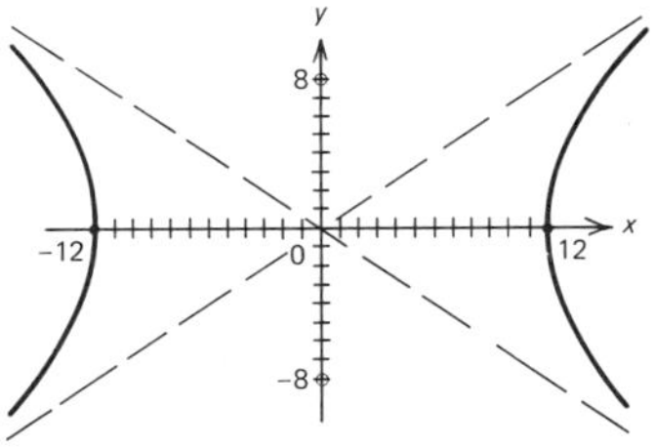


23.

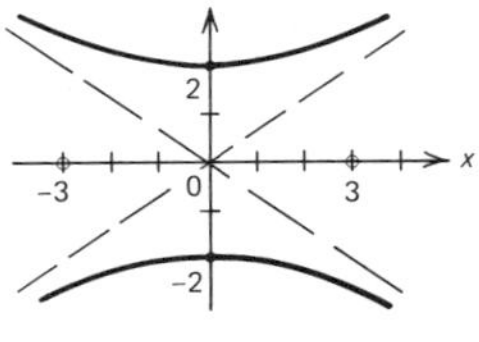

25.

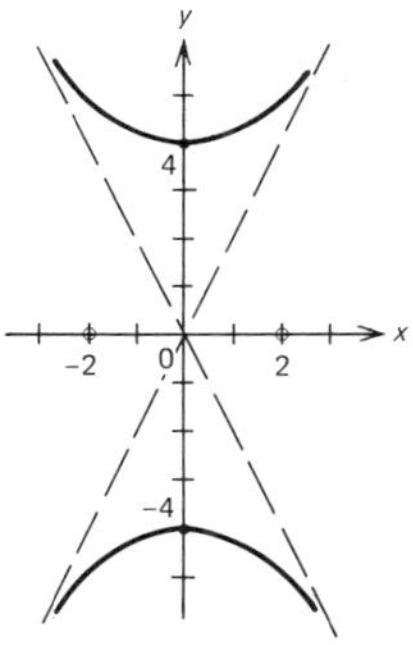


27.

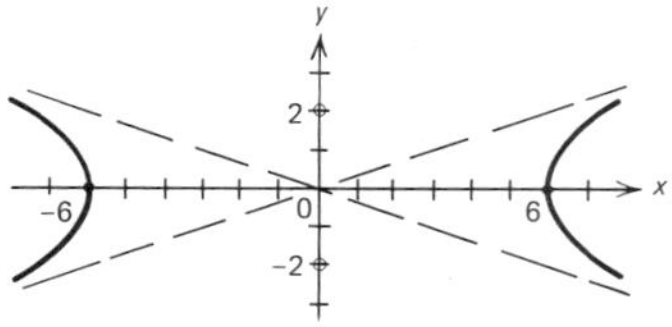


29.

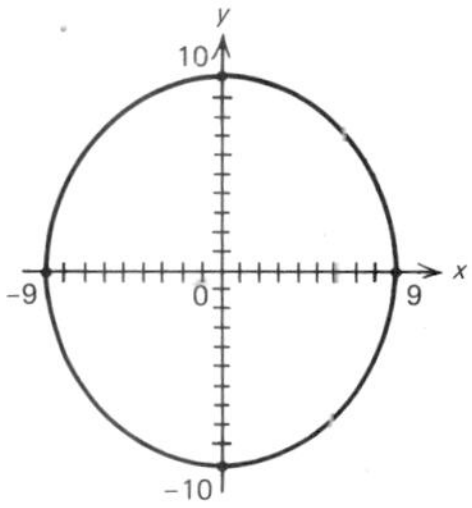


## PROBLEM SET 9.4

1.

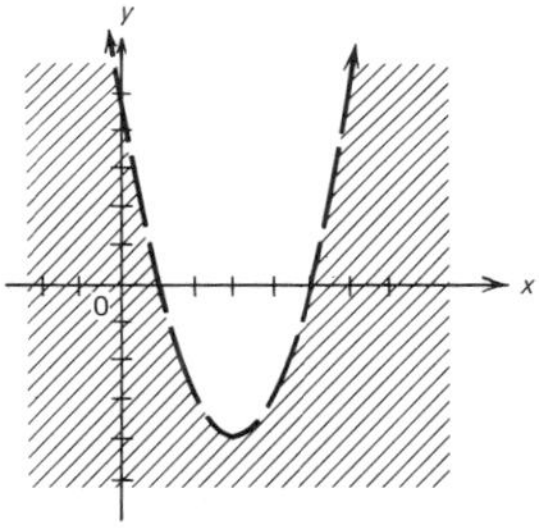


3.

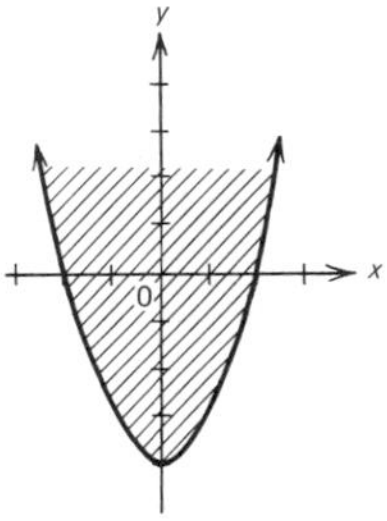


5.

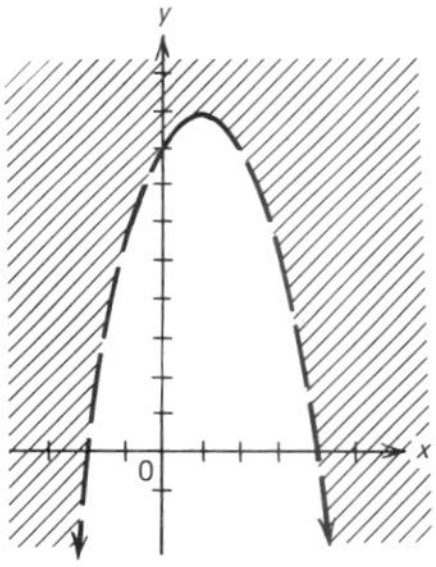


7.

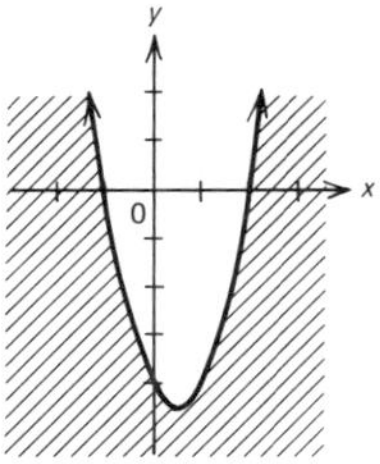


9.

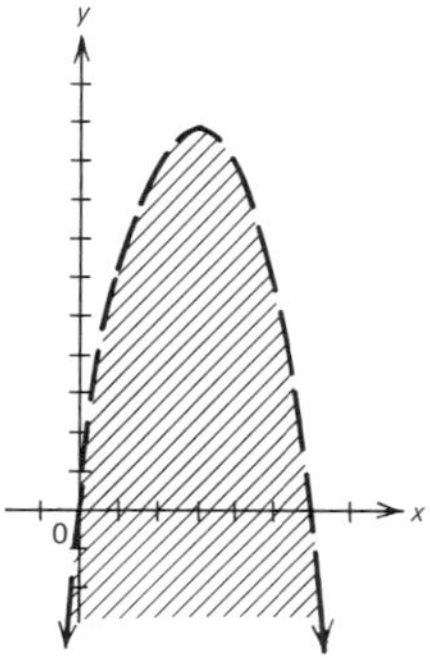


11.

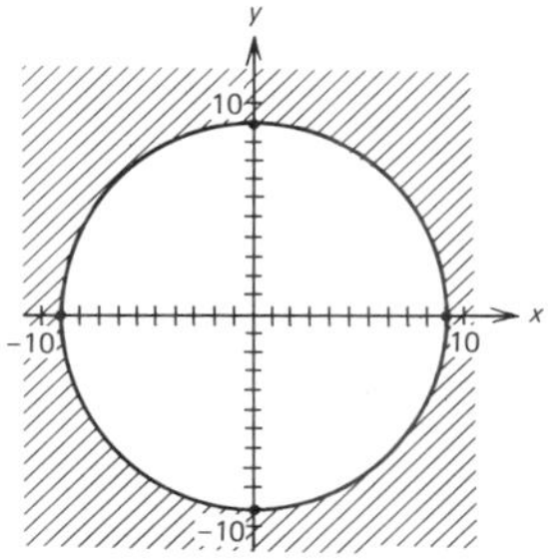


13.

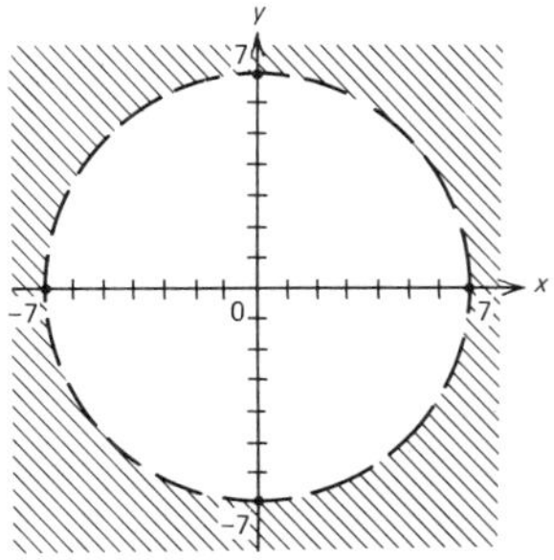


15.

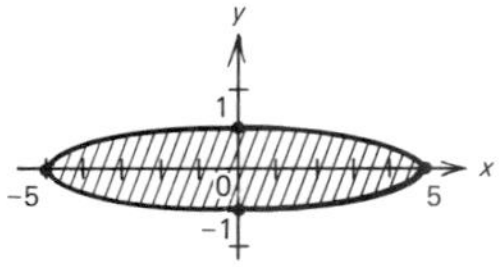


17.

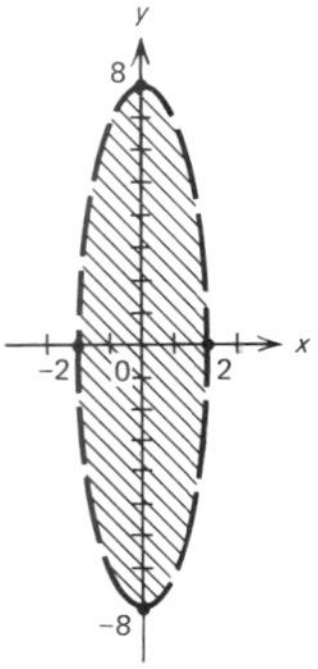

19.

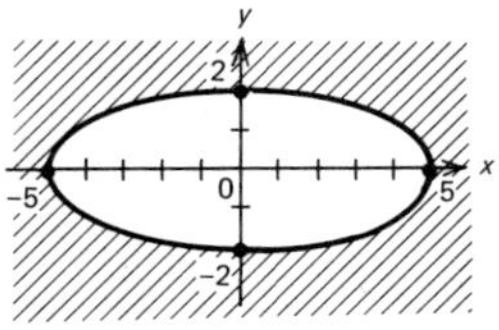


21.

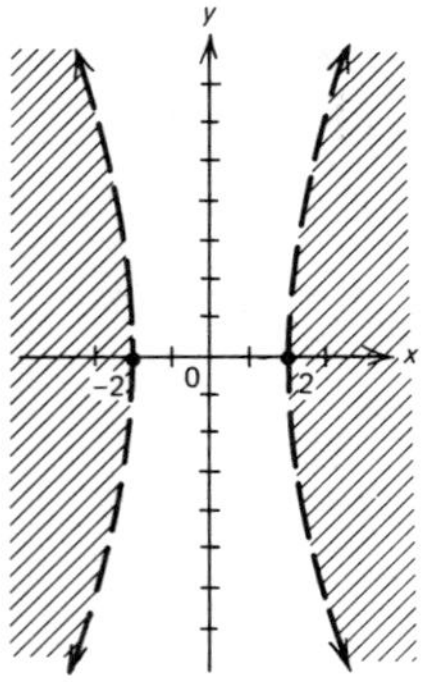


23.

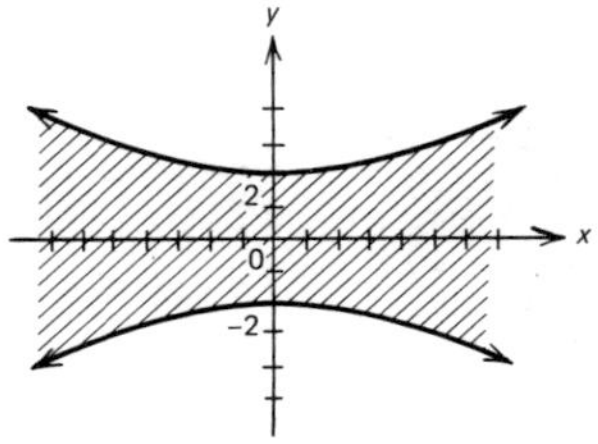


25.

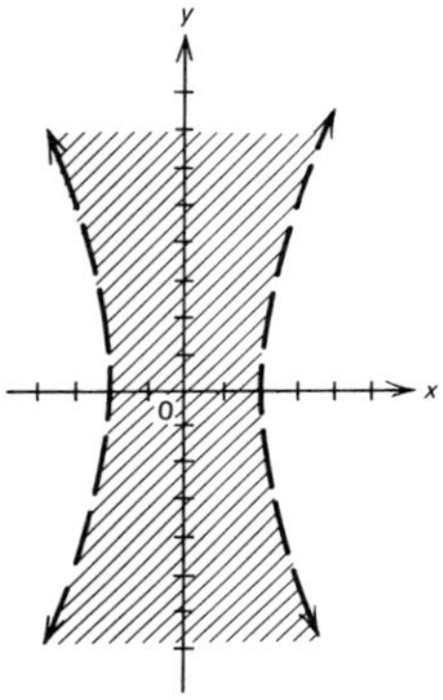


27.

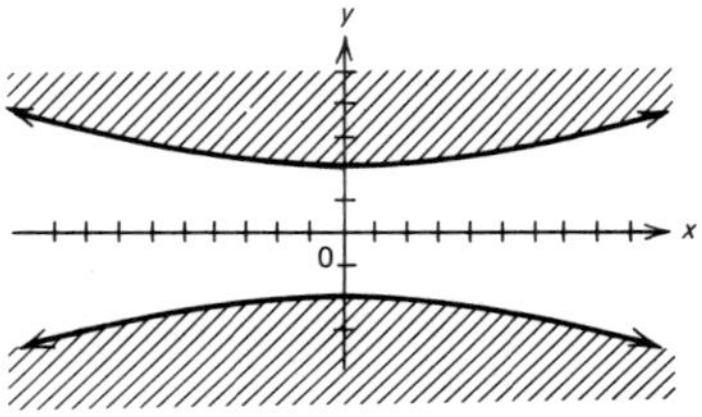


29.

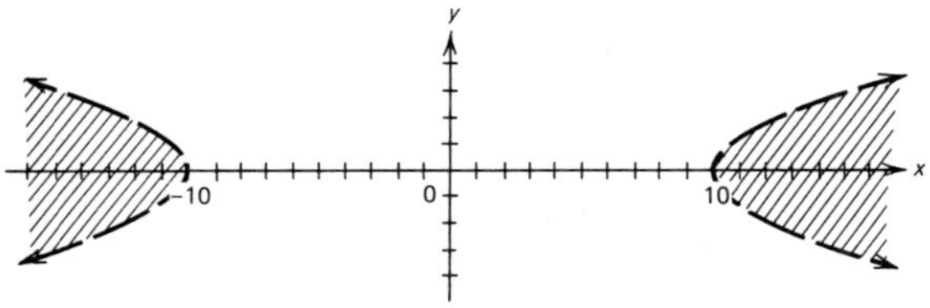


## PROBLEM SET 9.5

1. $f^{-1} = \{(5, 1), (7, 2), (10, 3), (14, 4), (19, 5), (25, 6)\}$ Both $f$ and $f^{-1}$ are functions.

3. $f^{-1} = \{(5, 1), (6, 1), (7, 1), (8, 1), (9, 1)\}$ $f$ is not a function; $f^{-1}$ is a function.

5. $f^{-1}$: $x = 3y - 1$ or $y = \dfrac{x+1}{3}$ Both are functions.

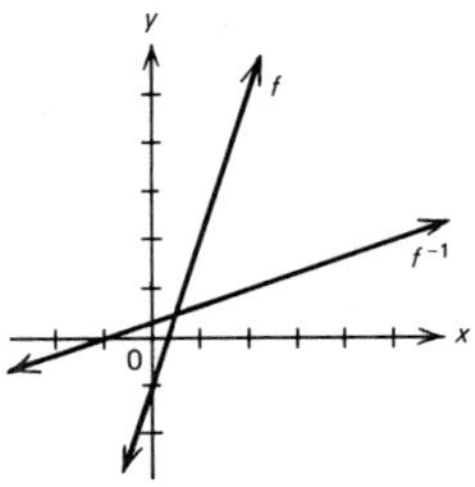

**7.** $f^{-1}$: $x = 5y - 2$ or $y = \dfrac{x+2}{5}$ Both are functions.

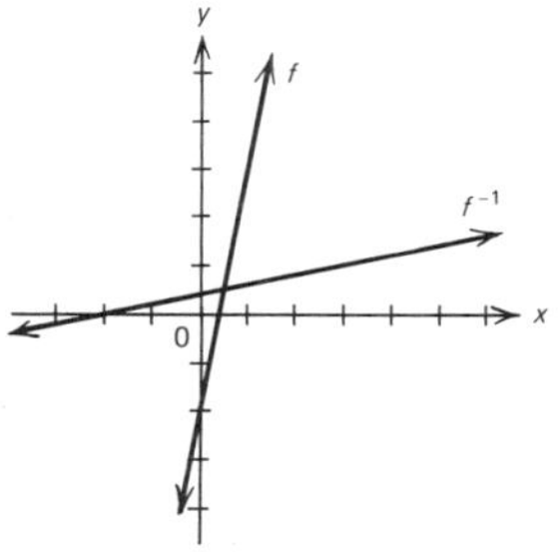


**9.** $f^{-1}$: $5x + 2y = 10$ Both are functions.

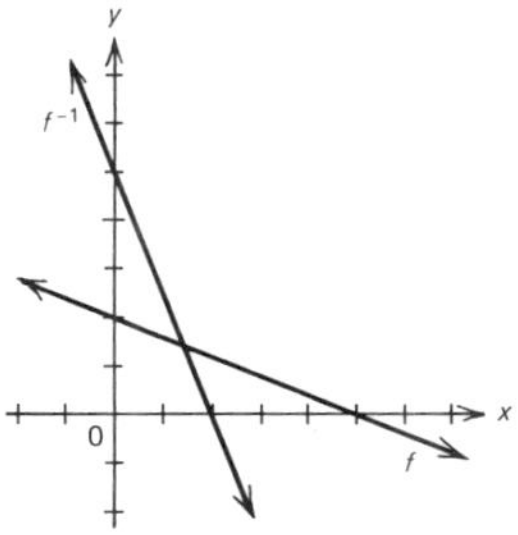


**11.** $f^{-1}$: $x = y^2$ $f$ is a function; $f^{-1}$ is not a function.

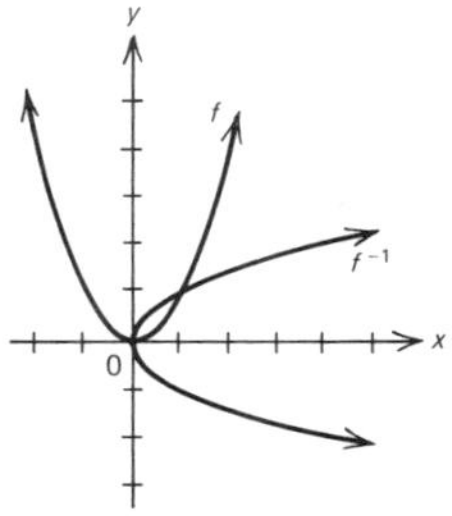


**13.** $f^{-1}$: $x = y^2 + 4$ $f$ is a function; $f^{-1}$ is not a function.

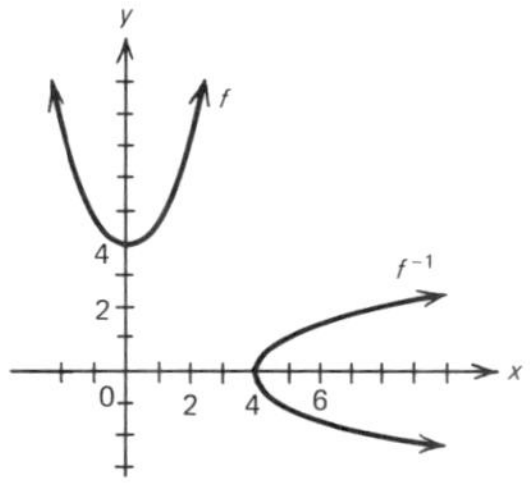


**15.** $f^{-1}$: $x = y^2 - 5y + 4$ $f$ is a function; $f^{-1}$ is not a function.

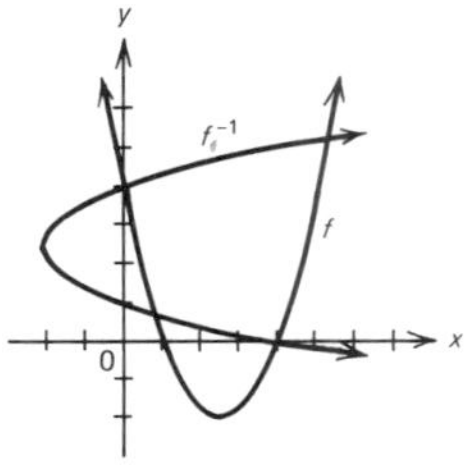


**17.** $f^{-1}$: $\dfrac{x^2}{25} + \dfrac{y^2}{4} = 1$ Neither is a function.

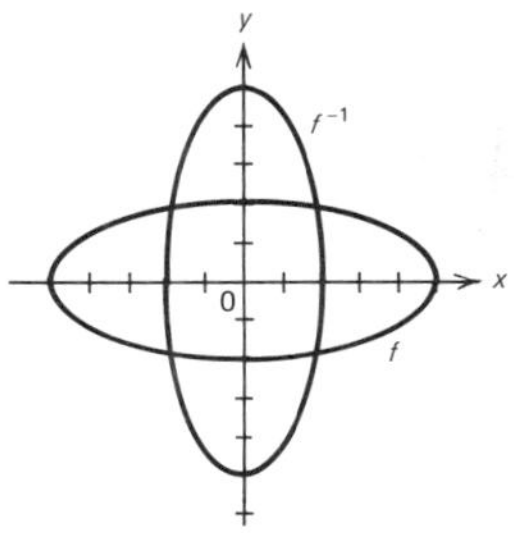


**19.** $f^{-1}$: $16x^2 + 16y^2 = 64$ Neither is a function.

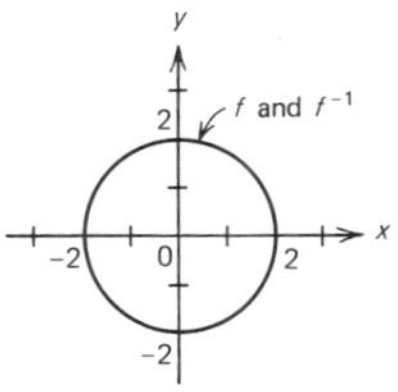

**21.** $f^{-1}$: $\dfrac{x^2}{9} - \dfrac{y^2}{1} = 1$ Neither is a function.

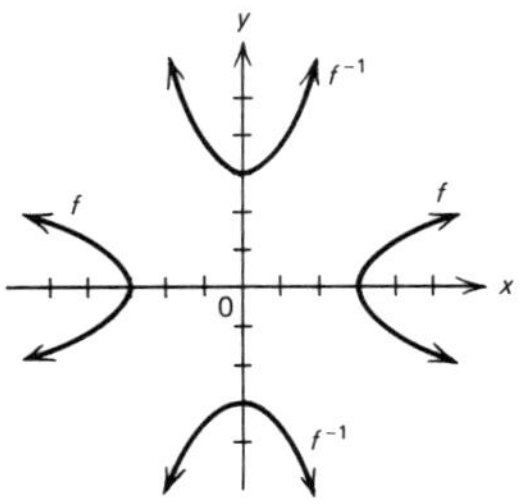


**23.** $f^{-1}$: $25x^2 - 9y^2 = 225$ Neither is a function.

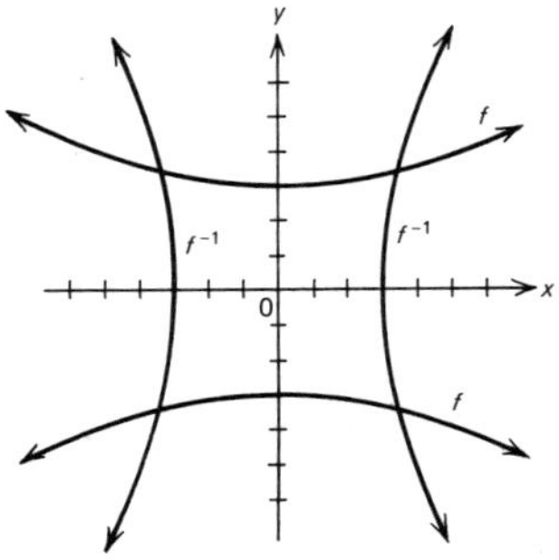


## CHAPTER 9 REVIEW EXERCISES

**1.** Domain = $\{1, 3, 8, 11, 13\}$ **2.** Range = $\{2, 3, 5, 10\}$ **3.** $f^{-1}$: = $\{(3, 1), (5, 3), (10, 8), (5, 11), (2, 13)\}$

**4.**

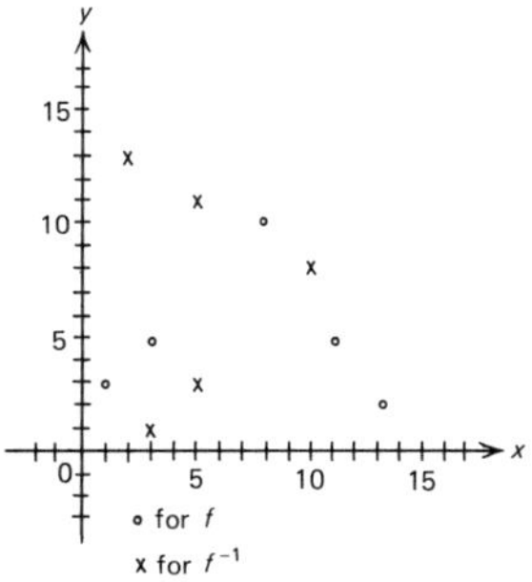


**5.** Yes **6.** No **7.** 1 **8.** 10 **9.** 101

**10.** $x^2 + 2xh + h^2 + 1$ **11.** $2xh + h^2$ **12.** $2x + h$ **13.**

| x | y |
|---|---|
| -3 | 17 |
| -2 | 3 |
| -1 | -5 |
| 0 | -7 |
| 1 | -3 |
| 2 | 7 |
| 3 | 23 |

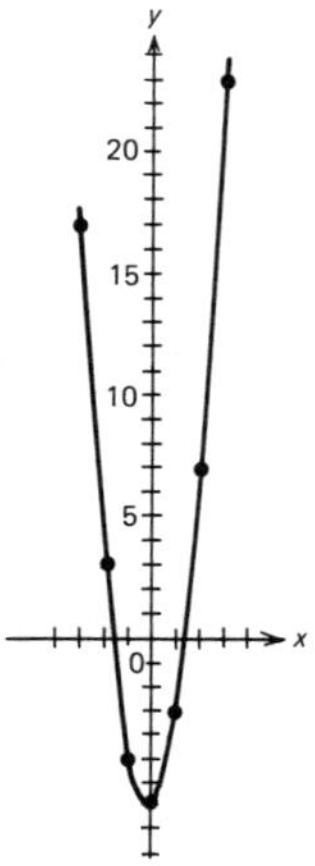

**14.** $(5, 0), (-2, 0); (0, -10); (\frac{3}{2}, -12\frac{1}{4})$

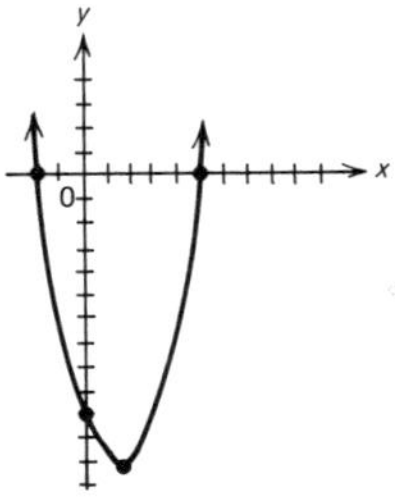


**15.**

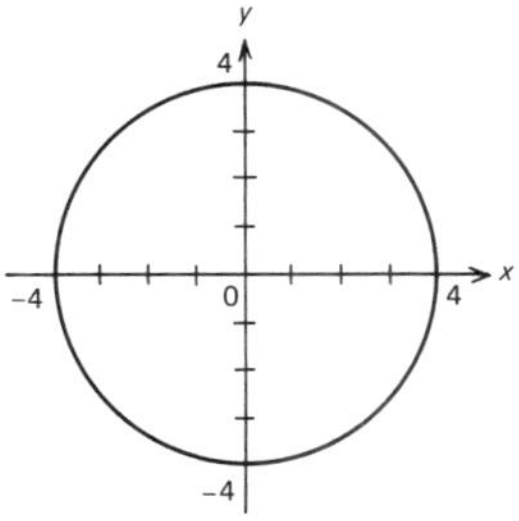


**16.**

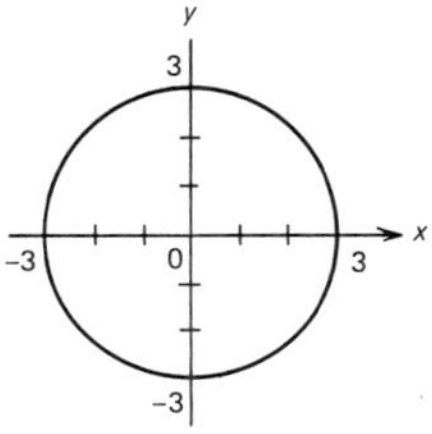


**17.**

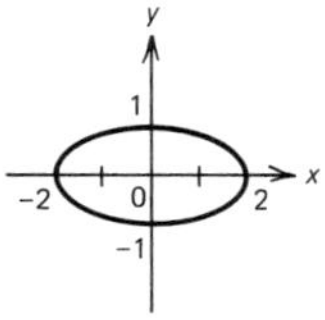


**18.**

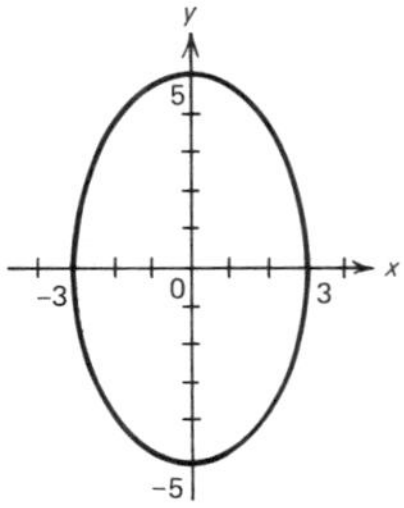


**19.**

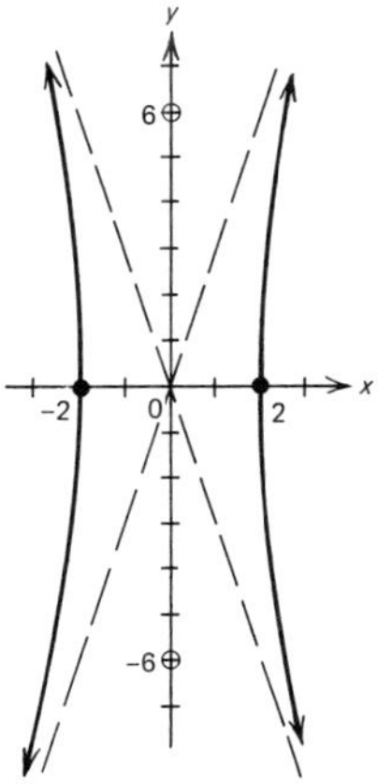


**20.**

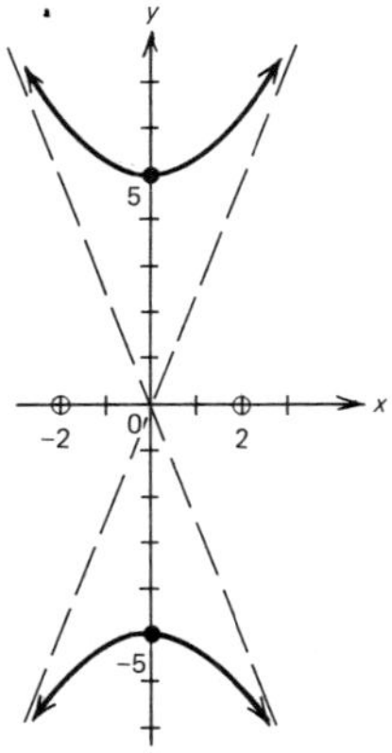


**21.**

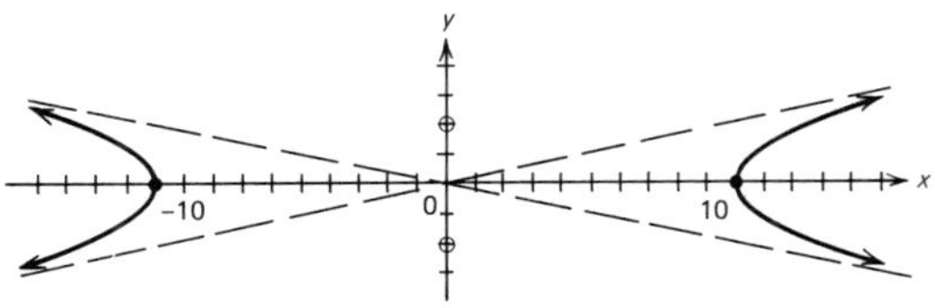


**22.**

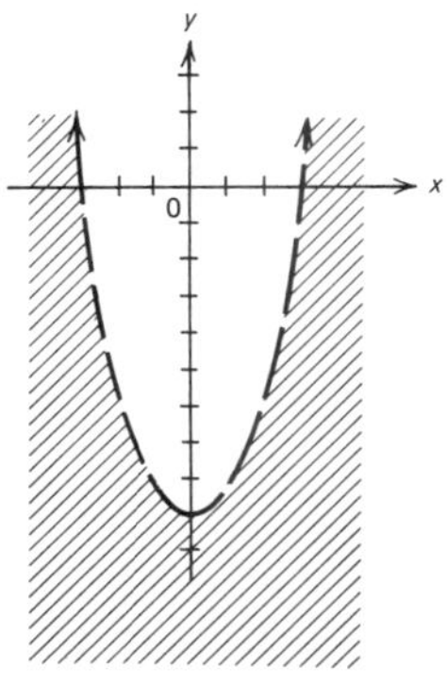

**23.**

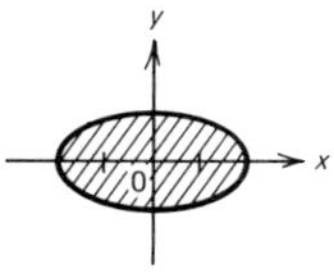


**24.**

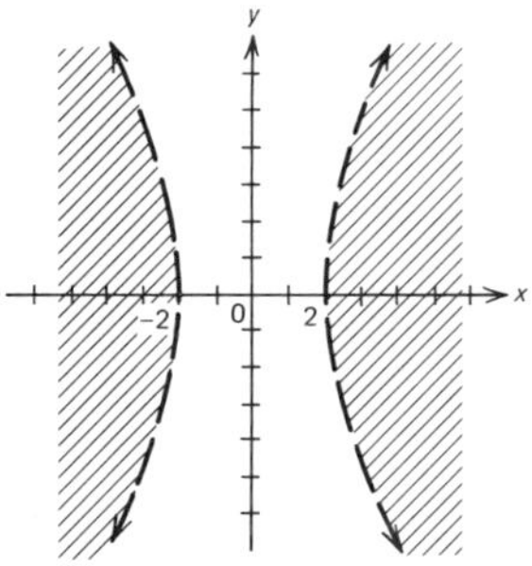


**25.** $f^{-1}$: $x = 2y - 5$ Both are functions.

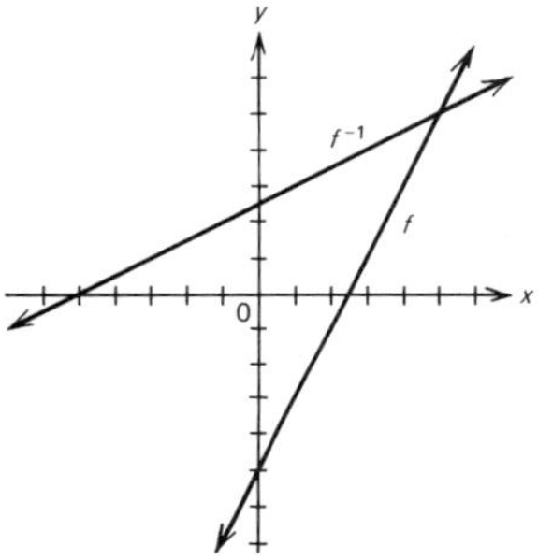


**26.** $f^{-1}$: $x = y^2 - 1$ $f$ is a function; $f^{-1}$ is not a function.

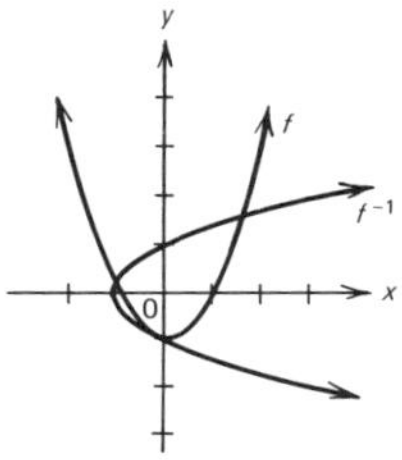


**27.** $f^{-1}$: $x^2 + 4y^2 = 36$ Neither is a function.

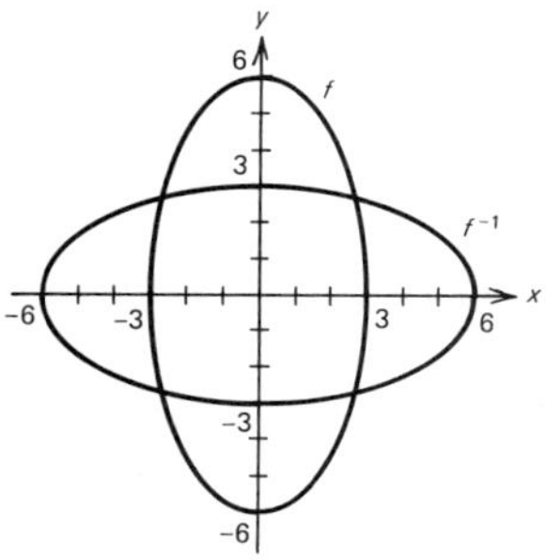


**28.** $f^{-1}$: $-4x^2 + 9y^2 = 36$ Neither is a function.

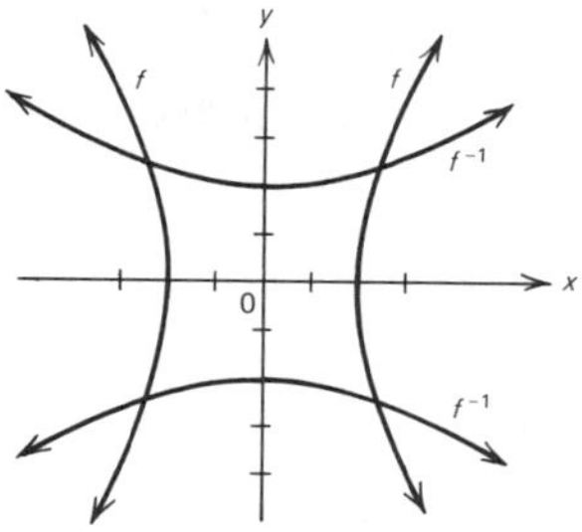


## PROBLEM SET 10.1

**1.** (5, 1) **3.** (0, −2) **5.** (1, 2) **7.** (−1, 3) **9.** (−2, 5) **11.** (3, 0) **13.** (2, 1) **15.** (0, 3) **17.** Either equation **19.** (−50, −36) **21.** $(\frac{5}{7}, \frac{2}{7})$ **23.** (1, 4) **25.** (−2, 0) **27.** ∅ **29.** $(\frac{17}{9}, \frac{1}{3})$ **31.** (1, 5) **33.** (3, −6) **35.** (2, −3) **37.** (1, 6) **39.** (−5, 2) **41.** (20,000 miles, \$2400) **43.** (500 items, \$50,000) **45.** (49,450, −165)

## PROBLEM SET 10.2

**1.** (11, 4) **3.** (5, 12) **5.** (5, 13) **7.** 45 oz of 20%; 15 oz of 40% **9.** 5 oz of 70%; 15 oz of 90%
**11.** \$6000 at 5%; \$2000 at 7% **13.** 8 nickels; 10 dimes **15.** 700 at \$3; 300 at \$5
**17.** 6 lbs of \$2; 14 lbs of \$3

## PROBLEM SET 10.3

**1.** $(3, 4, -2)$ **3.** $(1, 3, -2)$ **5.** $\varnothing$ **7.** $(3, 0, 5)$ **9.** $(1, -2, 3)$ **11.** $(1, 0, -3)$
**13.** $(8, 1, -5)$ **15.** $(1, -3, 8)$ **17.** $(2, -1, 4)$ **19.** $(1, 4, 3)$ **21.** $(10, 7, 8)$

## PROBLEM SET 10.4

**1.** $-2$ **3.** $-34$ **5.** $37$ **7.** $-26$ **9.** $6$ **11.** $0$ **13.** $-5$ **15.** $-66$ **17.** $1$
**19.** $-7$

## PROBLEM SET 10.5

**1.** $\left(\frac{46}{13}, \frac{9}{13}\right)$ **3.** $\left(\frac{83}{23}, \frac{1}{23}\right)$ **5.** $\left(\frac{124}{68}, \frac{-38}{68}\right)$ **7.** $\left(\frac{53}{83}, \frac{84}{83}\right)$ **9.** $\left(\frac{56}{32}, \frac{24}{32}\right)$ **11.** $\left(\frac{43}{68}, \frac{-10}{68}, \frac{40}{68}\right)$
**13.** $\left(\frac{222}{559}, \frac{-130}{559}, \frac{22}{559}\right)$ **15.** $\left(\frac{231}{75}, \frac{309}{75}, \frac{15}{75}\right)$ **17.** $\left(\frac{-296}{16}, \frac{-242}{16}, \frac{-216}{16}\right)$ **19.** $\left(\frac{-80}{82}, \frac{165}{82}, \frac{-99}{82}\right)$
**21.** $\left(\frac{54}{573}, \frac{-186}{573}, \frac{-243}{573}\right)$ **23.** $(12, 6)$

## PROBLEM SET 10.6

**1.** (2, 5) and (5, 2) **3.** (1, 7) and $\left(\frac{13}{3}, \frac{11}{3}\right)$ **5.** (2, 2) **7.** (4, 2) and $\left(\frac{-62}{13}, \frac{64}{13}\right)$ **9.** (4, 7) and (7, 4)
**11.** (5, 1) and $\left(\frac{-3}{2}, \frac{-10}{3}\right)$ **13.** (1, 10) and (6, 55) **15.** $(6, 1); (6, -1); (-6, 1); (-6, -1)$
**17.** $(2, 3); (2, -3); (-2, 3); (-2, -3)$ **19.** $(4, 5); (-4, 5); \left(\sqrt{\frac{51}{2}}, \frac{-9}{2}\right); \left(-\sqrt{\frac{51}{2}}, \frac{-9}{2}\right)$ **21.** (5, 7) and (7, 5)
**23.** (3, 8) and (8, 3) **25.** (50, 15,000) and (150, 45,000)

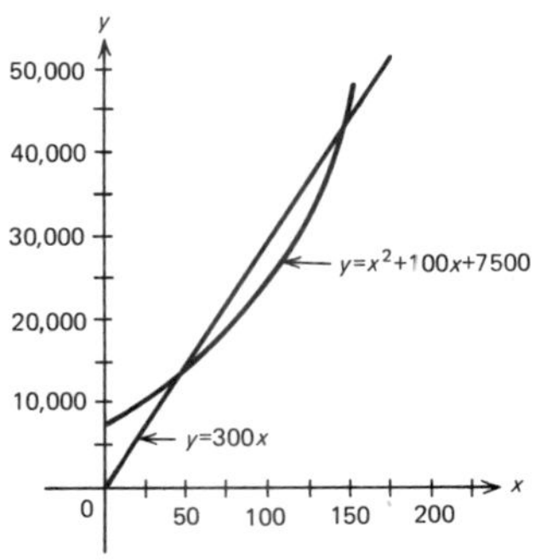

## PROBLEM SET 10.7

1.

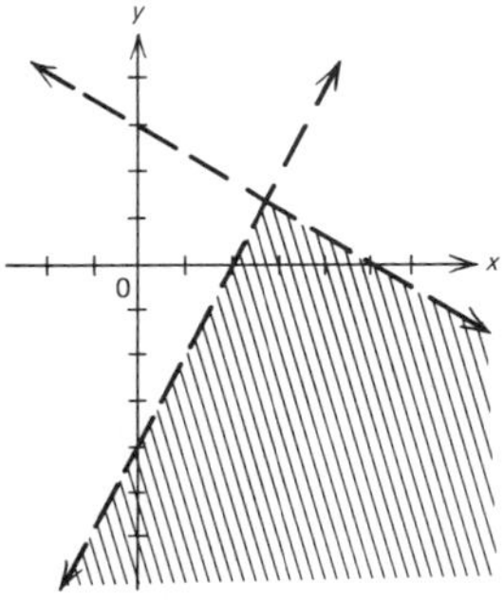


3.

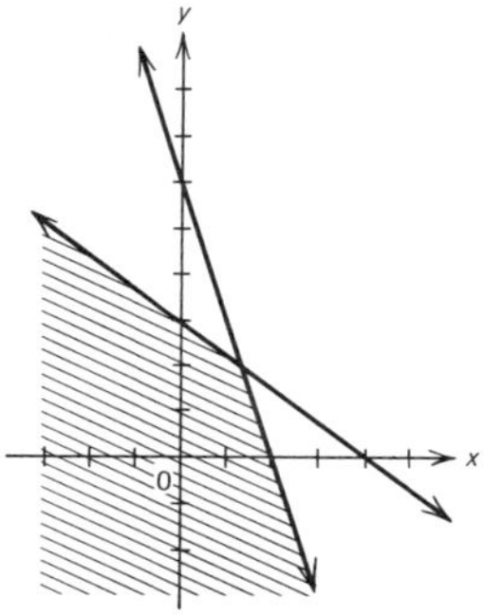


5.

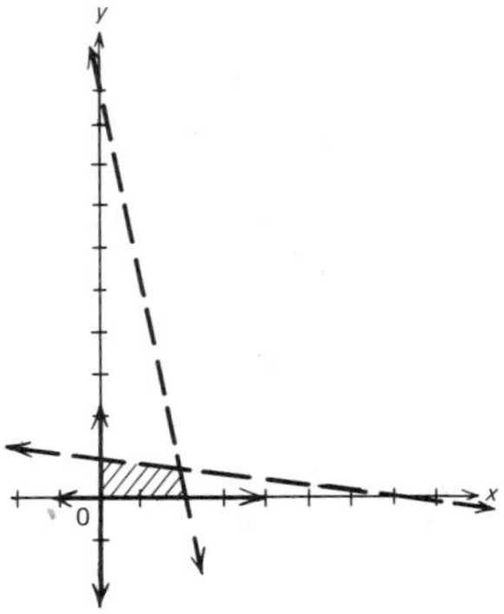


7.

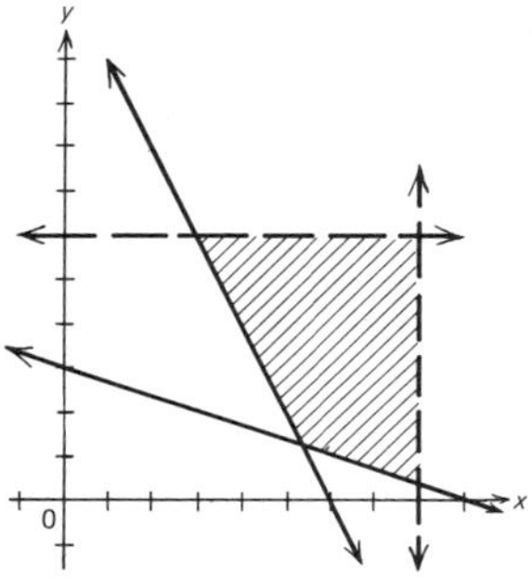


9.

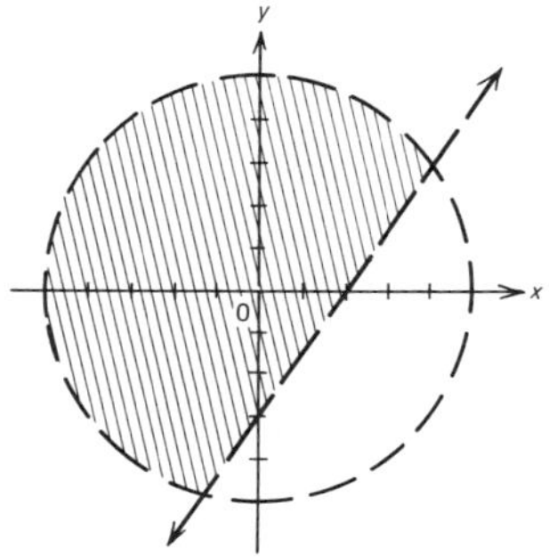


11.

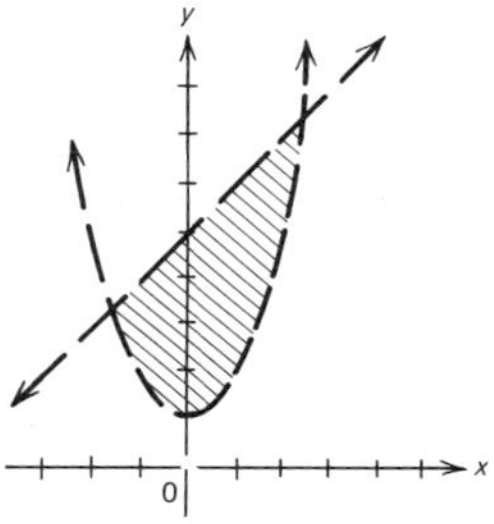


13.

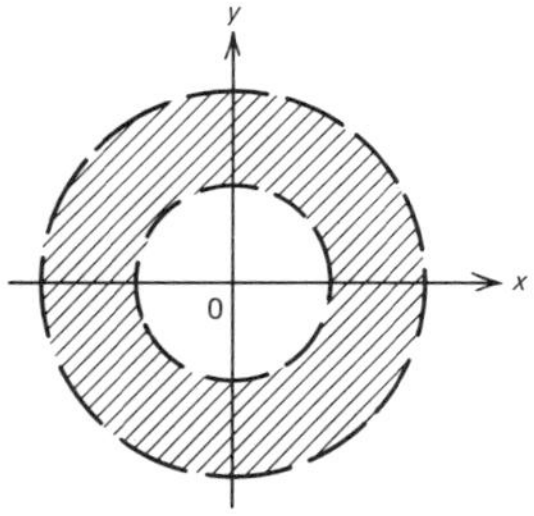


15.

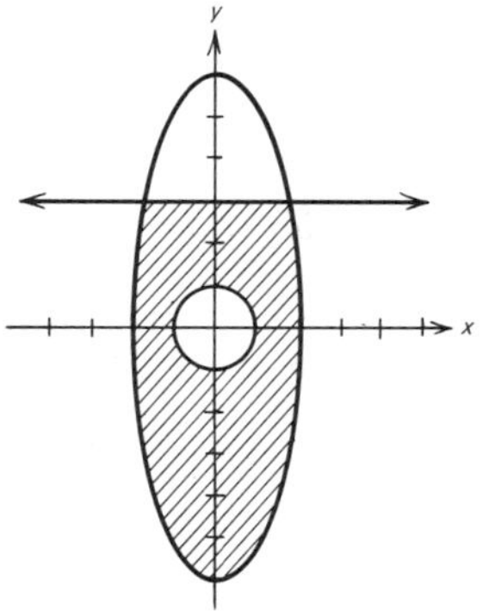


17.

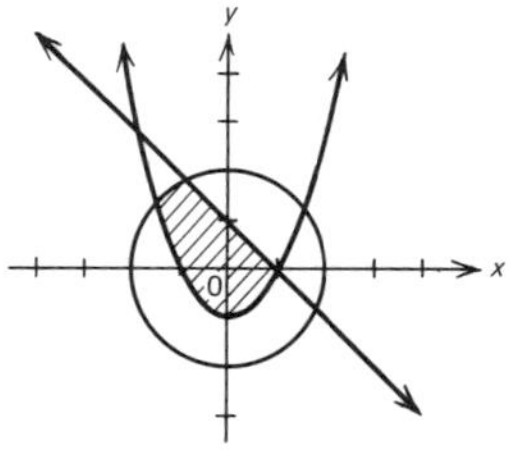


19.

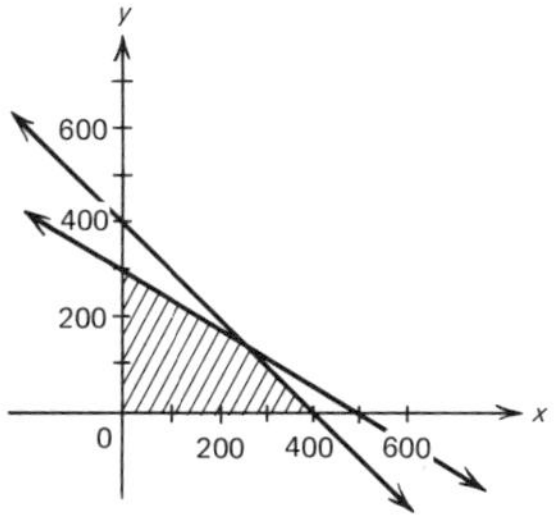


21.

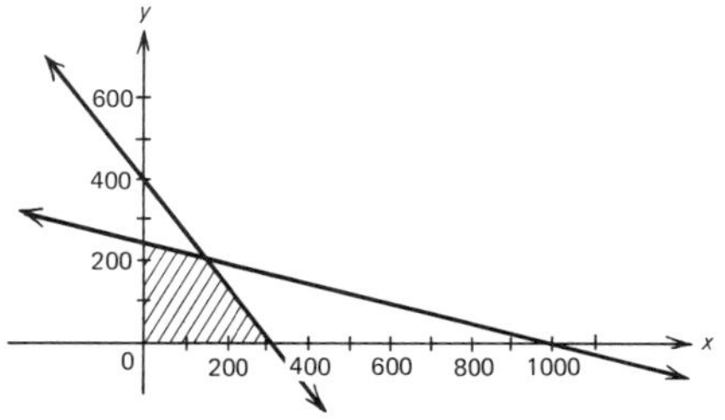

## CHAPTER 10 REVIEW EXERCISES

**1.** (4, 3) **2.** (5, −2) **3.** (4, −3) **4.** (7, 0, −6) **5.** (1, 0, −6) **6.** (1, 2, 3) **7.** 47

**8.** −140 **9.** $\left(\frac{57}{97}, \frac{13}{97}\right)$ **10.** $\left(\frac{437}{219}, \frac{105}{219}, \frac{-68}{219}\right)$ **11.** (10, 9) **12.** (8, 3) and $\left(\frac{3}{2}, 16\right)$

**13.** (6, 5); (6, −5); (−6, 5); (−6, −5)

**14.**

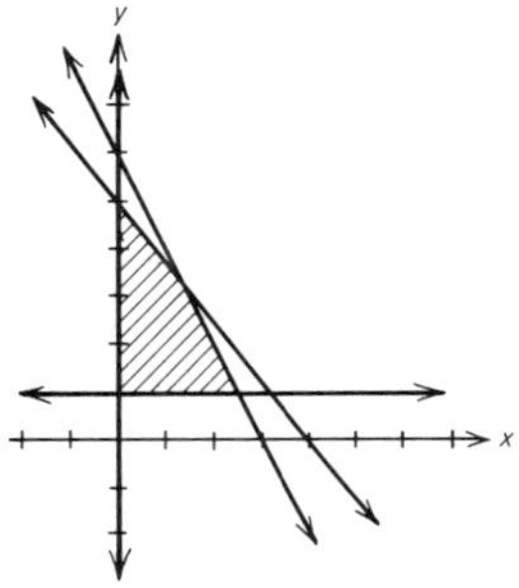


**15.**

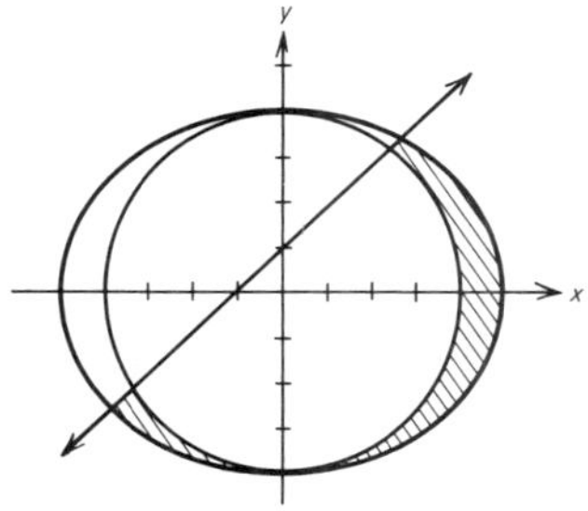


**16.** (7, 9) **17.** 8 qts of 20%; 4 qts of 50% **18.** 13 dimes, 5 quarters **19.** (7, 9); (−9, −7)

## PROBLEM SET 11.1

**1.** $\log_2 128 = 7$ **3.** $\log_{10} 1000 = 3$ **5.** $\log_{100} 10 = \frac{1}{2}$ **7.** $\log_5 \frac{1}{25} = -2$ **9.** $\log_{16} \frac{1}{4} = \frac{-1}{2}$

**11.** $2^3 = 8$ **13.** $5^4 = 625$ **15.** $64^{1/3} = 4$ **17.** $4^{-2} = \frac{1}{16}$ **19.** $4^{-3/2} = \frac{1}{8}$ **21.** 49 **23.** $\frac{1}{9}$

**25.** 2 **27.** 5 **29.** 3 **31.** 81 **33.** 3 **35.** −1 **37.** $\frac{-1}{2}$ **39.** 2 **41.** 10

**43.** 0 **45.** $\frac{3}{2}$ **47.** 5 **49.** 1 **51.** 0 **53.** 20 **55.** 60 **57.** 110 **59.** 1 **61.** 3

## PROBLEM SET 11.2

**1.** $\log_5 4 + \log_5 a$ **3.** $\log_{10} 7 + \log_{10} x + \log_{10} y + \log_{10} z$ **5.** $\log_2 6 - \log_2 r$ **7.** $\log_8 5 - \log_8 3$

**9.** $5 \log_2 x$ **11.** $r \log_{12} 2$ **13.** $\frac{1}{2} \log_7 y$ **15.** $2 \log_2 x + 5 \log_2 y$ **17.** $9 \log_4 u - 2 \log_4 v$

**19.** $\frac{1}{2} \log_{10} a - 3 \log_{10} b$ **21.** $\frac{1}{2}(\log_3 m + \log_3 n - \log_3 k)$

**23.** $\frac{1}{3}(\log_{10} x + 2 \log_{10} y) - \frac{1}{2}(\log_{10} u - 3 \log_{10} v)$ **25.** $\log_2 xy$ **27.** $\log_4 \frac{a}{b}$ **29.** $\log_{10} x^2$

**31.** $\log_{10} \frac{xy}{z}$ **33.** $\log_3 \frac{u^2 v^3}{w^4}$ **35.** $\log_{10} \sqrt[3]{\frac{x^2 y^4}{z^7}}$ **37.** 32 **39.** $\frac{9}{10}$ **41.** 135 **43.** 3

**45.** 9 **47.** 2 **49.** $\gamma = \log_{10} \left(\frac{\beta}{b}\right)^k$ **51.** $\log_e Q = \log_e \frac{bI^5 P_b^{0.9}}{P_a^{0.5}}$

## PROBLEM SET 11.3

**1.** 1 **3.** 6 **5.** $-5$ **7.** 0.8142 **9.** 0.7243 **11.** 0.6021 **13.** 4.6637 **15.** 6.2330 **17.** 3.9494 **19.** 2.3010 **21.** 6.8451 **23.** $9.7168 - 10$ **25.** $7.8633 - 10$ **27.** $5.9138 - 10$ **29.** $8.7782 - 10$ **31.** 3960 **33.** 298 **35.** 775 **37.** 0.00463 **39.** 0.00000107 **41.** 0.316 **43.** 25.7 **45.** 5.02 **47.** 1260 **49.** 20.792 **51.** 0.3010 **53.** 1.6990 **55.** 5.8 **57.** 4.1

## PROBLEM SET 11.4

**1.** 66 **3.** 1.45 **5.** 0.0000516 **7.** 38,900 **9.** 27.3 **11.** 0.0317 **13.** 9.12 **15.** 73.1 **17.** 6.19 **19.** 3.21 **21.** 71.6 **23.** 222 **25.** 0.397 **27.** 0.987 **29.** 7.92 **31.** $7.16 \times 10^9$ **33.** \$193,000 **35.** 0.955 **37.** 5,940,000,000

## PROBLEM SET 11.5

**1.** 1.54 **3.** 10.2 **5.** 0.856 **7.** $-0.356$ **9.** 2.17 **11.** 0.52 **13.** 0.46 **15.** 0.535 **17.** 1.12 **19.** 7 **21.** 1.98 **23.** 1.09 **25.** 1.40 **27.** 1.83 **29.** 0.585 **31.** 1.19 **33.** 1.946 **35.** 2.163 **37.** $-0.315$ **39.** 11.9 **41.** 109.9 **43.** 9.1 **45.** 20.5

## PROBLEM SET 11.6

**1.**

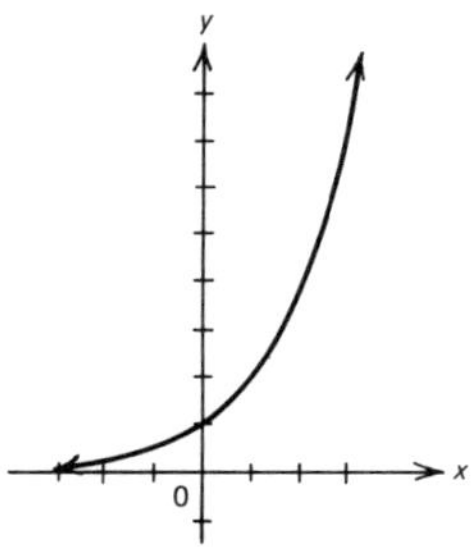


**3.**

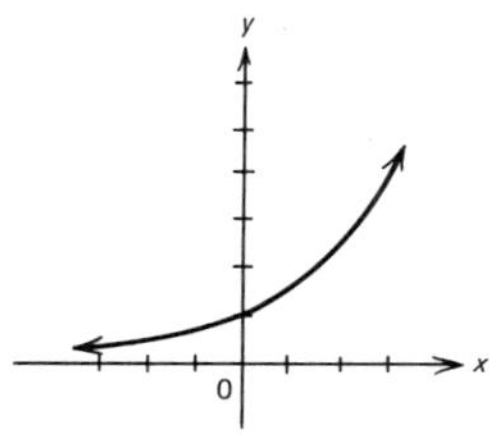


**5.**

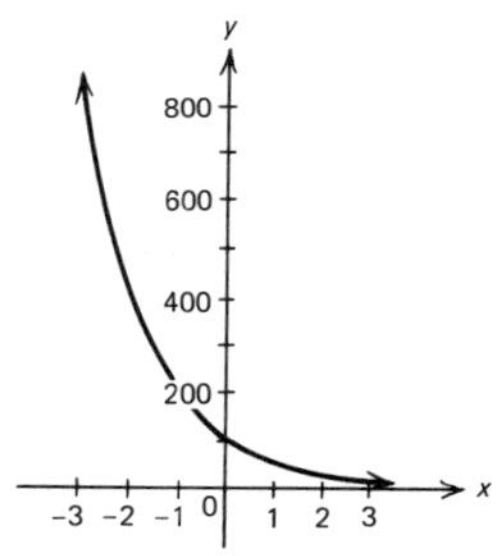


**7.**

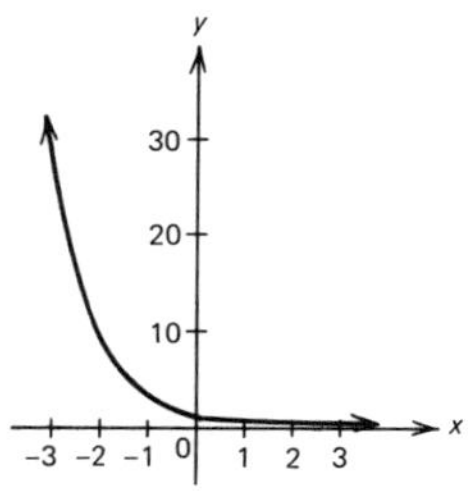


**9.**

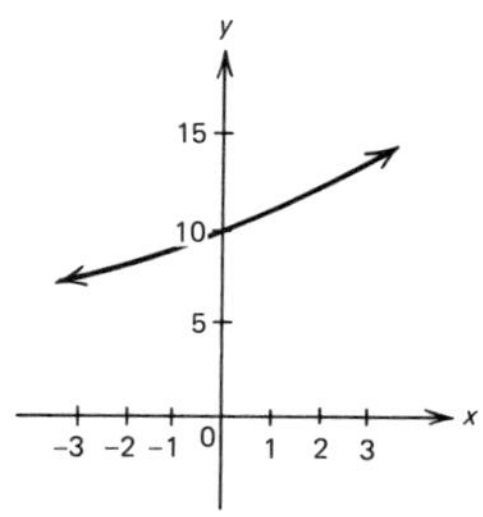


**11.**

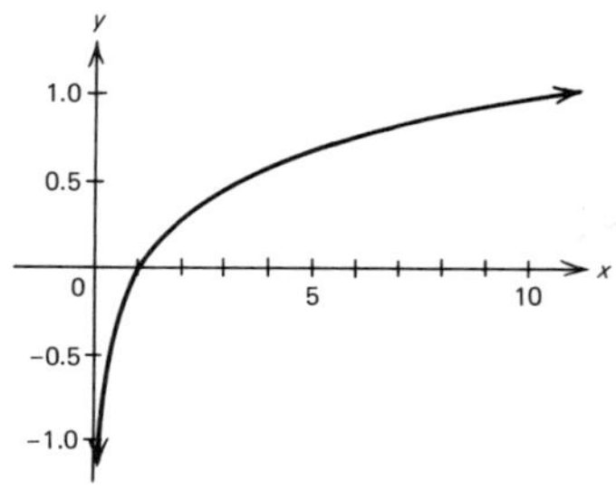


**13.**

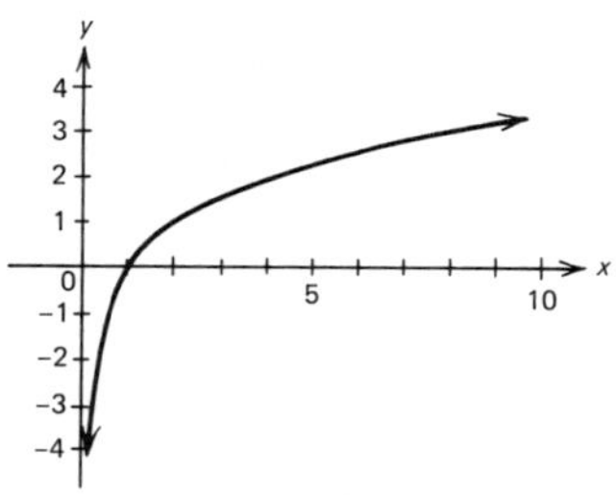


**15.**

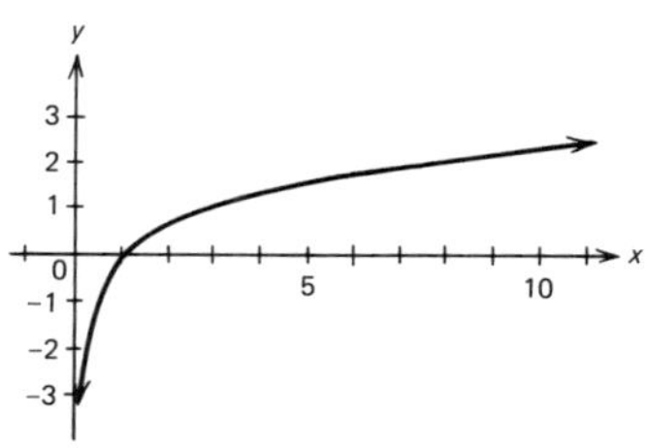

**17.** 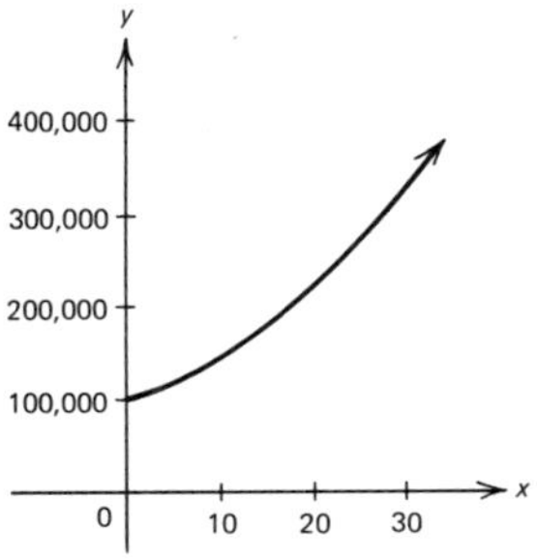


**19.** 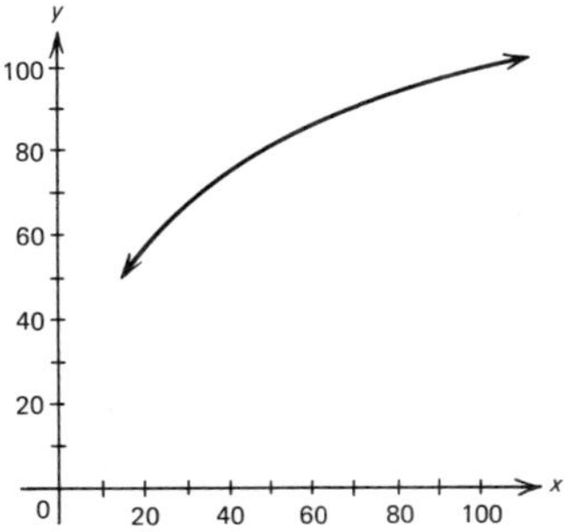


**21.** 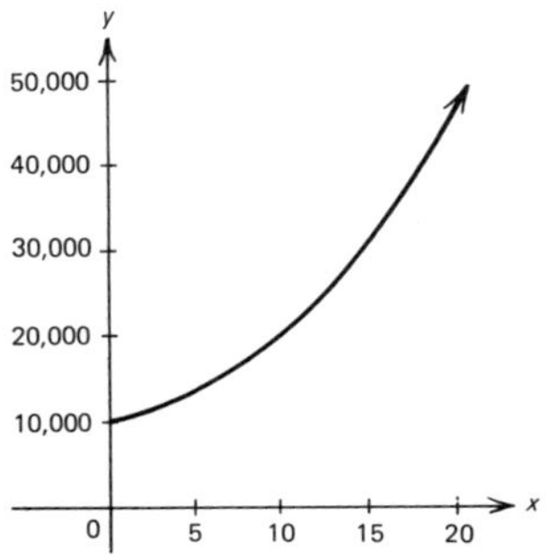


## CHAPTER 11 REVIEW EXERCISES

**1.** $\log_{16} 8 = \frac{3}{4}$ **2.** $5^3 = 125$ **3.** 16 **4.** 10 **5.** 4 **6.** 2

**7.** $\log_8 6 + \log_8 x + \log_8 y + \log_8 z$ **8.** $\log_7 T - \log_7 W$ **9.** $\frac{1}{2} \log_9 r$

**10.** $\frac{1}{4}(\log_{10} x - \log_{10} y - 3 \log_{10} z)$ **11.** $\log_{10} \frac{x^{2/5}}{y^{4/9}}$ **12.** 25 **13.** 1 **14.** 4.7160

**15.** 7.5440 − 10 **16.** 56.4 **17.** 0.00332 **18.** 251 **19.** 8210 **20.** 268 **21.** 18.5
**22.** 0.809 **23.** 1940 **24.** 0.494 **25.** 15.7 **26.** 3.51 **27.** 1.59 **28.** 2.996

**29.** 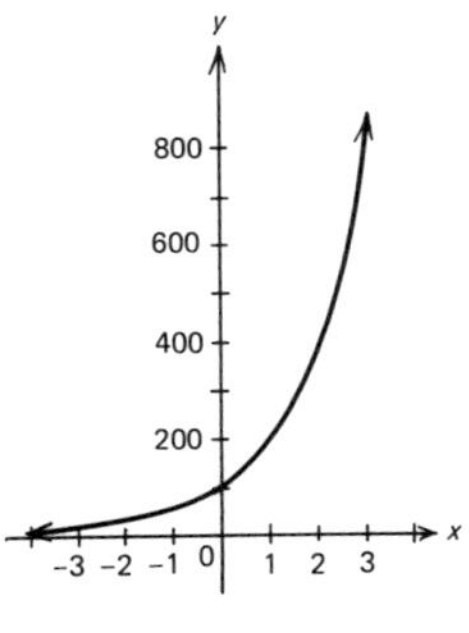


**30.** 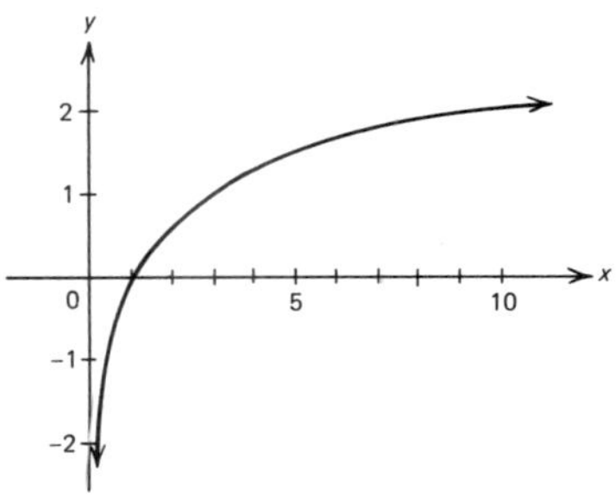

## PROBLEM SET 12.1

**1.** 3, 4, 5, 6, 7 **3.** 1, 5, 10, 17, 26 **5.** 8, 13, 18, 23, 28 **7.** $-1, 1, -1, 1, -1$ **9.** 2, 4, 8, 16, 32

**11.** $\frac{1}{1}, \frac{1}{2}, \frac{1}{3}, \frac{1}{4}, \frac{1}{5}$ **13.** $\frac{2}{1}, \frac{3}{4}, \frac{4}{9}, \frac{5}{16}, \frac{6}{25}$ **15.** $2, 1\frac{1}{2}, 1\frac{1}{3}, 1\frac{1}{4}, 1\frac{1}{5}$ **17.** $n^2$ **19.** $3n$ **21.** $3^n$

**23.** $\dfrac{1}{2n}$ **25.** $(-1)^n n$ **27.** $x^n$

**29.** 7,500,000; 3,750,000; 2,500,000; 1,875,000; 1,500,000; 1,250,000

**31.** $\dfrac{5}{36}Rc; \dfrac{3}{16}Rc; \dfrac{21}{100}Rc; \dfrac{2}{9}Rc$ **33.** $\dfrac{9}{400}Rc; \dfrac{5}{144}Rc; \dfrac{33}{784}Rc; \dfrac{3}{64}Rc$

## PROBLEM SET 12.2

**1.** 21 **3.** 30 **5.** 748 **7.** $\dfrac{137}{60}$ **9.** $\dfrac{63}{64}$ **11.** $\dfrac{61}{20}$ **13.** $x + x^2 + x^3 + x^4$

**15.** $1^x + 2^x + 3^x + 4^x + 5^x$ **17.** $\dfrac{-1}{x} + \dfrac{1}{x^2} - \dfrac{1}{x^3} + \dfrac{1}{x^4} - \dfrac{1}{x^5} + \dfrac{1}{x^6}$ **19.** $\sum_{n=1}^{5} 5n$ **21.** $\sum_{n=2}^{6} n^2$

**23.** $\sum_{n=3}^{8} \dfrac{1}{n}$ **25.** $\sum_{n=1}^{6} (-1)^n n$ **27.** $\sum_{n=1}^{5} x^n$ **29.** $\sum_{n=1}^{4} \dfrac{x^n}{n^2}$ **31.** $\sum_{n=1}^{3} (x_n - 5)^3$ **33.** 18,375,000

## PROBLEM SET 12.3

**1.** 1 **3.** 5 **5.** $-5$ **7.** $\frac{1}{2}$ **9.** 4, 9, 14, 19, 24; $a_n = 5n - 1$

**11.** 8, 5, 2, $-1$, $-4$; $a_n = -3n + 11$ **13.** $-2$, 2, 6, 10, 14; $a_n = 4n - 6$ **15.** $4, 4\frac{1}{3}, 4\frac{2}{3}, 5, 5\frac{1}{3}$; $a_n = \dfrac{n}{3} + \dfrac{11}{3}$

**17.** 13 **19.** 248 **21.** $-61$ **23.** $13\frac{1}{2}$ **25.** 90 **27.** 6275 **29.** $-390$ **31.** 175
**33.** $a_n = 60{,}000 - 5000n$; 55,000, 50,000, 45,000, 40,000 **35.** 12,000, 12,800, 13,600, 14,400 **37.** 156,000
**39.** 18,500

## PROBLEM SET 12.4

**1.** 3 **3.** 10 **5.** $\frac{1}{2}$ **7.** $-3$ **9.** $\frac{1}{2}$, 2, 8, 32, 128 **11.** 12, 4, $\frac{4}{3}, \frac{4}{9}, \frac{4}{27}$
**13.** 10, $-50$, 250, $-1250$, 6250 **15.** $\frac{1}{3}, \frac{1}{6}, \frac{1}{12}, \frac{1}{24}, \frac{1}{48}$ **17.** 729 **19.** $4 \times 10^8$ **21.** $1\frac{7}{8}$ **23.** 3645
**25.** 1093 **27.** 444,444,444 **29.** $118\frac{1}{8}$ **31.** 2735 **33.** $Lr^{12}$ **35.** 1.5
**37.** 49.5, 49.005, 48.515, 48.030; $a_n = 50(0.99)^n$ **39.** 5120
**41.** 160,000, 128,000, 102,400, 81,920, 65,536; $a_n = 200{,}000(0.8)^n$ **43.** 33,219.42 **45.** 137,285.89

## PROBLEM SET 12.5

**1.** 2 **3.** $\frac{2}{3}$ **5.** $\frac{18}{5}$ **7.** 27 **9.** No sum **11.** $\frac{4}{9}$ **13.** $\frac{1}{99}$ **15.** $\frac{174}{999}$ **17.** $\frac{1}{2}$
**19.** 25 **21.** 40

## PROBLEM SET 12.6

**1.** $x^4 + 4x^3y + 6x^2y^2 + 4xy^3 + y^4$ **3.** $m^6 + 6m^5n + 15m^4n^2 + 20m^3n^3 + 15m^2n^4 + 6mn^5 + n^6$
**5.** $u^6 - 6u^5v + 15u^4v^2 - 20u^3v^3 + 15u^2v^4 - 6uv^5 + v^6$ **7.** $x^{12} + 36x^{11} + 594x^{10} + 5940x^9 + \cdots$

**9.** $u^8 + 40u^7 + 700u^6 + 7000u^5 + \cdots$ **11.** $32a^5 + 240a^4b + 720a^3b^2 + 1080a^2b^3 + \cdots$
**13.** $x^9 - 9x^8y + 36x^7y^2 - 84x^6y^3 + \cdots$ **15.** $x^5 - 30x^4 + 360x^3 - 2160x^2 + \cdots$
**17.** $16x^4 - 96x^3y + 216x^2y^2 - 216xy^3 + \cdots$ **19.** four: $\frac{1}{256}$; three: $\frac{12}{256}$; two: $\frac{54}{256}$; one: $\frac{108}{256}$; none: $\frac{81}{256}$
**21.** six: $\frac{1}{4096}$; five: $\frac{18}{4096}$; four: $\frac{135}{4096}$; three: $\frac{540}{4096}$; two: $\frac{1215}{4096}$; one: $\frac{1458}{4096}$; none: $\frac{729}{4096}$

## CHAPTER 12 REVIEW EXERCISES

**1.** 4, 7, 12, 19, 28 **2.** $\frac{-1}{1}, \frac{1}{2}, \frac{-1}{3}, \frac{1}{4}, \frac{-1}{5}$ **3.** $\frac{1}{3^n}$ **4.** $\frac{1}{x^n}$ **5.** $\frac{137}{60}$

**6.** $\frac{-1}{x} + \frac{1}{x^2} - \frac{1}{x^3} + \frac{1}{x^4} - \frac{1}{x^5} + \frac{1}{x^6}$ **7.** $\sum_{n=1}^{5} n^3$ **8.** $\sum_{n=1}^{10} \frac{n^2}{x^n}$ **9.** $-3$ **10.** $-7$ **11.** 65

**12.** $3 + 3n$ **13.** 2 **14.** 2048 **15.** 4092 **16.** $6\left(\frac{1}{2}\right)^{n-1}$ **17.** 20 **18.** $\frac{17}{99}$

**19.** $a^{10} + 20a^9 + 180a^8 + 960a^7 + \cdots$ **20.** $128x^7 - 448x^6y + 672x^5y^2 - 560x^4y^3 + \cdots$

# INDEX OF MATHEMATICAL TERMS

# INDEX OF APPLICATIONS

## A



## B



## C



## D

## P



## R



## S



## T



## V



## W

# FOUR-PLACE COMMON LOGARITHMS

| *N* | 0 | 1 | 2 | 3 | 4 | 5 | 6 | 7 | 8 | 9 |
|---|---|---|---|---|---|---|---|---|---|---|
| **1.0** | .0000 | .0043 | .0086 | .0128 | .0170 | .0212 | .0253 | .0294 | .0334 | .0374 |
| **1.1** | .0414 | .0453 | .0492 | .0531 | .0569 | .0607 | .0645 | .0682 | .0719 | .0755 |
| **1.2** | .0792 | .0828 | .0864 | .0899 | .0934 | .0969 | .1004 | .1038 | .1072 | .1106 |
| **1.3** | .1139 | .1173 | .1206 | .1239 | .1271 | .1303 | .1335 | .1367 | .1399 | .1430 |
| **1.4** | .1461 | .1492 | .1523 | .1553 | .1584 | .1614 | .1644 | .1673 | .1703 | .1732 |
| **1.5** | .1761 | .1790 | .1818 | .1847 | .1875 | .1903 | .1931 | .1959 | .1987 | .2014 |
| **1.6** | .2041 | .2068 | .2095 | .2122 | .2148 | .2175 | .2201 | .2227 | .2253 | .2279 |
| **1.7** | .2304 | .2330 | .2355 | .2380 | .2405 | .2430 | .2455 | .2480 | .2504 | .2529 |
| **1.8** | .2553 | .2577 | .2601 | .2625 | .2648 | .2672 | .2695 | .2718 | .2742 | .2765 |
| **1.9** | .2788 | .2810 | .2833 | .2856 | .2878 | .2900 | .2923 | .2945 | .2967 | .2989 |
| **2.0** | .3010 | .3032 | .3054 | .3075 | .3096 | .3118 | .3139 | .3160 | .3181 | .3201 |
| **2.1** | .3222 | .3243 | .3263 | .3284 | .3304 | .3324 | .3345 | .3365 | .3385 | .3404 |
| **2.2** | .3424 | .3444 | .3464 | .3483 | .3502 | .3522 | .3541 | .3560 | .3579 | .3598 |
| **2.3** | .3617 | .3636 | .3655 | .3674 | .3692 | .3711 | .3729 | .3747 | .3766 | .3784 |
| **2.4** | .3802 | .3820 | .3838 | .3856 | .3874 | .3892 | .3909 | .3927 | .3945 | .3962 |
| **2.5** | .3979 | .3997 | .4014 | .4031 | .4048 | .4065 | .4082 | .4099 | .4116 | .4133 |
| **2.6** | .4150 | .4166 | .4183 | .4200 | .4216 | .4232 | .4249 | .4265 | .4281 | .4298 |
| **2.7** | .4314 | .4330 | .4346 | .4362 | .4378 | .4393 | .4409 | .4425 | .4440 | .4456 |
| **2.8** | .4472 | .4487 | .4502 | .4518 | .4533 | .4548 | .4564 | .4579 | .4594 | .4609 |
| **2.9** | .4624 | .4639 | .4654 | .4669 | .4683 | .4698 | .4713 | .4728 | .4742 | .4757 |
| **3.0** | .4771 | .4786 | .4800 | .4814 | .4829 | .4843 | .4857 | .4871 | .4886 | .4900 |
| **3.1** | .4914 | .4928 | .4942 | .4955 | .4969 | .4983 | .4997 | .5011 | .5024 | .5038 |
| **3.2** | .5051 | .5065 | .5079 | .5092 | .5105 | .5119 | .5132 | .5145 | .5159 | .5172 |
| **3.3** | .5185 | .5198 | .5211 | .5224 | .5237 | .5250 | .5263 | .5276 | .5289 | .5302 |
| **3.4** | .5315 | .5328 | .5340 | .5353 | .5366 | .5378 | .5391 | .5403 | .5416 | .5428 |
| **3.5** | .5441 | .5453 | .5465 | .5478 | .5490 | .5502 | .5514 | .5527 | .5539 | .5551 |
| **3.6** | .5563 | .5575 | .5587 | .5599 | .5611 | .5623 | .5635 | .5647 | .5658 | .5670 |
| **3.7** | .5682 | .5694 | .5705 | .5717 | .5729 | .5740 | .5752 | .5763 | .5775 | .5786 |
| **3.8** | .5798 | .5809 | .5821 | .5832 | .5843 | .5855 | .5866 | .5877 | .5888 | .5899 |
| **3.9** | .5911 | .5922 | .5933 | .5944 | .5955 | .5966 | .5977 | .5988 | .5999 | .6010 |
| **4.0** | .6021 | .6031 | .6042 | .6053 | .6064 | .6075 | .6085 | .6096 | .6107 | .6117 |
| **4.1** | .6128 | .6138 | .6149 | .6160 | .6170 | .6180 | .6191 | .6201 | .6212 | .6222 |
| **4.2** | .6232 | .6243 | .6253 | .6263 | .6274 | .6284 | .6294 | .6304 | .6314 | .6325 |
| **4.3** | .6335 | .6345 | .6355 | .6365 | .6375 | .6385 | .6395 | .6405 | .6415 | .6425 |
| **4.4** | .6435 | .6444 | .6454 | .6464 | .6474 | .6484 | .6493 | .6503 | .6513 | .6522 |
| **4.5** | .6532 | .6542 | .6551 | .6561 | .6571 | .6580 | .6590 | .6599 | .6609 | .6618 |
| **4.6** | .6628 | .6637 | .6646 | .6656 | .6665 | .6675 | .6684 | .6693 | .6702 | .6712 |
| **4.7** | .6721 | .6730 | .6739 | .6749 | .6758 | .6767 | .6776 | .6785 | .6794 | .6803 |
| **4.8** | .6812 | .6821 | .6830 | .6839 | .6848 | .6857 | .6866 | .6875 | .6884 | .6893 |
| **4.9** | .6902 | .6911 | .6920 | .6928 | .6937 | .6946 | .6955 | .6964 | .6972 | .6981 |
| **5.0** | .6990 | .6998 | .7007 | .7016 | .7024 | .7033 | .7042 | .7050 | .7059 | .7067 |
| **5.1** | .7076 | .7084 | .7093 | .7101 | .7110 | .7118 | .7126 | .7135 | .7143 | .7152 |
| **5.2** | .7160 | .7168 | .7177 | .7185 | .7193 | .7202 | .7210 | .7218 | .7226 | .7235 |
| **5.3** | .7243 | .7251 | .7259 | .7267 | .7275 | .7284 | .7292 | .7300 | .7308 | .7316 |
| **5.4** | .7324 | .7332 | .7340 | .7348 | .7356 | .7364 | .7372 | .7380 | .7388 | .7396 |
| *N* | 0 | 1 | 2 | 3 | 4 | 5 | 6 | 7 | 8 | 9 |